POTENTIAL THEORY IN GRAVITY AND MAGNETIC APPLICATIONS

The Stanford–Cambridge Program is an innovative publishing venture resulting from the collaboration between Cambridge University Press and Stanford University and its Press.

The Progam provides a new international imprint for the teaching and communication of pure and applied sciences. Drawing on Stanford's eminent faculty and associated institutions, books within the Program reflect the high quality of teaching and research at Stanford University.

The Program includes textbooks at undergraduate and graduate level, and research monographs, across a broad range of the sciences.

Cambridge University Press publishes and distributes books in the Stanford–Cambridge Program throughout the world.

POTENTIAL THEORY IN GRAVITY AND MAGNETIC APPLICATIONS

RICHARD J. BLAKELY

CAMBRIDGE
UNIVERSITY PRESS

CAMBRIDGE UNIVERSITY PRESS
Cambridge, New York, Melbourne, Madrid, Cape Town, Singapore, São Paulo, Delhi

Cambridge University Press
The Edinburgh Building, Cambridge CB2 8RU, UK

Published in the United States of America by Cambridge University Press, New York

www.cambridge.org
Information on this title: www.cambridge.org/9780521415088

First published 1995
Reprinted 1996
First paperback edition 1996

A catalogue record for this publication is available from the British Library

ISBN 978-0-521-41508-8 hardback
ISBN 978-0-521-57547-8 paperback

Transferred to digital printing 2009

To Diane

Contents

Introduction

Though this be madness, yet there is method in't.

(William Shakespeare)

I think I did pretty well, considering I started out with nothing but a bunch of blank paper.

(Steve Martin)

Pierre Simon, Marquis de Laplace, showed in 1782 that Newtonian potential obeys a simple differential equation. Laplace's equation, as it now is called, arguably has become the most universal differential equation in the physical sciences because of the wide range of phenomena that it describes. The theory of the potential spawned by Laplace's equation is the subject of this book, but with particular emphasis on the application of this theory to gravity and magnetic fields of the earth and in the context of geologic and geophysical investigations.

A Brief History of Magnetic and Gravity Methods

The geomagnetic field must surely rank as the longest studied of all the geophysical properties of the earth. Curiosity about the mutual attraction of lodestones can be traced back at least to the time of Thales, a philosopher of ancient Greece in the sixth century B.C. (Needham [194]). The tendency of lodestones to align preferentially in certain directions was known in China by the first century A.D., and perhaps as early as the second century B.C. This apparently was the first recognition that the earth is associated with a property that affects magnetic objects, thus paving the way for the advent of the magnetic compass in China and observations of magnetic declination.

The compass arrived in Europe much later, probably late in the twelfth century A.D., but significant discoveries were to follow. Petrus Peregrinus, a scholar of thirteenth-century Italy, performed several important experiments on spherical pieces of lodestone. His findings, written in 1269, described for the first time the concepts of magnetic polarity, magnetic meridians, and the idea that like poles repel but opposite poles attract. Georg Hartmann, Vicar of Nuremberg, was the first European to measure magnetic declination in about 1510. He also discovered magnetic inclination in 1544, but his writings went undiscovered until after Robert Norman, an English hydrographer, published his own careful experiments on inclination conducted in 1576. In 1600, William Gilbert, physician to Queen Elizabeth I, published his landmark treatise, *De Magnete*, culminating centuries of European and Chinese thought and experimentation on the geomagnetic field. Noting that the earth's magnetic field has a form much like that of a spherically shaped piece of lodestone, Gilbert proclaimed that *"magnus magnes ipse est globus terrestris"* ("the whole earth is a magnet"), and magnetism thus became the first physical property, other than roundness, attributed to the earth as a whole (Merrill and McElhinny [183]). In 1838, the German mathematician Carl Friederich Gauss gave geomagnetic observations their first global-scale mathematical formalism by applying spherical harmonic analysis to a systematic set of magnetic measurements available at the time.

The application of magnetic methods to geologic problems advanced in parallel with the development of magnetometers. Geologic applications began at least as early as 1630, when a sundial compass was used to prospect for iron ore in Sweden (Hanna [110]), thus making magnetic-field interpretation one of the oldest of the geophysical exploration techniques. Early measurements of the magnetic field for exploration purposes were made with land-based, balanced magnets similar in principle of operation to today's widely used gravity meters. Max Thomas Edelmann used such a device during the first decade of this century to make the first airborne magnetic measurements via balloon (Heiland [121]). It was soon recognized that measurements of the magnetic field via aircraft could provide superior uniform coverage compared to surface measurements because of the aircraft's ability to quickly cover remote and inaccessible areas, but balanced-magnet instruments were not generally amenable to the accelerations associated with moving platforms. It was military considerations, related to World War II, that spurred the development of a suitable magnetometer for routine aeromagnetic measurements. In 1941, Victor Vacquier, Gary Muffly, and R. D. Wyckoff, employees of Gulf Research and Development Company under contract with the U.S. government, modified 10-year-old flux-gate technology, combined it with suitable stabilizing equipment, and thereby developed a magnetometer for airborne detection of submarines. In 1944, James R. Balsley and Homer Jensen of the U.S. Geological Survey used a magnetometer of similar design in the first modern airborne geophysical survey near Boyertown, Pennsylvania (Jensen [143]).

A second major advance in magnetometer design was the development of the proton-precession magnetometer by Varian Associates in 1955. This relatively simple instrument measures the magnitude of the total field without the need for elaborate stabilizing or orienting equipment. Consequently, the proton-precession magnetometer is relatively inexpensive and easy to operate and has revolutionized land-based and shipborne measurements. Various other magnetometer designs have followed with greater resolution (Reford [240]) to be sure, but the proton-precession magnetometer remains a mainstay of field surveys.

Shipborne magnetic measurements were well under way by the 1950s. By the mid 1960s, ocean-surface measurements of magnetic intensity in the Northeast-Pacific (Raff and Mason [234]) had discovered curious anomalies lineated roughly north–south. Fred Vine and Drummond Matthews [286] and, independently, Lawrence Morley and Andre Larochelle [186] recognized that these lineations reflect a recording of the reversing geomagnetic field by the geologic process of seafloor spreading, and thus was spawned the plate-tectonic revolution.

The gravity method too has a formidable place in the history of science. The realization that the earth has a force of attraction surely must date back to our initial awareness that dropped objects fall to the ground, observations that first were quantified by the well-known experiments of Galileo Galilei around 1590. In 1687 Isaac Newton published his landmark treatise, *Philosophiae Naturalis Principia Mathematica*, in which he proposed (among other revolutionary concepts) that the force of gravity is a property of all matter, Earth included.

In 1672 a French scholar, Jean Richer, noted that a pendulum-based clock designed to be accurate in Paris lost a few minutes per day in Cayenne, French Guiana, and so pendulum observations were discovered as a way to measure the spatial variation of the geopotential. Newton correctly interpreted the discrepancy between these two measurements as reflecting the oblate shape of the earth. The French believed otherwise at the time, and to prove the point, the French Academy of Sciences sent two expeditions, one to the equatorial regions of Ecuador and the other to the high latitudes of Sweden, to carefully measure and compare the length of a degree of arc at both sites (Fernie [88, 89, 90]). The Ecuador expedition was led by several prominent French scientists, among them Pierre Bouguer, sometimes credited for the first careful observations of the shape of the earth and for whom the "Bouguer anomaly" is named.

The reversible pendulum was constructed by H. Kater in 1818, thereby facilitating absolute measurements of gravity. Near the end of the same century, R. Sterneck of Austria reported the first pendulum instrument and used it to measure gravity in Europe. Other types of pendulum instruments followed, including the first shipborne instrument developed by F. A. Vening Meinesz of The Netherlands in 1928, and soon gravity measurements were being recorded worldwide. The Hungarian geodesist, Roland von Eötvös, constructed the first torsional balance in 1910. Many gravity meters of various types were developed and patented during 1928 to 1930 as U.S. oil companies became interested in exploration applications. Most modern instruments suitable for field studies, such as the LaCoste and Romberg gravity meter and the Worden instrument, involve astatic principles in measuring the vertical displacement of a small mass suspended from a system of delicate springs and beams. Various models of the LaCoste and Romberg gravity meter are commonly used in land-based and shipborne studies and, more recently, in airborne surveys (e.g., Brozena and Peters [43]).

The application of gravity measurements to geological problems can be traced back to the rival hypotheses of John Pratt and George Airy published between 1855 and 1859 concerning the isostatic support of topography. They noted that plumb lines near the Himalayas were deflected from the vertical by amounts less than predicted by the topographic mass of the mountain range. Both Airy and Pratt argued that in the absence of forces other than gravity, the rigid part of the crust and mantle "floats" on a mobile, denser substratum, so the total mass in any vertical column down to some depth of compensation must balance from place to place. Elevated regions, therefore, must be compensated at depth by mass deficiencies, whereas topographic depressions are underlain by mass excesses. Pratt explained this observation in terms of lateral variations in density; that is, the Himalayas are elevated because they are less dense than surrounding crust. Airy proposed, on the other hand, that the crust has laterally uniform density but variable thickness, so mountain ranges rise above the surrounding landscape by virtue of underlying crustal roots.

The gravity method also has played a key role in exploration geophysics. Hugo V. Boeckh used an Eötvös balance to measure gravity over anticlines and domes and explained his observations in terms of the densities of rocks that form the structures. He thus was apparently the first to recognize the application of the gravity method in the exploration for petroleum (Jakosky [140]). Indeed the first oil discovered in the United States by geophysical methods was located in 1926 using gravity measurements (Jakosky [140]).

About This Book

Considering this long and august history of the gravity and magnetic methods, it might well be asked (as I certainly have done during the waning stages of this writing) why a new textbook on potential theory is needed now. I believe, however, that this book will fill a significant gap. As a graduate student at Stanford University, I quickly found myself involved in a thesis topic that required a firm foundation in potential theory. It seemed to me then, and I find it true today as a professional geophysicist, that no single textbook is available covering the topic of potential theory while emphasizing applications to geophysical problems. The classic texts on potential theory published during the middle of this century are still available today, notably those by Kellogg [146] and by Ramsey [235] (which no serious student of potential theory should be without). These books deal thoroughly with the fundamentals of potential theory, but they are not concerned particularly with geophysical applications. On the other hand, several good texts are available on the broad topics of applied geophysics (e.g., Telford, Geldart, and Sheriff [279]) and global geophysics (e.g., Stacey [270]). These books cover the wide range of geophysical methodologies, such as seismology, electromagnetism, and so forth, and typically devote a few chapters to gravity and magnetic methods; of necessity they do not delve deeply into the underlying theory.

This book attempts to fill the gap by first exploring the principles of potential theory and then applying the theory to problems of crustal and lithospheric geophysics. I have attempted to do this by structuring the book into essentially two parts. The first six chapters build the foundations of potential theory, relying heavily on Kellogg [146], Ramsey [235], and Chapman and Bartels [56]. Chapters 1 and 2 define the meaning of a potential and the consequences of Laplace's equation. Special attention is given therein to the all-important Green's identities, Green's functions, and Helmholtz theorem. Chapter 3 focuses these theoretical principles on Newtonian potential, that is, the gravitational potential of mass distributions in both two and three dimensions. Chapters 4 and 5 expand these discussions to magnetic fields caused by distributions of magnetic media. Chapter 6 then formulates the theory on a spherical surface, a topic of obvious importance to global representations of the earth's gravity and magnetic fields.

The last six chapters apply the foregoing principles of potential theory to gravity and magnetic studies of the crust and lithosphere. Chapters 7 and 8 examine the gravity and magnetic fields of the earth on a global and regional scale and describe the calculations and underlying theory by which measurements are transformed into "anomalies." These discussions set the stage for the remaining chapters, which provide a sampling of the myriad schemes in the literature for interpreting gravity and magnetic anomalies. These schemes are divided into the forward method (Chapter 9), the inverse method (Chapter 10), inverse and forward manipulations in the Fourier domain (Chapter 11), and methods of data enhancement (Chapter 12). Here I have concentrated on the mathematical rather than the technical side of the methodology, neglecting such topics as the nuts-and-bolts operations of gravity meters and magnetometers and the proper strategies in designing gravity or magnetic surveys.

Some of the methods discussed in Chapters 9 through 12 are accompanied by computer subroutines in Appendix B. I am responsible for the programming therein (user beware), but the methodologies behind the algorithms are from the literature. They include some of the "classic" techniques, such as the so-called Talwani method discussed in Chapter 9, and several more modern methods, such as the horizontal-gradient calculation first discussed by Cordell [66]. Those readers wishing to make use of these subroutines should remember that the programming is designed to instruct rather than to be particularly efficient or "elegant."

It would be quite beyond the scope of this or any other text to fully describe all of the methodologies published in the modern geophysical literature. During 1992 alone, *Geophysics* (the technical journal of the U.S.-based Society of Exploration Geophysicists) published 17 papers that arguably should have been covered in Chapters 9 through 12. Multiply that number by the several dozen international journals of similar stature and then times the 50 some-odd years that the modern methodology has been actively discussed in the literature, and it becomes clear that each technique could not be given its due. Instead, my approach has been to describe the various methodologies with key examples from the literature, including both classic algorithms and promising new techniques, and with apologies to all of my colleagues not sufficiently cited!

Acknowledgments

The seeds of this book began in graduate-level classes that I prepared and taught at Oregon State University and Stanford University between 1973 and 1990. The final scope of the book, however, is partly a reflection of interactions and discussions with many friends and colleagues. Foremost are my former professors at Stanford University during my graduate studies, especially Allan Cox, George Thompson, and Jon Claerbout, who introduced me to geological applications of potential theory and time-series analysis. My colleagues at the U.S. Geological Survey, Stanford University, Oregon State University, and elsewhere were always available for discussions and fomentation, especially Robert Jachens, Robert Simpson, Thomas Hildenbrand, Richard Saltus, Andrew Griscom, V. J. S. Grauch, Gerald Connard, Gordon Ness, and Michael McWilliams. I am grateful to Richard Saltus and Gregory Schreiber for carefully checking and critiquing all chapters, and to William Hinze, Tiki Ravat, Robert Langel, and Robert Jachens for reviewing and proofreading various parts of early versions of this manuscript. I am especially grateful to Lauren Cowles, my chief contact and editor at Cambridge University Press, for her patience, assistance, and flexible deadlines.

Finally, but at the top of the list, I thank my wife, Diane, and children, Tammy and Jason, for their unwavering support and encouragement, not just during the writing of this book, but throughout my career. This book is dedicated to Diane, who could care less about geophysics but always recognized its importance to me.

Richard J. Blakely

1

The Potential

Laplace's equation is the most famous and most universal of all partial differential equations. No other single equation has so many deep and diverse mathematical relationships and physical applications.

(G. F. D. Duff and D. Naylor)

Every arrow that flies feels the attraction of the earth.

(Henry Wadsworth Longfellow)

Two events in the history of science were of particular significance to the discussions throughout this book. In 1687, Isaac Newton put forth the Universal Law of Gravitation: Each particle of matter in the universe attracts all others with a force directly proportional to its mass and inversely proportional to the square of its distance of separation. Nearly a century later, Pierre Simon, Marquis de Laplace, showed that gravitational attraction obeys a simple differential equation, an equation that now bears his name. These two hallmarks have subsequently developed into a body of mathematics called *potential theory* that describes not only gravitational attraction but also a large class of phenomena, including magnetostatic and electrostatic fields, fields generated by uniform electrical currents, steady transfer of heat through homogeneous media, steady flow of ideal fluids, the behavior of elastic solids, probability density in random-walk problems, unsteady water-wave motion, and the theory of complex functions and conformal mapping.

The first few chapters of this book describe some general aspects of potential theory of most interest to practical geophysics. This chapter defines the meaning of a potential field and how it relates to Laplace's equation. Chapter 2 will delve into some of the consequences of this relationship, and Chapters 3, 4, and 5 will apply the principles of potential theory to gravity and magnetic fields specifically. Readers finding

1

these treatments too casual are referred to textbooks by Ramsey [235], Kellogg [146], and MacMillan [172].

1.1 Potential Fields

A few definitions are needed at the outset. We begin by building an understanding of the general term *field* and, more specifically, *potential field*. The cartesian coordinate system will be used in the following development, but any orthogonal coordinate system would provide the same results. Appendix A describes the vector notation employed throughout this text.

1.1.1 Fields

A *field* is a set of functions of space and time. We will be concerned primarily with two kinds of fields. *Material fields* describe some physical property of a material at each point of the material and at a given time. Density, porosity, magnetization, and temperature are examples of material fields. A *force field* describes the forces that act at each point of space at a given time. The gravitational attraction of the earth and the magnetic field induced by electrical currents are examples of force fields.

Fields also can be classed as either *scalar* or *vector*. A scalar field is a single function of space and time; displacement of a stretched string, temperature of a volume of gas, and density within a volume of rock are scalar fields. A vector field, such as flow of heat, velocity of a fluid, and gravitational attraction, must be characterized by three functions of space and time, namely, the components of the field in three orthogonal directions.

Gravitational and magnetic attraction will be the principal focus of later chapters. Both are vector fields, of course, but geophysical instruments generally measure just one component of the vector, and that single component constitutes a scalar field. In later discussions, we often will drop the distinction between scalar and vector fields. For example, gravity meters used in geophysical surveys measure the vertical component g_z (a scalar field) of the acceleration of gravity $\mathbf{g}$ (a vector field), but we will apply the word "field" to both $\mathbf{g}$ and g_z interchangeably.

A vector field can be characterized by its *field lines* (also known as *lines of flow* or *lines of force*), lines that are tangent at every point to the vector field. Small displacements along a field line must have x, y, and z

components proportional to the corresponding x, y, and z components of the field at the point of its displacement. Hence, if $\mathbf{F}$ is a continuous vector field, its field lines are described by integration of the differential equation

$$\frac{dx}{F_x} = \frac{dy}{F_y} = \frac{dz}{F_z}. \tag{1.1}$$

Exercise 1.1 We will find in Chapter 3 that the gravitational attraction of a uniform sphere of mass M, centered at point Q, and observed outside the sphere at point P is given by $\mathbf{g} = -\gamma M \hat{\mathbf{r}}/r^2$, where γ is a constant, r is the distance from Q to P, and $\hat{\mathbf{r}}$ is a unit vector directed from Q to P. Let Q be at the origin and use equation 1.1 to describe the gravitational field lines at each point outside of the sphere.

1.1.2 Points, Boundaries, and Regions

Regions and points are also part of the language of potential theory, and precise definitions are necessary for future discussions. A *set of points* refers to a group of points in space satisfying some condition. Generally, we will be dealing with *infinite* sets, sets that consist of a continuum of points which are infinite in number even though the entire set may fit within a finite volume. For example, if r represents the distance from some point Q, the condition that $r \leq 1$ describes an infinite set of points inside and on the surface of a sphere of unit radius. A set of points is *bounded* if all points of the set fit within a sphere of finite radius.

Consider a set of points ξ. A point P is said to be a *limit point* of ξ if every sphere centered about P contains at least one point of ξ other than P itself. A limit point does not necessarily belong to the set. For example, all points satisfying $r \leq 1$ are limit points of the set of points satisfying $r < 1$. A point P is an *interior point* of ξ if some sphere about P contains only points of ξ. Similarly, P is an *exterior point* of ξ if a sphere exists centered about P that contains no points of ξ.

The *boundary* of ξ consists of all limit points that are not interior to ξ. For example, any point satisfying $r = 1$ lies on the boundary of the set of points satisfying $r < 1$. A *frontier point* of ξ is a point that, although not an exterior point, is nevertheless a limit point of all exterior points. The set of all frontier points is the *frontier* of ξ. The distinction between boundary and frontier is a fine one but will be an issue in one derivation in Chapter 2.

A set of points is *closed* if it contains all of its limit points and *open* if it contains only interior points. Hence, the set of points described by

$r \leq 1$ is closed, whereas the set of points $r < 1$ is open. A *domain* is an open set of points such that any two points of the set can be connected by a finite set of connected line segments composed entirely of interior points. A *region* is a domain with or without some part of its boundary, and a *closed region* is a region that includes its entire boundary.

1.2 Energy, Work, and the Potential

Consider a test particle under the influence of a force field **F** (Figure 1.1). The test particle could be a small mass m acted upon by the gravitational field of some larger body or an electric charge moving under the influence of an electric field. Such physical associations are not considered until later chapters, so the present discussion is restricted to general force and energy relationships.

The kinetic energy expended by the force field in moving the particle from one point to another is defined as the *work* done by the force field. Newton's second law of motion requires that the momentum of the particle at any instant must change at a rate proportional to the magnitude of the force field and in a direction parallel to the direction taken by the force field at the location of the particle; that is,

$$\lambda \mathbf{F} = m \frac{d}{dt} \mathbf{v} \,, \tag{1.2}$$

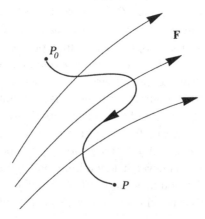

Fig. 1.1. While under the influence of force field **F**, a particle of mass m leaves point P_0 at time t_0 and moves by an arbitrary route to point P, arriving at time t.

where λ is a constant that depends on the units used, and $\mathbf{v}$ is the velocity of the particle. We select units in order to make $\lambda = 1$ and multiply both sides of equation 1.2 by $\mathbf{v}$ to obtain

$$\mathbf{F} \cdot \mathbf{v} = \frac{1}{2} m \frac{d}{dt} v^2$$

$$= \frac{d}{dt} E \,, \tag{1.3}$$

where E is the kinetic energy of the particle. If the particle moves from point P_0 to P during time interval t_0 to t (Figure 1.1), then the change in kinetic energy is given by integration of equation 1.3 over the time interval,

$$E - E_0 = \int_{t_0}^{t} \mathbf{F} \cdot \mathbf{v} \, dt'$$

$$= \int_{P_0}^{P} \mathbf{F} \cdot \mathbf{ds}$$

$$= W(P, P_0) \,, \tag{1.4}$$

where $\mathbf{ds}$ represents elemental displacement of the particle. The quantity $W(P, P_0)$ is the work required to move the particle from point P_0 to P. Equation 1.4 shows that the change in kinetic energy of the particle equals the work done by $\mathbf{F}$.

In general, the work required to move the particle from P_0 to P differs depending on the path taken by the particle. A vector field is said to be *conservative* in the special case that work is independent of the path of the particle. We assume now that the field is conservative and move the particle an additional small distance Δx parallel to the x axis, as shown in Figure 1.2. Then

$$W(P, P_0) + W(P + \Delta x, P) = W(P + \Delta x, P_0) \,,$$

and rearranging terms yields

$$W(P + \Delta x, P_0) - W(P, P_0) = W(P + \Delta x, P)$$

$$= \int_{P}^{P + \Delta x} F_x(x, y, z) \, dx \,.$$

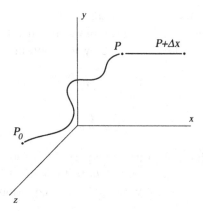

Fig. 1.2. A particle of mass travels through a conservative field; the particle moves first from P_0 to P, then parallel to the x axis an additional distance Δx.

The integral can be solved by dividing both sides of the equation by Δx and applying the law of the mean,

$$\frac{W(P + \Delta x, P_0) - W(P, P_0)}{\Delta x} = F_x(x + \epsilon\Delta x, y, z),$$

where $0 < \epsilon < 1$. As Δx becomes arbitrarily small, we have

$$\frac{\partial W}{\partial x} = F_x. \tag{1.5}$$

We can repeat this derivation for the y and z directions, multiply each equation by appropriate unit vectors, and add them to equation 1.5 to obtain

$$\mathbf{F}(x, y, z) = \left(\frac{\partial W}{\partial x}, \frac{\partial W}{\partial y}, \frac{\partial W}{\partial z}\right)$$

$$= \nabla W. \tag{1.6}$$

Hence, *the derivative of the work in any direction is equal to the component of force in that direction.* The vector force field $\mathbf{F}$ is completely specified by the scalar field W, which we call the *work function* of $\mathbf{F}$ (Kellogg [146]).

We have shown, therefore, that a conservative field is given by the gradient of its work function. With equations 1.4 and 1.6, we also can show the converse relationship. If the work function W has continuous derivatives, then we can integrate equation 1.6 as follows:

$$W(P, P_0) = \int_{P_0}^{P} \mathbf{F} \cdot \mathbf{ds}$$

$$= \int_{P_0}^{P} \left(\frac{\partial W}{\partial x} dx + \frac{\partial W}{\partial y} dy + \frac{\partial W}{\partial z} dz \right)$$

$$= \int_{P_0}^{P} dW$$

$$= W(P) - W(P_0). \tag{1.7}$$

Hence, the work depends only on the values of W at endpoints P and P_0, not on the path taken, and this is precisely the definition of a conservative field. Consequently, *any vector field that has a work function with continuous derivatives as described in equation 1.6 is conservative.* A corollary to equation 1.7 results if the path of the particle is a closed loop. Then P equals P_0, $W(P, P_0) = 0$, and no net work is required to move the particle around the closed loop.

The *potential* ϕ of vector field $\mathbf{F}$ is defined as the work function or as its negative depending on the convention used. Kellogg [146] summarizes these conventions as follows: If particles of like sign attract each other (e.g., gravity fields), then $\mathbf{F} = \nabla \phi$ and the potential equals the work done by the field. If particles of like sign repel each other (e.g., electrostatic fields), then $\mathbf{F} = -\nabla \phi$, and the potential equals the work done against the field by the particle. In the latter case, the potential ϕ is the potential energy of the particle; in the former case, ϕ is the negative of the particle's potential energy.

Note that any constant can be added to ϕ without changing the important result that

$$\mathbf{F} = \nabla \phi.$$

This constant is chosen generally so that ϕ approaches 0 at infinity. In other words, the potential at point P is given by

$$\phi(P) = \int_{\infty}^{P} \mathbf{F} \cdot \mathbf{ds}. \tag{1.8}$$

The value of the potential at a specific point, therefore, is not nearly so important as the *difference* in potential between two separated points.

1.2.1 Equipotential Surfaces

As its name implies, an *equipotential surface* is a surface on which the potential remains constant; that is,

$$\phi(x, y, z) = \text{constant}.$$

If $\hat{s}$ is a unit vector lying tangent to an equipotential surface of $\mathbf{F}$, then $\hat{s} \cdot \mathbf{F} = \frac{\partial \phi}{\partial s}$ at any point and must vanish according to the definition of an equipotential surface. It follows that field lines at any point are always perpendicular to their equipotential surfaces and, conversely, any surface that is everywhere perpendicular to all field lines must be an equipotential surface. Hence, no work is done in moving a test particle along an equipotential surface. Only one equipotential surface can exist at any point in space. The distance between equipotential surfaces is a measure of the density of field lines; that is, a force field will have greatest intensity in regions where its equipotential surfaces are separated by smallest distances.

Exercise 1.2 Prove that equipotential surfaces never intersect.

1.3 Harmonic Functions

To summarize the previous section, a conservative field $\mathbf{F}$ has a scalar potential ϕ given by $\mathbf{F} = \nabla\phi$ (or $\mathbf{F} = -\nabla\phi$, depending on sign convention). Moreover, if $\mathbf{F} = \nabla\phi$, then $\mathbf{F}$ is conservative and is said to be a potential field. In the following we discuss another property of potential fields: The potential ϕ of field $\mathbf{F}$, under certain conditions to be discussed in Chapter 2, satisfies an important second-order differential equation called *Laplace's equation*,

$$\nabla^2\phi = 0, \qquad (1.9)$$

at points not occupied by sources of $\mathbf{F}$. Several surprising and illustrative results follow from this statement. We start by discussing the physical meaning of Laplace's equation, first with the trivial one-dimensional case and then the general equation.

1.3.1 Laplace's Equation

Consider a stretched rubber band subject to a static force directed in the y direction, as shown in Figure 1.3. The displacement ϕ of the rubber band in the y direction is described by the differential equation

$$\alpha \frac{d^2 \phi}{dx^2} = -F(x),$$

where α is a constant and $F(x)$ is the force in the y direction per unit length in the x direction. If $F(x) = 0$, the rubber band lies along a straight line, and

$$\frac{d^2 \phi}{dx^2} = 0.$$

The second-order derivative of a function is a measure of the function's curvature, and the previous equation illustrates the obvious result: The stretched rubber band has no curvature in the absence of external forces. This is simply the one-dimensional case of Laplace's equation, but it illustrates an important property of harmonic functions that will extend to two- and three-dimensional cases. Laplace's equation is not satisfied along any part of the band containing a local minimum or maximum. Indeed, if $\phi(x)$ in this example is to satisfy Laplace's equation, the maximum and minimum displacements must occur at the two end points of the rubber band, as shown in Figure 1.3.

Now consider a membrane stretched over an uneven frame, such as

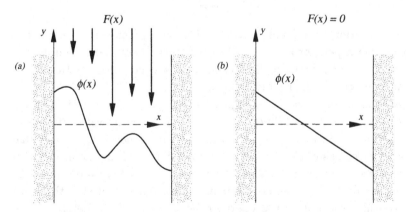

Fig. 1.3. Displacement in the y direction of a stretched rubber band due to an applied force $F(x)$. (a) Static force $F(x)$ is nonzero and varies along the x axis. (b) $F(x) = 0$ so that $\phi(x)$ has no maxima or minima except at the ends of the band.

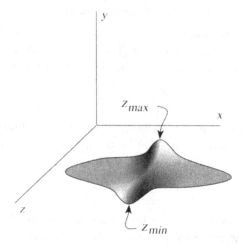

Fig. 1.4. Stretched membrane attached to an uneven loop of wire. Note that the membrane reaches maximum and minimum values of z at the wire.

might be formed by a thin film of soap spread across a twisted loop of wire (Figure 1.4). Let $\phi(x, y)$ represent the displacement of the membrane in the z direction. In the absence of external forces, the displacement of the membrane satisfies Laplace's equation in two dimensions,

$$\frac{\partial^2 \phi}{\partial x^2} + \frac{\partial^2 \phi}{\partial y^2} = 0 \,.$$

This condition would not be satisfied at any point of the membrane containing a peak or a trough. Hence, Laplace's equation requires that maximum and minimum displacements can occur only on the frame, that is, on the boundary of the membrane.

As a three-dimensional example, let $\phi(x, y, z)$ represent the concentration of a solute in a fluid, such as salt dissolved in water. If the salt is concentrated at some point within the fluid, then $\nabla^2 \phi \neq 0$ at that point. In regions of the fluid where $\nabla^2 \phi = 0$, the salt apparently has arranged itself so that no localized zones of excess concentration (maxima) or excess dilution (minima) occur in the region. It is useful, therefore, to consider the differential operator ∇^2 as a means to determine the variations in the concentration of a distribution; if $\nabla^2 \phi = 0$ throughout a region, then ϕ at each point of the region is never more (or less) concentrated than all surrounding parts of the region.

With these preliminary remarks, we define a *harmonic function* as any function that (1) satisfies Laplace's equation; (2) has continuous, single-valued first derivatives; and (3) has second derivatives. We might expect from the previous examples (and soon will prove) that a function that is harmonic throughout a region R must have all maxima and minima on the boundary of R and none within R itself. The converse is not necessarily true, of course; a function with all maxima and minima on its boundary is not necessarily harmonic because it may not satisfy the three criteria listed.

The definition of the second derivative of a one-dimensional function demonstrates another important property of a harmonic function. The second derivative of $\phi(x)$ is given by

$$\lim_{\Delta x \to 0} \frac{1}{\Delta x^2} \left\{ \phi(x) - \frac{1}{2} \left[\phi(x - \Delta x) + \phi(x + \Delta x) \right] \right\} = -\frac{1}{2} \frac{d^2 \phi}{dx^2} \, .$$

If ϕ satisfies the one-dimensional case of Laplace's equation, then the right-hand side of the previous equation vanishes and

$$\phi(x) = \frac{1}{2} \lim_{\Delta x \to 0} \left[\phi(x - \Delta x) + \phi(x + \Delta x) \right] \, .$$

Hence, the value of a harmonic function ϕ at any point is the average of ϕ at its neighboring points. This is simply another way of stating the now familiar property of a potential: A function can have no maxima or minima within a region in which it is harmonic. We will discuss a more rigorous proof of this statement in Chapter 2.

1.3.2 An Example from Steady-State Heat Flow

Consider a temperature distribution T specified throughout some region R of a homogeneous material. All heat sources and sinks are restricted from the region. According to *Fourier's law*, the flow of heat $\mathbf{J}$ at any point of R is proportional to the change in temperature at that point, that is,

$$\mathbf{J} = -k\nabla T \, , \qquad (1.10)$$

where k is thermal conductivity, a constant of the medium. Equation 1.10 tells us, on the basis of the discussion in Section 1.2, that temperature is a potential, and flow of heat $\mathbf{J}$ is a potential field analogous to a force field. Field lines for $\mathbf{J}$ describe the pattern and direction of heat transfer,

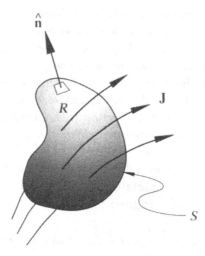

Fig. 1.5. Heat flow **J** through a region R containing no heat sources or sinks. Region R is bounded by surface S, and $\hat{\mathbf{n}}$ is the unit vector normal to S.

in the same sense that gravitational field lines describe the gravitational forces acting upon a particle of mass.

Consider the free flow of heat in and out of a region R bounded by surface S, as shown by Figure 1.5. The total heat in region R is given by

$$H = c\rho \int_R T\, dv\,, \tag{1.11}$$

where c is the specific heat and ρ is the density of the material. The change in total heat within R must equal the net flow of heat across boundary S; that is,

$$\frac{dH}{dt} = -\int_S \mathbf{J}\cdot\hat{\mathbf{n}}\, dS$$

$$= k\int_S \nabla T\cdot\hat{\mathbf{n}}\, dS\,,$$

where $\hat{\mathbf{n}}$ is the unit vector normal to S. The divergence theorem (Appendix A) can be used to convert the surface integral into a volume integral; that is,

$$\frac{dH}{dt} = k \int_R \nabla \cdot \nabla T \, dv$$

$$= k \int_R \nabla^2 T \, dv \,. \tag{1.12}$$

From equation 1.11, the change in heat over time is also given by

$$\frac{dH}{dt} = c\rho \int_R \frac{\partial T}{\partial t} dv \,, \tag{1.13}$$

and combining equations 1.13 and 1.12 provides

$$\int_R \left(c\rho \frac{\partial T}{\partial t} - k\nabla^2 T \right) dv = 0 \,. \tag{1.14}$$

This integral vanishes for every choice of the region R. If the integrand, which we assume is continuous, is not zero throughout R, then we could choose some portion of R so as to contradict equation 1.14. Hence, the integrand itself must be zero throughout R, or

$$\kappa\nabla^2 T = \frac{\partial T}{\partial t} \,, \tag{1.15}$$

where $\kappa = k/c\rho$ is thermal diffusivity. Equation 1.15 is the equation of conductive heat transfer. If all heat sources and sinks lie outside of region R and do not change with time, then steady-state conditions eventually will be obtained and equation 1.15 becomes

$$\nabla^2 T = 0 \tag{1.16}$$

throughout R. Hence, temperature under steady-state conditions satisfies Laplace's equation and is harmonic.

The temperature distribution accompanying steady-state transfer of heat is an easily visualized example of a harmonic function, one which clarifies some of the theoretical results discussed earlier. For example, imagine a volume of rock with no internal heat sources or heat sinks (Figure 1.5) and in steady-state condition. On the basis of previous discussions, we can state a number of characteristics of the temperature distribution within the volume of rock. (1) We conclude that the temperature within the volume of rock cannot reach any maximum or minimum values. This is a reasonable result; if temperature is in a steady-state condition, all maxima and minima should occur at heat sources and sinks, and these have been restricted from this part of the rock. (2) Furthermore, maximum and minimum temperatures must occur on the boundary of the volume and not within the volume. After all, *some* point of the boundary must be closer to any external heat sources (or sinks) than all interior points; likewise, *some* point of the boundary will be farther from any external heat source (or sink) than all interior points. (3) It also is reasonable that the temperature at any point is the average of the temperatures in a small region around that point.

Exercise 1.3 One end of a glass rod is kept in boiling water and the opposite end in ice water until the temperature of the rod reaches equilibrium. Suddenly the two ends are switched so that the hot end is in ice water and the cold end is in boiling water. Describe how the temperature of the rod changes with time. Is the temperature harmonic?

Finding a solution to Laplace's equation, if indeed one exists, is a boundary-value problem of, in this case, the *Dirichlet type*; that is, find a representation for ϕ throughout a region R, given that $\nabla^2 \phi = 0$ within R and given specified values of ϕ on the surface that bounds R. For example, the steady-state temperature can be calculated, in principle, throughout a spherically shaped region of homogeneous matter by solving equation 1.16 subject to specified boundary conditions. We will have considerably more to say about this subject in later chapters.

1.3.3 Complex Harmonic Functions

This section provides a very brief review of complex functions sufficient to draw one important conclusion: The real and imaginary parts of a complex function are harmonic in regions where the complex function is analytic. For additional information about complex functions, the interested reader is referred to the textbook by Churchill [59].

First we need some definitions. In the following, x and y are real variables describing a two-dimensional cartesian coordinate system. The coordinate system represents the complex plane, and any point of the plane is identified by the complex number $z = x + iy$, where $i = \sqrt{-1}$. As discussed in Section 1.1.2 for general cases, we can consider *sets of points* of the complex plane. A *neighborhood* of a point z_0 of the complex plane is the set of all points such that $|z - z_0| < \epsilon$, where ϵ is a positive constant. An *interior point* of a set of points has some neighborhood containing only points of the set. Sets that contain only interior points are called *open regions*. Open, connected regions of the complex plane are called *domains*.

If a complex number w is prescribed for each value of a set of complex numbers z, usually a domain, then w is a complex function of the complex variable z, written $w(z)$. Complex functions can be written in terms of their real and imaginary parts,

$$w(z) = u(z) + iv(z),$$

where $u(z)$ and $v(z)$ are real functions of the complex variable z. For example, if

$$w(z) = z^2$$
$$= x^2 - y^2 + 2ixy,$$

then $u(x, y) = x^2 - y^2$ and $v(x, y) = 2xy$, and the domain of definition in this case is the entire complex plane.

The derivative of a complex function requires special consideration. In order for a real function $f(x)$ to have a derivative, the ratio of the change in f to a change in x, $\Delta f / \Delta x$, must have a limit as Δx approaches 0. Similarly, for the derivative of complex function $w(z)$ to exist in a domain, it is necessary that the ratio

$$r(z) = \lim_{\Delta z \to 0} \frac{\Delta w}{\Delta z},$$

have a limit. In the complex plane, however, there are different paths along which Δz can approach zero, and the value of the ratio $r(z)$ may depend on that path. For example, the complex function $w(z) = z^2$ has $r(z) = 2(x_0 + iy_0)$ at point (x_0, y_0), independent of how Δz approaches zero, whereas the complex function $w(z) = x^2 + y^2 - i\,2xy$ is dependent on the path taken as $\Delta z \to 0$. In this latter case, the ratio has no limit and the derivative does not exist.

A function $w(z) = u(x, y) + iv(x, y)$ is said to be *analytic* in a domain of the complex plane if the real functions $u(x, y)$ and $v(x, y)$ have continuous partial derivatives and if $w(z)$ has a derivative with respect to z at every point of the domain. The *Cauchy–Riemann conditions* provide an easy way to determine whether such conditions are met. If $u(x, y)$ and $v(x, y)$ have continuous derivatives of first order, then the Cauchy–Riemann conditions,

$$\frac{\partial u}{\partial x} = \frac{\partial v}{\partial y}, \tag{1.17}$$

$$\frac{\partial u}{\partial y} = -\frac{\partial v}{\partial x}. \tag{1.18}$$

are necessary and sufficient conditions for the analyticity of $w(z)$. The derivative of the complex function is given by

$$\frac{dw}{dz} = \frac{\partial u}{\partial x} + i\frac{\partial v}{\partial x}$$

$$= \frac{\partial v}{\partial y} - i\frac{\partial u}{\partial y}.$$

Now consider a complex function analytic within some domain T. The two-dimensional Laplacian of its real part $u(x, y)$ is given by

$$\nabla^2 u = \frac{\partial^2 u}{\partial x^2} + \frac{\partial^2 u}{\partial y^2}.$$

The Cauchy–Riemann conditions are applicable here because $w(z)$ is analytic; hence, we assume that derivatives of second order exist and employ the Cauchy–Riemann conditions to get

$$\nabla^2 u = \frac{\partial^2 v}{\partial x \partial y} - \frac{\partial^2 v}{\partial x \partial y}$$

$$= 0.$$

Consequently, the real part of a complex function satisfies the two-dimensional case of Laplace's equation in domains in which the function is analytic, and since the necessary derivatives exist, the real part of $w(z)$ must be harmonic. Hence, *if a complex function is analytic in domain T, it has a real part that is harmonic in T. Likewise, it can be shown that the imaginary part of an analytic complex function also is harmonic in domains of analyticity.*

Exercise 1.4 Use the Cauchy-Riemann conditions to show that the imaginary part of an analytic function satisfies Laplace's equation.

Conversely, if u is harmonic in T, there must exist a function v such that $u + iv$ is analytic in T, and v is given by

$$v = \int_{z_0}^{z} \left[-\frac{\partial u}{\partial y} dx + \frac{\partial u}{\partial x} dy \right].$$

We will have occasion later in this text to use these rather abstract properties of complex numbers in some practical geophysical applications.

1.4 Problem Set

1. The potential of $\mathbf{F}$ is given by $(x^2 + y^2)^{-1}$.

   (a) Find $\mathbf{F}$.
   (b) Describe the field lines of $\mathbf{F}$.
   (c) Describe the equipotential surfaces of $\mathbf{F}$.
   (d) Demonstrate by integration around the perimeter of a rectangle in the x, y plane that $\mathbf{F}$ is conservative. Let the rectangle extend from x_1 to x_2 in the x direction and from y_1 to y_2 in the y direction, and let $x_1 > 0$.

2. Prove that the intensity of a conservative force field is inversely proportional to the distance between its equipotential surfaces.

3. If all mass lies interior to a closed equipotential surface S on which the potential takes the value C, prove that in all space outside of S the value of the potential is between C and 0.

4. If the lines of force traversing a certain region are parallel, what may be inferred about the intensity of the force within the region?

5. Two distributions of matter lie entirely within a common closed equipotential surface C. Show that all equipotential surfaces outside of C also are common.

6. For what integer values of n is the function $(x^2 + y^2 + z^2)^{\frac{n}{2}}$ harmonic?

7. You are monitoring the magnetometer aboard an interstellar spacecraft and discover that the ship is approaching a magnetic source described by

$$\mathbf{B} = \frac{\hat{\mathbf{r}}}{r}.$$

   (a) Remembering Maxwell's equation for $\mathbf{B}$, will you report to Mission Control that the magnetometer is malfunctioning, or is this a possible source?
   (b) What if the magnetometer indicates that $\mathbf{B}$ is described by

$$\mathbf{B} = \frac{\hat{\mathbf{r}}}{r^2} ?$$

8. The physical properties of a spherical body are homogeneous. Describe the temperature at all points of the sphere if the temperature is harmonic throughout the sphere and depends only on the distance from its center.

9. As a crude approximation, the temperature of the interior of the earth depends only on distance from the center of the earth. Based

on the results of the previous exercise, would you expect the temperature of the earth to be harmonic everywhere inside? Explain your
answer?

10. Assume a spherical coordinate system and let $\mathbf{r}$ be a vector directed
from the origin to a point P with magnitude equal to the distance
from the origin to P. Prove the following relationships:

$$\nabla \cdot \mathbf{r} = 3,$$

$$\nabla r = \frac{\mathbf{r}}{r},$$

$$\nabla \cdot \left(\frac{\mathbf{r}}{r^3}\right) = 0, \quad r \neq 0,$$

$$\nabla \frac{1}{r} = -\frac{\mathbf{r}}{r^3}, \quad r \neq 0,$$

$$\nabla \times \mathbf{r} = 0,$$

$$\mathbf{A} \cdot \nabla \frac{1}{r} = -\frac{\mathbf{A} \cdot \mathbf{r}}{r^3}, \quad r \neq 0,$$

$$(\mathbf{A} \cdot \nabla) \mathbf{r} = A_r \frac{\mathbf{r}}{r}.$$

2

Consequences of the Potential

It may be no surprise that human minds can deduce the laws of
falling objects because the brain has evolved to devise strategies
for dodging them.

(Paul Davies)

Only mathematics and mathematical logic can say as little as the
physicist means to say.

(Bertrand Russell)

In Chapter 1, we learned that a conservative vector field $\mathbf{F}$ can be expressed as the gradient of a scalar ϕ, called the potential of $\mathbf{F}$, and
conversely $\mathbf{F}$ is conservative if $\mathbf{F} = \nabla\phi$. It was asserted that such potentials satisfy Laplace's equation at places free of all sources of $\mathbf{F}$ and
are said to be harmonic. This led to several important characteristics of
the potential. In the same spirit, this chapter investigates a number of
additional consequences that follow from Laplace's equation.

2.1 Green's Identities

Three identities can be derived from vector calculus and Laplace's equation, and these lead to several important theorems and additional insight into the nature of potential fields. They are referred to as *Green's
identities.*†

---

† The name Green, appearing repeatedly in this and subsequent chapters, refers to
George Green (1793–1841), a British mathematician of Caius College, Cambridge,
England. He is perhaps best known for his paper, *Essay on the Application of
Mathematical Analysis to the Theory of Electricity and Magnetism,* and was ap-
parently the first to use the term "potential."

2.1.1 Green's First Identity

Green's first identity is derived from the divergence theorem (Appendix A). Let U and V be continuous functions with continuous partial derivatives of first order throughout a closed, regular region R, and let U have continuous partial derivatives of second order in R. The boundary of R is surface S, and $\hat{\mathbf{n}}$ is the outward normal to S. If $\mathbf{A} = V\nabla U$, then

$$\int_R \nabla \cdot \mathbf{A}\, dv = \int_R \nabla \cdot (V\nabla U)\, dv$$

$$= \int_R [\nabla V \cdot \nabla U + V\nabla^2 U]dv\,.$$

Using the divergence theorem yields

$$\int_R [\nabla V \cdot \nabla U + V\nabla^2 U]dS = \int_S \mathbf{A} \cdot \hat{\mathbf{n}}\, Sv$$

$$= \int_S V\nabla U \cdot \hat{\mathbf{n}}\, dS$$

$$= \int_S V\frac{\partial U}{\partial n}\, dS,$$

that is,

$$\int_R V\nabla^2 U\, dv + \int_R \nabla U \cdot \nabla V\, dv = \int_S V\frac{\partial U}{\partial n}\, dS\,. \qquad (2.1)$$

Equation 2.1 is *Green's first identity* and is true for all functions U and V that satisfy the differentiability requirements stated earlier.

Several very interesting theorems result from Green's first identity if U and V are restricted a bit further. For example, if U is harmonic and continuously differentiable in R, and if $V = 1$, then $\nabla^2 U = 0$, $\nabla V = 0$, and equation 2.1 becomes

$$\int_S \frac{\partial U}{\partial n}\, dS = 0\,. \qquad (2.2)$$

Thus *the normal derivative of a harmonic function must average to zero on any closed boundary surrounding a region throughout which the function is harmonic and continuously differentiable* (Figure 2.1). It also can be shown (Kellogg [146, p. 227]) that the converse of equation 2.2 is

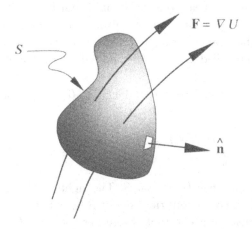

$$\mathbf{F} = \nabla U$$

Fig. 2.1. Region R subject to force field $\mathbf{F}$. Surface S bounds region R. Unit vector $\hat{\mathbf{n}}$ is outward normal at any point on S.

true; that is, if U and its derivatives of first order are continuous in R, and $\frac{\partial U}{\partial n}$ integrates to zero over its closed boundary, then U must be harmonic throughout R. Hence, equation 2.2 is a necessary and sufficient condition for U to be harmonic throughout the region.

Equation 2.2 provides an important boundary condition for many geophysical problems. Suppose that vector field $\mathbf{F}$ has a potential U which is harmonic throughout some region. Because $\frac{\partial U}{\partial n} = \mathbf{F} \cdot \hat{\mathbf{n}}$ on the surface of the region, equation 2.1 can be written as

$$\int_S \mathbf{F} \cdot \hat{\mathbf{n}} \, dS = 0 \,, \tag{2.3}$$

and applying the divergence theorem (Appendix A) yields

$$\int_R \nabla \cdot \mathbf{F} \, dv = 0 \,.$$

In words, the normal component of a conservative field must average to zero on the closed boundary of a region in which its potential is harmonic. Hence, the flux of $\mathbf{F}$ into the region exactly equals the flux leaving the region, implying that no sources of $\mathbf{F}$ exist in the region. Moreover, the condition that $\nabla \cdot \mathbf{F} = 0$ throughout the region is sufficient to conclude that no sources lie within the region.

Steady-state heat flow, for example, is harmonic (as discussed in Chapter 1) in regions without heat sources or sinks and must satisfy equation 2.3. If region R is in thermal equilibrium and contains no heat sources or sinks, the heat entering R must equal the heat leaving R. Equation 2.3 is often called *Gauss's law* and will prove useful in subsequent chapters.

Now let U be harmonic in region R and let $V = U$. Then, from Green's first identity,

$$\int\limits_R (\nabla U)^2 \, dv = \int\limits_S U \frac{\partial U}{\partial n} \, dS. \qquad (2.4)$$

Consider equation 2.4 when $U = 0$ on S. The right-hand side vanishes and, because $(\nabla U)^2$ is continuous throughout R by hypothesis, $(\nabla U)^2 = 0$. Therefore, U must be a constant. Moreover, because $U = 0$ on S and because U is continuous, the constant must be zero. Hence, *if U is harmonic and continuously differentiable in R and if U vanishes at all points of S, U also must vanish at all points of R.* This result is intuitive from steady-state heat flow. If temperature is zero at all points of a region's boundary and no sources or sinks are situated within the region, then clearly the temperature must vanish throughout the region once equilibrium is achieved.

Green's first identity leads to a statement about uniqueness, sometimes referred to as *Stokes's theorem.* Let U_1 and U_2 be harmonic in R and have identical boundary conditions, that is,

$$U_1(S) = U_2(S).$$

The function $U_1 - U_2$ also must be harmonic in R. But $U_1 - U_2$ vanishes on S and the previous theorem states that $U_1 - U_2$ also must vanish at every point of R. Therefore, U_1 and U_2 are identical. Consequently, *a function that is harmonic and continuously differentiable in R is uniquely determined by its values on S*, and the solution to the Dirichlet boundary-value problem is unique. Stokes's theorem makes intuitive sense when applied to steady-state heat flow. A region will eventually reach thermal equilibrium if heat is allowed to flow in and out of the region. It seems reasonable that, for any prescribed set of boundary temperatures, the region will always attain the same equilibrium temperature distribution throughout the region regardless of the initial temperature distribution. In other words, the steady-state temperature of the region is uniquely determined by the boundary temperatures.

The surface integral in equation 2.4 also vanishes if $\frac{\partial U}{\partial n} = 0$ on S. A similar proof could be developed to show that *if U is single-valued, harmonic, and continuously differentiable in R and if $\frac{\partial U}{\partial n} = 0$ on S, then U is a constant throughout R.* Again, steady-state heat flow provides some insight. If the boundary of R is thermally insulated, equilibrium temperatures inside R must be uniform. Moreover, *a single-valued harmonic function is determined throughout R, except for an additive constant, by the values of its normal derivatives on the boundary.*

Exercise 2.1 Prove the previous two theorems.

These last theorems relate to the Neumann boundary-value problem and show that such solutions are unique to within an additive constant.

The uniqueness of harmonic functions also extends to mixed boundary-value problems. *If U is harmonic and continuously differentiable in R and if*

$$\frac{\partial U}{\partial n} + hU = g$$

on S, where h and g are continuous functions of S, and h is never negative, then U is unique in R.

Exercise 2.2 Prove the previous theorem.

We have shown that under many conditions Laplace's equation has only one solution in a region, thus describing the uniqueness of harmonic functions. But can we say that even that one solution always exists? The answer to this interesting question requires a set of "existence theorems" for harmonic functions that are beyond the scope of this chapter. Interested readers are referred to Chapter XI of Kellogg [146, p. 277] for a comprehensive discussion.

2.1.2 Green's Second Identity

If we interchange U and V in equation 2.1 and subtract the result from equation 2.1, we obtain *Green's second identity:*

$$\int_R [U\nabla^2 V - V\nabla^2 U]\, dv = \int_S \left[U\frac{\partial V}{\partial n} - V\frac{\partial U}{\partial n} \right] dS, \qquad (2.5)$$

where it is understood that U and V are continuously differentiable and have continuous partial derivatives of first and second order in R.

A corollary results if U and V are both harmonic:

$$\int\limits_S \left[U \frac{\partial V}{\partial n} - V \frac{\partial U}{\partial n} \right] dS = 0 \,. \qquad (2.6)$$

This relationship will prove useful later in this chapter in discussing certain kinds of boundary-value problems. Also notice that if $V = 1$ in equation 2.5 and if U is the potential of $\mathbf{F}$, then

$$\int\limits_R \nabla^2 U \, dv = \int\limits_S \mathbf{F} \cdot \hat{\mathbf{n}} \, dS \,.$$

In regions of space where U is harmonic, we have the same result as in Section 2.1.1,

$$\int\limits_S \mathbf{F} \cdot \hat{\mathbf{n}} \, dS = 0 \,,$$

namely, that the normal component of a conservative field averages to zero over any closed surface.

2.1.3 Green's Third Identity

The third identity is a bit more difficult to derive. We begin by letting $V = \frac{1}{r}$ in Green's second identity (equation 2.5), where r is the distance between points P and Q inside region R (Figure 2.2):

$$\int\limits_R \left[U \nabla^2 \frac{1}{r} - \frac{1}{r} \nabla^2 U \right] dv = \int\limits_S \left[U \frac{\partial}{\partial n} \frac{1}{r} - \frac{1}{r} \frac{\partial U}{\partial n} \right] dS, \quad P \neq Q. \qquad (2.7)$$

Integration is with respect to point Q. It is easily shown that $\nabla^2 \frac{1}{r} = 0$ so long as $P \neq Q$. To insure that $P \neq Q$, we surround P with a small sphere σ and exclude it from R. Equation 2.7 becomes

$$-\int\limits_R \frac{\nabla^2 U}{r} \, dv = \int\limits_S \left[U \frac{\partial}{\partial n} \frac{1}{r} - \frac{1}{r} \frac{\partial U}{\partial n} \right] dS + \int\limits_\sigma \left[U \frac{\partial}{\partial n} \frac{1}{r} - \frac{1}{r} \frac{\partial U}{\partial n} \right] dS \,.$$

$$(2.8)$$

Exercise 2.3 Show that $\frac{1}{r}$ is harmonic for any region where $r \neq 0$. What happens at $r = 0$?

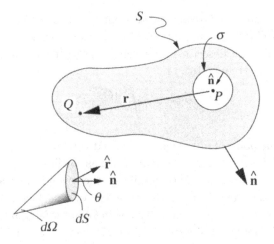

Fig. 2.2. Derivation of Green's third identity. Point P is inside surface S but is excluded from region R. Angle $d\Omega$ is the solid angle subtended by dS at point P.

First consider the integral over σ (Figure 2.2). We use the relationships,

$$\frac{\partial}{\partial n} = -\frac{\partial}{\partial r},$$

$$\frac{\partial}{\partial r}\frac{1}{r} = -\frac{1}{r^2},$$

$$\cos\theta\, dS = r^2\, d\Omega,$$

where $d\Omega$ is the solid angle subtended at P by dS and $\theta = 0$. The last integral of equation 2.8 becomes

$$\int_\sigma \left[\frac{1}{r^2}U + \frac{1}{r}\frac{\partial U}{\partial r}\right] r^2 d\Omega = \int_\sigma U\, d\Omega + \int_\sigma r\frac{\partial U}{\partial r}d\Omega$$

$$= 4\pi\bar{U} + \int_\sigma r\frac{\partial U}{\partial r}d\Omega.$$

As the sphere becomes arbitrarily small, the right-hand side of the previous expression approaches $4\pi U(P)$, and equation 2.8 becomes

$$U(P) = -\frac{1}{4\pi}\int_R \frac{\nabla^2 U}{r}dv + \frac{1}{4\pi}\int_S \frac{1}{r}\frac{\partial U}{\partial n}dS - \frac{1}{4\pi}\int_S U\frac{\partial}{\partial n}\frac{1}{r}dS. \quad (2.9)$$

Equation 2.9 is *Green's third identity*.

The significance of Green's third identity will become clear in later chapters. In Chapter 3 (equation 3.5), for example, we will show that a mass distribution described by a density $\rho(Q)$ has a gravitational potential at point P given by

$$U(P) = \gamma \int_R \frac{\rho(Q)}{r}\, dv,$$

where Q is the point of integration, r is the distance from P to Q, and γ is a constant. This integral has the same form as the first integral of Green's third identity if $\rho = -\frac{1}{4\pi\gamma}\nabla^2 U$. Similarly, the second integral of Green's identity has the same form as the potential of a surface distribution of mass σ, where $\sigma = \frac{1}{4\pi\gamma}\frac{\partial U}{\partial n}$. We will show in Chapter 5 (equation 5.2) that the magnetic potential of a distribution of magnetization $\mathbf{M}$ is given by

$$V(P) = C_{\mathrm{m}} \int_R \mathbf{M} \cdot \nabla_Q \frac{1}{r}\, dv,$$

where C_{m} is a constant, and this has the same form as the third integral of Green's third identity if $\mathbf{M}$ is spread over S and directed normal to S. But remember that no physical meanings were attached to U in deriving Green's third identity; that is, U was only required to have a sufficient degree of continuity. Green's third identity shows, therefore, that any function with sufficient differentiability can be expressed as the sum of three potentials: the potential of a volume distribution with density proportional to $-\nabla^2 U$, the potential of a surface distribution with density proportional to $\frac{\partial U}{\partial n}$, and the potential of a surface of magnetization proportional to $-U$. Hence, we have the surprising result that *any function with sufficient differentiability is a potential.*

An important consequence follows from Green's third identity when U is harmonic. Then equation 2.9 becomes

$$U(P) = \frac{1}{4\pi} \int_S \left[\frac{1}{r}\frac{\partial U}{\partial n} - U\frac{\partial}{\partial n}\frac{1}{r} \right] dS. \tag{2.10}$$

This important result shows that *a harmonic function can be calculated at any point of a region in which it is harmonic simply from the values of the function and its normal derivatives over the region's boundary.* This equation is called the *representation formula* (Strauss [274]), and we will return to it later in this chapter and again in Chapter 12.

Green's third identity demonstrates an important limitation that faces any interpretation of a measured potential field in terms of its causative sources. It was shown earlier that a harmonic function satisfying a given set of Dirichlet boundary conditions is unique, but the converse is not true. If U is harmonic in a region R, it also must be harmonic in each subregion of R. Likewise equation 2.10 must apply to the boundary of each subregion. It follows that the potential within any subregion of R can be related to an infinite variety of surface distributions. Hence, no unique boundary conditions exist for a given harmonic function. This property of nonuniqueness will be a common theme in following chapters.

2.1.4 Gauss's Theorem of the Arithmetic Mean

Another consequence of Green's third identity occurs when U is harmonic, the boundary S is the surface of a sphere, and point P is at the center of the sphere. If a is the radius of the sphere, then equation 2.10 becomes

$$U(P) = \frac{1}{4\pi a} \int\limits_{S} \frac{\partial U}{\partial n} \, dS - \frac{1}{4\pi} \int\limits_{S} U \left(-\frac{1}{a^2} \right) \, dS.$$

The first integral vanishes according to Green's first identity, so that

$$U(P) = \frac{1}{4\pi a^2} \int\limits_{S} U \, dS. \tag{2.11}$$

Hence, *the value of a harmonic function at any point is simply the average of the harmonic function over any sphere concentric about the point*, so long as the function is harmonic throughout the sphere. This relationship is called *Gauss's theorem of the arithmetic mean*.

We discussed the *maximum principle* by example in Section 1.3: *If U is harmonic in region R, a closed and bounded region of space, then U attains its maximum and minimum values on the boundary of R*, except in the trivial case where U is constant. Now we are in a position to prove it. The proof is by contradiction. Let Σ represent a set of points of R at which U attains a maximum M (Figure 2.3). Hence,

$$U(\Sigma) = M.$$

Σ cannot equal the total of R because we have stated that U is not constant, and Σ must be closed because of the continuity of U. Suppose that Σ contains at least one interior point of R. It can be shown that if Σ has one point interior to R it also has a frontier point interior to R, which we call P_0. Because P_0 is interior to R, a sphere can be constructed centered about P_0 that lies entirely within R. By the definition of a

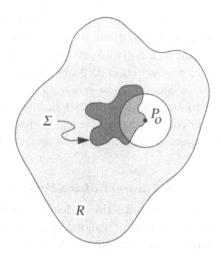

Fig. 2.3. Region R includes a set of points Σ at which U attains a maximum. Point P_0 is a frontier point of Σ. Any sphere centered on P_0 must contain at least one point of Σ and one point of R not in Σ.

frontier point (Section 1.1.2), such a sphere must pass outside of Σ where $U < M$. Gauss's theorem of the arithmetic mean and the continuity of U therefore imply that $U(P_0) < M$. But P_0 is also a member of Σ, and this requires that $U(P_0) = M$. A contradiction arises, and our original suppositions, that P_0 and at least one point of Σ lie interior to R, must be in error. Hence, no maxima of U can exist interior to R. An analogous proof can be constructed to show that the same is true of all minima of U.

2.2 Helmholtz Theorem

We said in Chapter 1 that a vector field $\mathbf{F}$ is conservative if the work required to move a particle through the field is independent of the path of the particle, in which case $\mathbf{F}$ can be represented as the gradient of a scalar ϕ,

$$\mathbf{F} = \nabla\phi \,,$$

called the potential of $\mathbf{F}$. Conversely, if $\mathbf{F}$ has a scalar potential, then $\mathbf{F}$ is conservative. These concepts are a subset of the *Helmholtz theorem* (Duff and Naylor [81]) which states that *any vector field* $\mathbf{F}$ *that is continuous and zero at infinity can be expressed as the gradient of a scalar and the curl of a vector, that is,*

$$\mathbf{F} = \nabla\phi + \nabla \times \mathbf{A}, \qquad (2.12)$$

where $\nabla\phi$ and $\nabla \times \mathbf{A}$ are orthogonal in the integral norm. The quantity ϕ is the scalar potential of $\mathbf{F}$, and $\mathbf{A}$ is the *vector potential.*

2.2.1 Proof of the Helmholtz Theorem

Given that $\mathbf{F}$ is continuous and vanishes at infinity, we can construct the integral

$$\mathbf{W}(P) = \frac{1}{4\pi} \int \frac{\mathbf{F}(Q)}{r} \, dv, \qquad (2.13)$$

where Q is the point of integration, r is the distance between P and Q, and the integral is taken over all space. Each of the three cartesian components of $\mathbf{W}$ has a form like

$$W_x = \frac{1}{4\pi} \int \frac{F_x}{r} \, dv. \qquad (2.14)$$

At this point, we borrow a result from Chapter 3: Equation 2.14 is a solution to a very important differential equation, *Poisson's equation*:

$$\nabla^2 W_x = -F_x. \qquad (2.15)$$

The relationship between equations 2.14 and 2.15 follows from Green's third identity because the integration in equation 2.14 is over all space, and we have stipulated that $\mathbf{F}$ and, therefore, the three components of $\mathbf{F}$ vanish at infinity.

Exercise 2.4 Show that equations 2.14 and 2.15 are consistent with Green's third identity.

With $\mathbf{W}$ defined as in equation 2.13, the relationship between equations 2.14 and 2.15 suggests that

$$\nabla^2 \mathbf{W} = -\mathbf{F}, \qquad (2.16)$$

where each component of $\mathbf{F}$ leads to an example of Poisson's equation. A vector identity (Appendix A) shows that $\nabla^2 \mathbf{W}$ can be represented by a gradient plus a curl, that is,

$$-\nabla^2 \mathbf{W} = -\nabla(\nabla \cdot \mathbf{W}) + \nabla \times (\nabla \times \mathbf{W}), \qquad (2.17)$$

and hence $\mathbf{F}$ is represented as the gradient of a scalar $(\nabla \cdot \mathbf{W})$ plus the curl of a vector $(\nabla \times \mathbf{W})$. We define $\phi = -\nabla \cdot \mathbf{W}$ and $\mathbf{A} = \nabla \times \mathbf{W}$ and substitute these definitions along with equation 2.16 into equation 2.17 to get the Helmholtz theorem,

$$\mathbf{F} = \nabla\phi + \nabla \times \mathbf{A}.$$

Exercise 2.5 Prove the vector identity $\nabla(\nabla \cdot \mathbf{W}) - \nabla \times \nabla \times \mathbf{W} = \nabla^2\mathbf{W}$.

Hence, the Helmholtz theorem is proven: If $\mathbf{F}$ is continuous and vanishes at infinity, it can be represented as the gradient of a scalar potential plus the curl of a vector potential.

The Helmholtz theorem is useful, however, only if the scalar and vector potentials can be derived directly from $\mathbf{F}$. This should be possible because of the way ϕ and $\mathbf{A}$ were defined, and the relationships can be seen by taking the divergence and curl of both sides of equation 2.12. The divergence yields

$$\nabla^2\phi = \nabla \cdot \mathbf{F},$$

which, comparing with equations 2.14 and 2.15, has the solution

$$\phi = -\frac{1}{4\pi} \int \frac{\nabla \cdot \mathbf{F}}{r}\, dv. \qquad (2.18)$$

The curl of equation 2.12 provides

$$\nabla^2\mathbf{A} = \nabla(\nabla \cdot \mathbf{A}) - \nabla \times \mathbf{F}.$$

For convenience, we define $\mathbf{A}$ so that it has no divergence, and consequently

$$\nabla^2\mathbf{A} = -\nabla \times \mathbf{F}.$$

Comparing this result with equations 2.13 and 2.16 leads to

$$\mathbf{A} = \frac{1}{4\pi} \int \frac{\nabla \times \mathbf{F}}{r}\, dv. \qquad (2.19)$$

Exercise 2.6 Show that equation 2.19 implies that $\nabla \cdot \mathbf{A} = 0$.

Consequently, the scalar potential ϕ and vector potential $\mathbf{A}$ can be derived from integral equations taken over all space and involving the divergence and curl, respectively, of $\mathbf{F}$ itself.

Exercise 2.7 Prove the last statement of the Helmholtz theorem; that is, show that $\nabla\phi$ and $\nabla \times \mathbf{A}$, both vanishing at infinity, are orthogonal under integration over three-dimensional space.

2.2.2 Consequences of the Helmholtz Theorem

The Helmholtz theorem shows that a vector field vanishing at infinity is completely specified by its divergence and its curl if they are known throughout space. If both the divergence and curl vanish at all points, then the field itself must vanish or be constant everywhere.

In addition to this statement, the following important observations follow directly from the Helmholtz theorem and from the integral representation for scalar and vector potentials.

Irrotational Fields

A vector field is *irrotational* in a region if its curl vanishes at each point of the region; that is, $\mathbf{F}$ is irrotational in a region if $\nabla \times \mathbf{F} = 0$ throughout the region. Such fields have no vorticity or "eddies." For example, if the flow of a fluid can be represented as an irrotational field, then a small paddlewheel placed within the fluid will not rotate. Examples of irrotational fields are common and include gravitational attraction, of considerable importance to future chapters.

Consider any surface S entirely within a region where $\nabla \times \mathbf{F} = 0$. Integration of the curl over the surface provides

$$\int_S (\nabla \times \mathbf{F}) \cdot \hat{\mathbf{n}} \, dS = 0 \,,$$

and applying Stokes's theorem (Appendix A) provides

$$\oint \mathbf{F} \cdot \mathbf{ds} = 0 \,,$$

where the closed line integral is taken around the perimeter of S. Because the integral holds for any closed surface within the region, no net work is done in moving around any closed loop that lies within an irrotational field, that is, work is independent of path, and $\mathbf{F}$ is conservative, a sufficient condition for the existence of a scalar potential such that $\mathbf{F} = \nabla \phi$. Hence, the condition that $\nabla \times \mathbf{F} = 0$ at each point of a region is sufficient to say that $\mathbf{F} = \nabla \phi$. Furthermore, a field that has a scalar potential has no curl because $\nabla \times \mathbf{F} = \nabla \times \nabla \phi$ vanishes identically (Appendix A). Hence, *the property that $\nabla \times \mathbf{F} = 0$ at every point of a region is a necessary and sufficient condition for the existence of a scalar potential such that $\mathbf{F} = \nabla \phi$.*

Solenoidal Fields

A vector field $\mathbf{F}$ is said to be *solenoidal* in a region if its divergence vanishes at each point of the region. A physical meaning for solenoidal fields can be had by integrating the divergence of $\mathbf{F}$ over any volume V within the region,

$$\int_V \nabla \cdot \mathbf{F}\, dv = 0\,,$$

and applying the divergence theorem (Appendix A) to get

$$\int_S \mathbf{F} \cdot \hat{\mathbf{n}}\, dS = 0\,, \qquad (2.20)$$

where S is the closed boundary of V. Hence, if the divergence of $\mathbf{F}$ vanishes in a region, the normal component of the field vanishes when integrated over any closed surface within the region. Or put another way, the "number" of field lines entering a region equals the number that exit the region, and sources or sinks of $\mathbf{F}$ do not exist in the region. For example, gravitational attraction is solenoidal in regions not occupied by mass.

It was stated in Section 2.1.1 that if a function ϕ can be found such that $\mathbf{F} = \nabla\phi$, then the condition expressed by equation 2.20 is necessary and sufficient to say that ϕ is harmonic throughout the region. From the Helmholtz theorem, $\nabla \cdot \mathbf{F} = \nabla^2\phi + \nabla \cdot \nabla \times \mathbf{A}$. The last term of this equation vanishes identically (Appendix A), and $\nabla \cdot \mathbf{F} = \nabla^2\phi$. Hence, if the divergence of a conservative field vanishes in a region, the potential of the field is harmonic in the region.

Note that if $\mathbf{F} = \nabla \times \mathbf{A}$, then $\nabla \cdot \mathbf{F} = 0$; that is, the divergence of a vector field vanishes if the vector can be expressed purely as the curl of another vector. Furthermore, the converse can be shown to be true by taking the curl of both sides of equation 2.19. Hence, *the property that $\nabla \cdot \mathbf{F} = 0$ is a necessary and sufficient condition for $\mathbf{F} = \nabla \times \mathbf{A}$.*

2.2.3 Example

Equation 2.18 is important to the geophysical interpretation of gravity and magnetic anomalies caused by crustal masses and magnetic sources, respectively. To see this, we use the magnetic field as an example and anticipate the results of future chapters.

A set of differential equations, called Maxwell's equations, describes the spatial and temporal relationships of electromagnetic fields and their sources. One of Maxwell's equations relates magnetic induction $\mathbf{B}$ and magnetization $\mathbf{M}$ in the absence of macroscopic currents:

$$\nabla \times \mathbf{B} = \mu_0 \nabla \times \mathbf{M},$$

where μ_0 is the permeability of free space. Magnetic field intensity $\mathbf{H}$ is related to magnetic induction and magnetization by the equation

$$\mathbf{B} = \mu_0(\mathbf{H} + \mathbf{M}). \tag{2.21}$$

Hence, in the absence of macroscopic currents,

$$\nabla \times \mathbf{H} = 0,$$

and it follows from the Helmholtz theorem that magnetic field intensity is irrotational and can be expressed in terms of a scalar potential, that is, $\mathbf{H} = -\nabla V$, where the minus sign is a matter of convention as discussed in Chapter 1. Moreover, equation 2.18 provides an expression for that scalar potential:

$$V = \frac{1}{4\pi} \int \frac{\nabla \cdot \mathbf{H}}{r} dv, \tag{2.22}$$

where it is understood that the integration is over all space. Another of Maxwell's equations states that magnetic induction has no divergence, that is, $\nabla \cdot \mathbf{B} = 0$. This fact plus equation 2.21 yields

$$\nabla \cdot \mathbf{H} = -\nabla \cdot \mathbf{M}, \tag{2.23}$$

and substituting into equation 2.22 provides

$$V = -\frac{1}{4\pi} \int \frac{\nabla \cdot \mathbf{M}}{r} dv. \tag{2.24}$$

Again integration is over all space. Equation 2.24 provides a way to calculate magnetic potential and magnetic field from a known (or assumed) spatial distribution of magnetic sources. This is called the *forward problem* when applied to the geophysical interpretation of measured magnetic fields. Equation 2.24 also is a suitable starting point for discussions of the *inverse problem*: the direct calculation of the distribution of magnetization from observations of the magnetic field. We will return to this equation in Chapter 5 and subsequent chapters.

2.3 Green's Functions

We now turn to *Green's functions,* important tools for solving certain classes of problems in potential theory. A heuristic approach will be used, first considering a mechanical system and then extending this result to Laplace's equation.

2.3.1 Analogy with Linear Systems

We begin with the differential equation describing motion of a particle subject to both a resistance R and an external force $f(t)$,

$$m \frac{d}{dt} v(t) = -R v(t) + f(t), \qquad (2.25)$$

where $v(t)$ is the velocity and m is the mass of the particle, respectively. One conceptual way to solve equation 2.25 is to abruptly strike the particle and observe its response; that is, we let the force be zero except over a short time interval $\Delta\tau$,

$$f(t) = \begin{cases} \frac{I}{\Delta\tau}, & \text{if } \tau < t < \tau + \Delta\tau\,; \\ 0, & \text{otherwise}\,. \end{cases} \qquad (2.26)$$

As soon as the force returns to zero, the velocity of the particle behaves like a decaying exponential, and the solution has the form

$$v(t) = A \exp\left[-\frac{R}{m} (t - (\tau + \Delta\tau)) \right], \quad t > \tau + \Delta\tau\,. \qquad (2.27)$$

The coefficient A can be found if the velocity of the particle is known at the moment that the force returns to zero; that is, $v(\tau + \Delta\tau) = A$. To find this velocity, we integrate both sides of equation 2.25 over the duration of the force

$$m[v(\tau + \Delta\tau) - v(\tau)] = -R \int_{\tau}^{\tau+\Delta\tau} v(t)\, dt + \frac{I}{\Delta\tau} \int_{\tau}^{\tau+\Delta\tau} dt\,.$$

The first integral can be ignored if $\Delta\tau$ is small and the particle has some mass. Also $v(\tau) = 0$. Hence,

$$m\, v(\tau + \Delta\tau) = I\,,$$

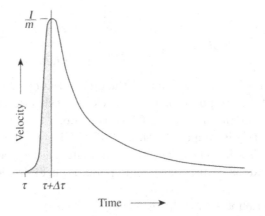

Fig. 2.4. Velocity of a particle of mass m resulting from an impulsive force of magnitude I.

and $A = I/m$ for small $\Delta\tau$. Combining this result with equation 2.27 provides

$$v(t) = \begin{cases} \dfrac{I}{m}\, e^{-\frac{R}{m}(t-\tau)}, & \text{if } t > \tau\,; \\ 0, & \text{if } t \leq \tau\,. \end{cases} \tag{2.28}$$

Equation 2.28 represents the response of the particle to a single abrupt blow (Figure 2.4). Now suppose that the particle suffers a series of blows I_k at time τ_k, $k = 1, 2, \ldots, N$. The response of the particle to each blow should be independent of all other blows, and the velocity becomes

$$v(t) = \sum_{k=1}^{N} \frac{I_k}{m}\, e^{-\frac{R}{m}(t-\tau_k)}, \quad t > \tau_N\,. \tag{2.29}$$

If the blows become sufficiently rapid, the particle is subjected to a continuous force. Then $I_k \to f(\tau)d\tau$ and

$$v(t) = \frac{1}{m} \int_{\tau_0}^{t} f(\tau)\, e^{-\frac{R}{m}(t-\tau)}\, d\tau, \quad t > \tau_0\,,$$

which can be rewritten as

$$v(t) = \int_{-\infty}^{t} \psi(t,\tau) f(\tau)\, d\tau, \tag{2.30}$$

where

$$\psi(t,\tau) = \begin{cases} 0, & \text{if } t < \tau; \\ \frac{1}{m}e^{-\frac{R}{m}(t-\tau)}, & \text{if } t \geq \tau. \end{cases}$$

Equation 2.30 is the solution to the differential equation 2.25. It presumes that the response of the particle at each instant of impact is independent of all other times. Given this property, the response of the particle to $f(t)$ is simply the sum of all the instantaneous forces, and the particle is said to be a *linear system*. Many mechanical and electrical systems (and, as it turns out, many potential-field problems) have this property.

The function $\psi(t,\tau)$ is the response of the particle at time t due to an impulse at time τ; it is called the *impulse response* or *Green's function* of the linear system. The Green's function, therefore, satisfies the initial conditions and is the solution to the differential equation 2.25 subject to the initial conditions when the forcing function is an impulse.

Equation 2.26 is a heuristic description of an impulse. In the limit as $\Delta\tau$ approaches zero, the impulse of equation 2.26 becomes arbitrarily large in amplitude and short in duration while its integral over time remains the same. The limiting case is called a Dirac delta function $\delta(t)$, which has the properties

$$\int_{-\infty}^{\infty} \delta(t)\,dt = 1\,;$$

$$\delta(t) = 0\,,\ t \neq 0\,;$$

$$\int_{-\infty}^{\infty} f(t)\delta(t)\,dt = f(0)\,;$$

$$\int_{-\infty}^{\infty} f(t)\delta(\tau - t)\,dt = f(\tau)\,. \tag{2.31}$$

These definitions and properties are meaningless if $\delta(t)$ is viewed as an ordinary function. It should be considered rather as a "generalized function" characterized by the foregoing properties.

Green's functions are very useful tools; equation 2.30 shows that if the Green's function ψ is known for a particular linear system, then the state of the linear system due to any forcing function can be derived for any time.

2.3.2 Green's Functions and Laplace's Equation

The previous mechanical example provides an analogy for potential theory. In Chapter 3, we will derive Poisson's equation

$$\nabla^2 U = -4\pi\gamma\rho. \tag{2.32}$$

This second-order differential equation describes the Newtonian potential U throughout space due to a mass distribution with density ρ. Clearly $\nabla^2 U = 0$ and U is harmonic in regions where $\rho = 0$. We seek a solution for U that satisfies the differential equation and the boundary condition that U is zero at infinity.

The density distribution in Poisson's equation is obviously the source of U and in this sense is analogous to the forcing function $f(t)$ of the previous section. We know from the previous section that the response to an impulsive forcing function $f(t) = \delta(t)$ is the Green's function, so we could try representing the density distribution in R as an "impulse" and see what happens to U. An impulsive source in three dimensions can be written as $\delta(P, Q)$, where

$$\int \delta(P, Q)\, dv = 1,$$

$$\delta(P, Q) = 0 \quad \text{if} \quad P \neq Q,$$

and where Q is the point of integration as in Section 2.2.1.

Hence, we let the density be $\delta(P, Q)$ and the potential be ψ_1 in equation 2.32,

$$\nabla^2 \psi_1 = -4\pi\gamma\delta(P, Q).$$

Then from the Helmholtz theorem and equations 2.14 and 2.15,

$$\psi_1(P, Q) = \gamma \int \frac{\delta(P, Q)}{r}\, dv$$

$$= \frac{\gamma}{r}, \tag{2.33}$$

where r is the distance between P and Q. This is a very interesting result. We see that γ/r is the solution to Poisson's equation when ρ is an "impulsive" density distribution located at Q. Indeed, we will show in Chapter 3 that γ/r is the Newtonian potential at P due to a point mass at Q. Hence, γ/r is the "impulse response" for Poisson's relation; with it, the potential due to any density distribution can be determined with an integral equation analogous to equation 2.30:

$$U(P) = \int \psi_1(P, Q)\rho(Q)\, dv \tag{2.34}$$

$$= \gamma \int \frac{\rho(Q)}{r}\, dv\,, \tag{2.35}$$

where it is understood that integration is over all space. This fundamental equation relating gravitational potential to causative density distributions will be derived in a different way in Chapter 3. The important point to be made here is that the function $\psi_1 = \gamma/r$ is analogous to the Green's function of the mechanical example in the previous section: It satisfies the required boundary condition, that ψ_1 is zero at infinity, and is the solution to Poisson's differential equation when the density is an "impulse."

Half-Space Regions

The representation formula, which followed from Green's third identity, shows that the value of a function harmonic in R can be found at any point within R strictly from the behavior of U and its normal derivative on the boundary of R, that is,

$$U(P) = -\frac{1}{4\pi} \int_S \left[U \frac{\partial}{\partial n} \frac{1}{r} - \frac{1}{r} \frac{\partial U}{\partial n} \right] dS\,, \tag{2.36}$$

where Q is the point of integration and r is the distance from P to Q. In practical situations, we are unlikely to have both the potential and its normal derivative at our disposal, and elimination of $\frac{\partial U}{\partial n}$ would make this equation much more useful. We should expect that such a simplification is possible because earlier results have shown that the potential is uniquely determined by its boundary conditions.

To eliminate $\frac{\partial U}{\partial n}$ from the third identity, we begin with Green's second identity. Let both U and V be harmonic in equation 2.5 so that

$$0 = -\frac{1}{4\pi} \int_S \left[U \frac{\partial V}{\partial n} - V \frac{\partial U}{\partial n} \right] dS\,.$$

Adding this equation to equation 2.36 provides

$$U(P) = -\frac{1}{4\pi} \int_S \left[U \frac{\partial}{\partial n} \frac{1}{r} - \frac{1}{r} \frac{\partial U}{\partial n} + U \frac{\partial V}{\partial n} - V \frac{\partial U}{\partial n} \right] dS$$

$$= -\frac{1}{4\pi} \int_S \left[U \frac{\partial}{\partial n} \left(V + \frac{1}{r} \right) - \left(V + \frac{1}{r} \right) \frac{\partial U}{\partial n} \right] dS\,.$$

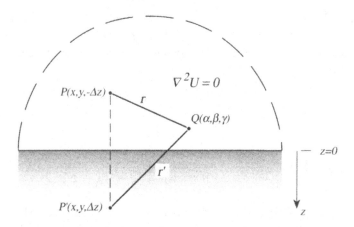

Fig. 2.5. Function U is harmonic throughout the half-space $z < 0$ and assumes known values on the surface $z = 0$. Parameter r is the distance between points P and Q; r' is the distance between P' and Q. Point P' is the image of point P such that $r = r'$ when Q is on the surface $z = 0$.

If we select a harmonic V such that $V + \frac{1}{r} = 0$ at each point of S, then

$$U(P) = -\frac{1}{4\pi} \int_S U \frac{\partial}{\partial n} \left(V + \frac{1}{r} \right) dS. \tag{2.37}$$

Hence, if for a particular geometry we can find a function V such that (1) V is harmonic throughout R and (2) $V + \frac{1}{r} = 0$ at each point of S, then U can be found throughout the region, and only values of U on the boundary will be required. The function $V + \frac{1}{r}$ is called the Green's function for Laplace's equation in restricted regions. It satisfies Laplace's equation throughout the region (except when $P = Q$) and is zero on the boundary.

In principle, equation 2.37 provides a simple way to solve Laplace's equation from specified boundary conditions. Unfortunately, the function V is very difficult to derive analytically except for the simplest sorts of geometrical situations, such as half-spaces and spheres. As an example, consider the half-space problem, where U is harmonic for all $z < 0$ and is known on the planar surface $z = 0$ (Figure 2.5). Boundary S then consists of the $z = 0$ plane plus the $z < 0$ hemisphere, as shown in Figure 2.5. We construct a point P' below the $z = 0$ plane that is the image of point P. The necessary properties are satisfied if we let $V = -\frac{1}{r'}$, where r' is the distance from P' to Q: namely, V is always harmonic since Q is always above or on the $z = 0$ plane, $V + \frac{1}{r} = 0$ when Q is on the $z = 0$ plane, and $V + \frac{1}{r} = 0$ when Q is on the infinite hemisphere. Hence, V defined in this way satisfies the necessary requirements to be used in equation 2.37; that is,

$$U(P) = -\frac{1}{4\pi} \int\limits_{S} U \frac{\partial}{\partial n} \left(\frac{1}{r} - \frac{1}{r'} \right) dS, \qquad (2.38)$$

$$U(x, y, -\Delta z) = \frac{\Delta z}{2\pi} \int\limits_{-\infty}^{\infty} \int\limits_{-\infty}^{\infty} \frac{U(\alpha, \beta, 0)}{[(x - \alpha)^2 + (y - \beta)^2 + \Delta z^2]^{\frac{3}{2}}} \, d\alpha \, d\beta, \qquad (2.39)$$

where $\Delta z \geq 0$. Equation 2.39 provides a way to calculate the potential at any point above a planar surface on which the potential is known. Such calculations are called *upward continuation*, a subject that will be revisited at some length in Chapter 12.

Terminology

The Green's function for Poisson's equation throughout space is usually derived from a general form of Poisson's equation, $\nabla^2 \phi = -f$, and thus is given by $G = \frac{1}{4\pi r}$ (e.g., Duff and Naylor [81], Strauss [274]). Here, we started with the gravitational case of Poisson's equation, $\nabla^2 U = -4\pi\gamma\rho$, and derived a slightly different form for the Green's function, $\psi_1 = \frac{\gamma}{r}$. The present approach led to equation 2.34, that is,

$$U(P) = \int \psi_1(P, Q) \, \rho(Q) \, dv \,,$$

where integration is over all space. This simple integral expression for the potential in terms of density and the Green's function will prove useful in following chapters.

We also showed that if a function V can be found satisfying just two properties (V is harmonic throughout a region, and $V + \frac{1}{r}$ is zero on the boundary of the region), then the representation formula reduces to a very simple form,

$$U(P) = -\frac{1}{4\pi} \int\limits_{S} U \frac{\partial}{\partial n} \left(V + \frac{1}{r} \right) dS \,.$$

The function $V + \frac{1}{r}$ is the Green's function for Laplace's equation in restricted regions. In future chapters, however, we will use a somewhat looser terminology. If we let $\psi_2(P,Q) = -\frac{1}{4\pi}\frac{\partial}{\partial n}(V+\frac{1}{r})$ in this equation, then

$$U(P) = \int_S U(Q)\,\psi_2(P,Q)\,dS\,,$$

which has a form similar to equation 2.34. For this reason, we also will refer to $\psi_2(P,S) = -\frac{1}{4\pi}\frac{\partial}{\partial n}(V+\frac{1}{r})$ as a kind of Green's function, one which provides U at points away from boundaries on which U is known.

2.4 Problem Set

1. In the following, T is temperature in region R bounded by surface S, and $\hat{\mathbf{n}}$ is the unit vector normal to S. If $\nabla^2 T = -f(P)$ in region R and $\frac{\partial T}{\partial n} = g(S)$ on surface S, show that

$$\int_R f\,dv + \int_S g\,dS = 0\,,$$

   and interpret the meaning of this equation.

2. Show, starting with Green's first identity, that if U is harmonic throughout all space, it must be zero everywhere.

3. Show that if two harmonic functions U_1 and U_2 satisfy $U_1 < U_2$ at each point of the boundary of a region, then $U_1 < U_2$ throughout the region.

4. A radial field is described by $\mathbf{F} = ar^n\hat{\mathbf{r}}$.

   (a) In regions where $r \neq 0$, find the values of n for which $\mathbf{F}$ is solenoidal.

   (b) For what values of n ($r \neq 0$) is the field irrotational?

5. Maxwell's equations state that magnetic induction $\mathbf{B}$ in the absence of moving charge is both solenoidal *and* irrotational, that is,

$$\nabla \cdot \mathbf{B} = 0,$$
$$\nabla \times \mathbf{B} = 0\,.$$

   Show that the three cartesian components of $\mathbf{B}$ are each harmonic in such situations.

6. Function U satisfies the two-dimensional Laplace's equation at every point of a circle. Find a Green's function that will provide the value of

U at any point inside the circle from the values of U on the boundary of the circle.

7. Function U is harmonic everywhere inside a sphere of radius a. Find a Green's function that will provide the value of U at any point inside the sphere from values of U on the surface of the sphere.

3
Newtonian Potential

And the Newtonian scheme was based on a set of assumptions, so few and so simple, developed through so clear and so enticing a line of mathematics that conservatives could scarcely find the heart and courage to fight it.

(Isaac Asimov)

The airplane stays up because it doesn't have the time to fall.

(Orville Wright)

The previous chapters reviewed Laplace's differential equation and its implications for conservative fields and scalar potentials in general. In this chapter, we become more specific and focus on the most important application of Laplace's equation, the force of gravity. As before, much of the discussion herein relies heavily on developments presented by Kellogg [146], Ramsey [235], and MacMillan [172].

3.1 Gravitational Attraction and Potential

In 1687, Newton published *Philosophiae Naturalis Principia Mathematica*, which, among other profundities, stated Newton's *law of gravitational attraction: The magnitude of the gravitational force between two masses is proportional to each mass and inversely proportional to the square of their separation.* In cartesian coordinates (Figure 3.1), the mutual force between a particle of mass m centered at point $Q = (x', y', z')$ and a particle of mass m_o at $P = (x, y, z)$ is given by

$$F = \gamma \frac{mm_o}{r^2},$$

where

$$r = [(x - x')^2 + (y - y')^2 + (z - z')^2]^{\frac{1}{2}},$$

43

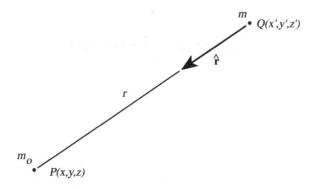

Fig. 3.1. Masses m and m_o experience a mutual gravitational force which is proportional to m, m_o, and r^{-2}. By convention, unit vector $\hat{\mathbf{r}}$ is directed from the gravitational source to the observation point, which in this case is located at test mass m_o.

and where γ is Newton's gravitational constant discussed subsequently. If we let mass m_o be a test particle with unit magnitude, then dividing the force of gravity by m_o provides the *gravitational attraction* produced by mass m at the location of the test particle:

$$\mathbf{g}(P) = -\gamma \frac{m}{r^2}\hat{\mathbf{r}}, \tag{3.1}$$

where $\hat{\mathbf{r}}$ is a unit vector directed from the mass m to the observation point P, that is, in cartesian coordinates,

$$\hat{\mathbf{r}} = \frac{1}{r}\left[(x - x')\hat{\mathbf{i}} + (y - y')\hat{\mathbf{j}} + (z - z')\hat{\mathbf{k}}\right].$$

The minus sign in equation 3.1 is necessary because of the convention, followed throughout this text, that $\hat{\mathbf{r}}$ is directed from the source to the observation point, opposite in sense to the gravitational attraction. Because $\mathbf{g}$ is force divided by mass, it has units of acceleration and is sometimes called *gravitational acceleration*. We will use the terms attraction and acceleration interchangeably in reference to $\mathbf{g}$.

Gravitational attraction, as described by equation 3.1, is an irrotational field because

$$\nabla \times \mathbf{g} = 0. \tag{3.2}$$

Exercise 3.1 Prove equation 3.2. Hint: Use equation 3.1 in the spherical coordinate system.

Hence, from the Helmholtz theorem and equation 2.12 (Chapter 2), gravitational acceleration is a conservative field and can be represented as the gradient of a scalar potential

$$\mathbf{g}(P) = \nabla U(P),\tag{3.3}$$

where

$$U(P) = \gamma \frac{m}{r}.\tag{3.4}$$

The function U is called the *gravitational potential* or *Newtonian potential*, and gravitational acceleration $\mathbf{g}$ is a potential field.

Exercise 3.2 Prove that the gradient of equation 3.4 yields equation 3.1.

Some textbooks (e.g., Grant and West [99]) consider gravitational potential to be *the work done by the test particle*, so that equation 3.3 is written $\mathbf{g} = -\nabla U$. Equation 3.3, however, follows the convention of Kellogg [146]: The gravitational potential is *the work done by the field on a test particle* and is the negative of the particle's potential energy. Because gravity is a conservative field, no net work is required to move a mass around a closed loop. Cross-country skiers and bicycle riders will appreciate that this statement pertains only to an ideal (frictionless, windless, etc.) world.

Units and the Gravitational Constant

In the *Système Internationale* (International System, abbreviated SI) and mksa system of units, m and m_o have units of kilograms, distance is in meters, and gravitational attraction is reported in m·sec^{-2}. In the cgs system of units, mass has units of grams, distance is in centimeters, and gravitational attraction is reported in units of cm·sec^{-2}. The cgs unit of acceleration is often referred to as the Gal (short for "Galileo"), where $1 \text{ Gal} = 1 \text{ cm·sec}^{-2}$, and the geophysical literature commonly reports gravitational attraction in units of mGal ($1 \text{ mGal} = 10^{-3} \text{ Gal}$). The conversion from cgs to SI units is $1 \text{ mGal} = 10^{-5} \text{ m·sec}^{-2}$.

Newton's gravitational constant γ is $6.67 \times 10^{-11} \text{ m}^3 \cdot \text{kg}^{-1} \cdot \text{sec}^{-2}$ in SI units and $6.67 \times 10^{-8} \text{ cm}^3 \cdot \text{g}^{-1} \cdot \text{sec}^{-2}$ in cgs units. Some texts, especially those dealing primarily with Newtonian attraction, use *astronomical units* for force, arranged so that $\gamma = 1$.

Refinements to Newton's gravitational constant have progressed ever since Newton proposed his law of gravitational attraction three centuries ago. Some concern still remains, however, as to just how constant Newton's gravitational constant really is. Recent work has reported very small deviations in the inverse-square relationship between gravitational force and distance (e.g., Stacey et al. [271]), but questions remain as to whether these deviations reflect a nonconstant gravitational constant or are caused by other physical processes in effect during the experiments (Zumberge et al. [297]). The question has not been resolved as of the writing of this book. Although of considerable interest for scientific and philosophical reasons, the issue of the constancy of Newton's gravitational constant has little impact on the applications for which this book is intended, and the problem will be ignored henceforth.

3.2 The Potential of Distributions of Mass

Gravitational potential obeys the *principle of superposition: The gravitational potential of a collection of masses is the sum of the gravitational attractions of the individual masses.* Hence, the net force on a test particle is simply the vector sum of the forces due to all masses in space. The superposition principle can be applied to find the gravitational attraction in the limit of a continuous distribution of matter. A continuous distribution of mass m is simply a collection of a great many, very small masses $dm = \rho(x, y, z)\, dv$, where $\rho(x, y, z)$ is the density distribution. Applying the principle of superposition yields

$$U(P) = \gamma \int_V \frac{dm}{r}$$

$$= \gamma \int_V \frac{\rho(Q)}{r}\, dv, \qquad (3.5)$$

where integration is over V, the volume actually occupied by mass. As usual, P is the point of observation, Q is the point of integration, and r is distance between P and Q. Density ρ has units of kilogram·meter^{-3} in SI units and gram·centimeter^{-3} in the cgs system. The conversion between the two systems is 1 kg·m^{-3}=10^{-3} g·cm^{-3}.

First consider observation points located outside of a mass distribution (Figure 3.2). If density is well behaved, integral 3.5 converges for all P outside of the mass (Kellogg [146]), and differentiation with respect to x, y, and z can be moved inside the integral. For example, the partial derivative of U with respect to x is

$$\frac{\partial U(P)}{\partial x} = -\gamma \int_V \frac{(x - x')}{r^3} \rho(Q)\, dv.$$

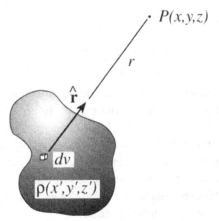

Fig. 3.2. Gravitational attraction at point P due to a density distribution ρ.

Repeating the differentiation of equation 3.5, once with respect to y and once with respect to z, and adding the three components will provide the attraction outside of any distribution of mass:

$$\mathbf{g}(P) = \nabla U(P)$$

$$= -\gamma \int_V \rho(Q) \frac{\hat{\mathbf{r}}}{r^2} \, dv. \tag{3.6}$$

Second-order derivatives can be derived in similar fashion; for example, the x component is

$$\frac{\partial^2 U}{\partial x^2} = \gamma \int_V \left[-\frac{\rho}{r^3} + \frac{3\rho(x - x')^2}{r^5} \right] dv.$$

Repeating for the y and z components and adding the three results yields

$$\nabla^2 U(P) = \frac{\partial^2 U}{\partial x^2} + \frac{\partial^2 U}{\partial y^2} + \frac{\partial^2 U}{\partial z^2}$$

$$= 0, \tag{3.7}$$

and the gravitational potential is harmonic at all points outside of the mass.

What about the potential inside distributions of mass? If P is inside the mass, the integrand in equation 3.5 is singular, and the integral is improper. Nevertheless, the integral can be shown to converge. In fact,

Kellogg [146] shows that the integral

$$I(P) = \int_V \frac{\rho}{r^n}\, dv$$

is convergent for P inside V and is continuous throughout V if $n < 3$, V is bounded, and ρ is piecewise continuous. Hence, $U(P)$ and $\mathbf{g}(P)$ exist and are continuous everywhere, both inside and outside the mass; so long as the density is well behaved. Kellogg [146] also shows that $\mathbf{g}(P) = \nabla U(P)$ for P inside the mass. This last point is not obvious because derivatives cannot be moved inside improper integrals.

The Helmholtz theorem (Section 2.2.2) tells us that if $\mathbf{g}$ satisfies $\mathbf{g} = \nabla U$ and vanishes strongly at infinity, then

$$U = \frac{1}{4\pi} \int \frac{\nabla \cdot \mathbf{g}}{r}\, dv . \tag{3.8}$$

Comparing the integrand of equation 3.8 with the integrand of equation 3.5 suggests that

$$\nabla^2 U(P) = -4\pi\gamma\rho(P) . \tag{3.9}$$

Equation 3.9 is *Poisson's equation*, which describes the potential at all points, even inside the mass distribution. Laplace's equation is simply a special case of Poisson's equation, valid for mass-free regions of space. Although the foregoing is not a rigorous proof of the relationship between equations 3.5 and 3.9, the example in Section 3.2.2 will demonstrate the validity of Poisson's equation.

The following theorems can be stated in summary:

1. The Newtonian potential U and the acceleration of gravity $\mathbf{g}$ exist and are continuous throughout space if caused by a bounded distribution of piecewise-continuous density.
2. The potential U is everywhere differentiable so equation $\mathbf{g} = \nabla U$ is true throughout space.
3. Poisson's equation $\nabla^2 U = -4\pi\gamma\rho$ describes the relationship between mass and potential throughout space. Laplace's equation $\nabla^2 U = 0$ is a special case of Poisson's equation valid in regions of space not occupied by mass.

Surface and Line Distributions

It is sometimes useful, as will be seen in the next sections, to consider the gravitational attraction and potential of mass distributions that are spread over vanishingly thin surfaces and along vanishingly narrow lines. The potential of a mass distribution spread over surface S and viewed at a point P not on the surface is given by

$$U(P) = \gamma \int_S \frac{\sigma(S)}{r} \, dS \,, \tag{3.10}$$

where σ is the *surface density* with units of mass per unit area. The potential of a mass concentrated along a line l is given by

$$U(P) = \gamma \int_l \frac{\lambda(l)}{r} \, dl \,, \tag{3.11}$$

where λ is the *line density* with units of mass per unit length. The gravitational attractions of these hypothetical distributions are easily derived from $\mathbf{g} = \nabla U$.

3.2.1 Example: A Spherical Shell

To investigate some of the points of the previous sections, consider the gravitational effects of a thin-walled, spherical shell of radius a and uniform surface density σ. We simplify the task (Figure 3.3) by arranging the coordinate system in order to take advantage of the symmetry of the problem: The origin is placed at the center of the sphere, and one axis is oriented so that it passes through P.

For P outside the shell, the potential is given by equation 3.10,

$$U(P) = \gamma \int_S \frac{\sigma(S)}{r} \, dS$$

$$= \gamma \sigma a^2 \int_0^{2\pi} \int_0^{\pi} \frac{\sin\theta}{r} \, d\theta \, d\phi \,.$$

The distance from P to any point on the sphere is

$$r = [R^2 + a^2 - 2aR\cos\theta]^{\frac{1}{2}} \,,$$

so

$$\frac{dr}{d\theta} = \frac{aR\sin\theta}{r} \,.$$

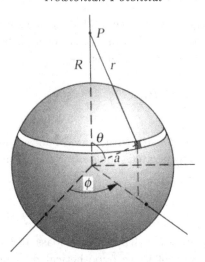

Fig. 3.3. Thin-walled, spherical shell with radius a observed at point P.

Substituting yields

$$U(P) = \frac{2\pi\gamma\sigma a}{R} \int\limits_{R-a}^{R+a} dr$$

$$= \gamma\frac{4\pi a^2 \sigma}{R} \tag{3.12}$$

$$= \gamma\frac{M}{R},$$

where M is the total mass of the shell. Therefore, *the gravitational potential at any point outside a uniform shell is equivalent to the potential of a point source located at the center of the shell with mass equal to the total mass of the shell.* It follows, therefore, that the gravitational attraction at points outside the shell is equivalent to the attraction of a point mass,

$$\mathbf{g}(P) = \nabla U(P)$$

$$= -\gamma\frac{M}{R^2}\,\hat{\mathbf{r}},$$

and that

$$\nabla^2 U(P) = 0.$$

Now consider P *inside* the shell. The previous derivation can be repeated but with slightly different limits of integration; that is,

$$U(P) = \frac{2\pi\gamma\sigma a}{R} \int_{a-R}^{a+R} dr$$

$$= \gamma\sigma 4\pi a \qquad (3.13)$$

$$= \gamma\frac{M}{a} .$$

All quantities in equation 3.13 are constant, so *the gravitational potential is constant everywhere inside a uniform shell.* Consequently, no gravitational forces exist inside the hollow shell because

$$\mathbf{g} = \nabla\left(\gamma\frac{M}{a}\right)$$

$$= 0 .$$

Obviously, $\nabla^2 U = 0$ within the shell because U is uniform throughout its interior.

Exercise 3.3 Equation 3.13 is easy to understand when P is located at the center of the shell. Observed at the center, the attraction due to any patch of the shell is exactly canceled by the attraction of an identical patch on the opposite side, so it seems reasonable that $\mathbf{g} = 0$ at the center. Less obvious is the fact that $\mathbf{g} = 0$ at points away from the center. Explain in terms of geometry and solid angles why all forces cancel at any point inside the shell.

3.2.2 Example: Solid Sphere

Equations 3.12 and 3.13 provide an easy way to investigate the gravitational effects of a solid sphere. For P outside the sphere, the problem is simple. A solid sphere of radius a is just a collection of concentric, thin-walled shells with radii ranging from 0 to a. The superposition principle states that the gravitational potential of the entire set of concentric shells is the sum of their individual potentials, which, according to the previous section, are each equivalent to a point mass at their centers. Consequently, *the potential of a solid sphere appears at all external points as a single point of mass located at the center of the sphere with magnitude equal to the total mass of the sphere*; that is,

$$U(P) = \gamma\frac{\frac{4}{3}\pi a^3\rho}{R}, \qquad (3.14)$$

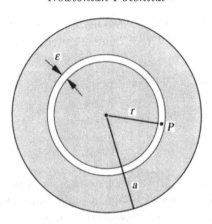

Fig. 3.4. Observation point P inside a sphere. Point P lies within a narrow spherical cavity between radius $r - \frac{\epsilon}{2}$ and $r + \frac{\epsilon}{2}$.

$$\mathbf{g}(P) = -\gamma \frac{\frac{4}{3}\pi a^3 \rho}{R^2} \hat{\mathbf{r}},$$

and $\nabla^2 U(P) = 0$ everywhere outside the sphere. Computer subroutine B.1 in Appendix B provides a Fortran subroutine that calculates the gravitational attraction at external points due to a sphere with homogeneous density.

To investigate the potential at points inside the sphere, we place P in a narrow, spherical cavity of radius r and thickness ϵ concentric about the center of the sphere (Figure 3.4). The potential at P is due to two sources: (1) That part of the sphere with radius less than $r - \frac{\epsilon}{2}$ and (2) the concentric shell with radius greater than $r + \frac{\epsilon}{2}$. Equation 3.14 gives the potential of the inner sphere:

$$U_{\mathrm{I}}(P) = \gamma \frac{\frac{4}{3}\pi(r - \frac{\epsilon}{2})^3 \rho}{r}.$$

We know from equation 3.13 that the potential of the outer shell must be constant because each concentric, thin-walled shell is a constant. Equation 3.13 can be integrated to provide the potential of the entire outer shell:

$$U_{\mathrm{O}}(P) = 4\pi\gamma\rho \int_{r+\frac{\epsilon}{2}}^{a} a' \, da'$$

$$= 2\pi\gamma\rho \left[a^2 - \left(r + \frac{\epsilon}{2}\right)^2 \right].$$

Adding $U_I(P)$ and $U_O(P)$ and letting $\epsilon \to 0$ provide the potential inside a spherical mass:

$$U(P) = \pi\gamma\rho \left[\frac{4(r - \frac{\epsilon}{2})^3}{3r} + 2\left[a^2 - \left(r + \frac{\epsilon}{2}\right)^2\right] \right]$$

$$= \tfrac{2}{3}\pi\gamma\rho\left[3a^2 - r^2\right]. \tag{3.15}$$

The gravitational attraction is given by

$$\mathbf{g}(P) = \nabla U(P)$$

$$= \frac{\partial}{\partial r}\frac{2}{3}\pi\gamma\rho\left[3a^2 - r^2\right]\hat{\mathbf{r}}$$

$$= -\tfrac{4}{3}\pi\gamma\rho r\,\hat{\mathbf{r}},$$

and *the attraction at internal points of a uniform sphere is proportional to the distance from the center.* The Laplacian (in spherical coordinates) of equation 3.15 yields

$$\nabla^2 U(P) = \frac{1}{r^2}\frac{\partial}{\partial r}\left[r^2\frac{\partial}{\partial r}U(P)\right]$$

$$= -4\pi\gamma\rho,$$

which is Poisson's equation.

Exercise 3.4 Show that $U(P)$ and $\mathbf{g}(P)$ are continuous across the surface of the sphere.

The last result shows that Poisson's differential equation describes the potential inside a uniformly dense sphere, and this result can be used to show that Poisson's equation holds inside all continuous distributions of mass. Within any well-behaved mass, we simply surround P with a small sphere and consider the potential as the sum of two parts, that is,

$$U(P) = U_S(P) + U_O(P),$$

where $U_S(P)$ is the potential at P due to the sphere and $U_O(P)$ is the potential caused by everything outside of the sphere. But $\nabla^2 U_O(P) = 0$ because no mass exists inside the spherical cavity. Furthermore, if the density is continuous around P, the sphere can be reduced in radius until its density is essentially uniform. Hence, $\nabla^2 U_S(P) = -4\pi\gamma\rho(P)$, and

$$\nabla^2 U(P) = -4\pi\gamma\rho(P)$$

for P inside a continuous density distribution.

Exercise 3.5 Graphically describe the potential and attraction of a uniform, thick-walled shell (inner radius a_1 and outer radius a_2) along a line extending from the center of the shell to infinity.

3.2.3 Example: Straight Wire of Finite Length

Consider the gravitational acceleration due to a straight wire extended along the z axis from $z = -a$ to $z = +a$ and observed at P on the x axis (Figure 3.5). The component of gravity in the y direction must be zero at P. The component of attraction in the direction parallel to the wire is also zero because the mass between $0 \le z \le a$ is equal to the mass between $-a \le z \le 0$. Hence, starting with equation 3.6,

$$\mathbf{g}(P) = -\gamma \int_R \rho \frac{\hat{\mathbf{r}}}{r^2} \, dv$$

$$= -\hat{\mathbf{i}} \gamma \lambda x \int_{-a}^{a} \frac{1}{r^3} \, dz' \,,$$

where λ is mass per unit length of the wire. Now make the following substitutions

$$\angle OPQ = \theta,$$
$$r = x \sec \theta,$$
$$z' = x \tan \theta,$$
$$dz' = x \sec^2 \theta \, d\theta,$$
$$\angle OPa = \theta_o$$

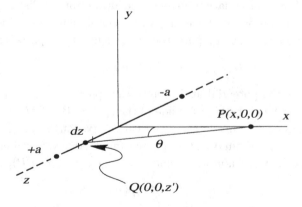

Fig. 3.5. Gravity at point P due to wire along z axis.

to derive

$$\mathbf{g}(P) = -\hat{\mathbf{i}}\,\frac{\gamma\lambda}{x} \int\limits_{-\theta_o}^{\theta_o} \cos\theta\,d\theta$$

$$= -\hat{\mathbf{i}}\,\frac{2\gamma\lambda}{x}\sin\theta_o$$

$$= -\hat{\mathbf{i}}\,2\gamma\lambda\frac{a}{x\sqrt{x^2+a^2}}\,. \tag{3.16}$$

Hence, the gravitational attraction of a finite length of wire viewed along a line perpendicular to the midpoint of the wire is directed toward the center of the wire.

3.3 Potential of Two-Dimensional Distributions

Masses that are infinitely extended in one dimension are said to be *two-dimensional*, for reasons that soon will become obvious. We begin by investigating the potential and attraction of infinite wires and apply these results to bodies of arbitrary cross-sectional shape.

3.3.1 Potential of an Infinite Wire

First consider the attraction of an infinitely long wire. As $a \to \infty$ in equation 3.16, the attraction becomes

$$\mathbf{g}(P) = -\frac{2\gamma\lambda\hat{\mathbf{i}}}{x}\,.$$

Hence, *the attraction of an infinitely long wire is inversely proportional to and in the direction of the perpendicular distance to the wire.* A general relationship is seen if P is moved to an arbitrary point of the x, y plane:

$$\mathbf{g}(P) = -\frac{2\gamma\lambda}{r}\hat{\mathbf{r}}\,, \tag{3.17}$$

where $\hat{\mathbf{r}}$ is directed from the wire to P and is understood to lie in the x, y plane; that is, $r^2 = (x - x')^2 + (y - y')^2$.

Although the gravitational attraction of an infinite wire is straightforward, the potential of an infinite wire is something of a problem. First consider the potential of a finite wire of length $2a$ (Figure 3.5):

$$U(P) = \gamma\lambda \int_{-a}^{a} \frac{1}{r} \, dz'$$

$$= \gamma\lambda \int_{-\theta_o}^{\theta_o} \sec\theta \, d\theta$$

$$= \gamma\lambda \log\left(\frac{1 + \sin\theta_o}{1 - \sin\theta_o}\right)$$

$$= \gamma\lambda \log \frac{\sqrt{x^2 + a^2} + a}{\sqrt{x^2 + a^2} - a}.$$

As $a \to \infty$, the potential also approaches infinity and obviously violates our requirements that the potential should vanish at infinity. This inconvenience is handled by redefining the meaning of the potential for infinitely extended bodies. The potential of an infinite wire is *defined* so that it vanishes at a unit distance from the wire. This is accomplished by adding a constant to the previous equation:

$$U(P) = \gamma\lambda\left[\log \frac{\sqrt{x^2 + a^2} + a}{\sqrt{x^2 + a^2} - a} - \log \frac{\sqrt{1 + a^2} + a}{\sqrt{1 + a^2} - a}\right].$$

Now, as $a \to \infty$,

$$U(P) = 2\gamma\lambda \log \frac{1}{x},$$

and moving P to an arbitrary point of the x, y plane provides the general result

$$U(P) = 2\gamma\lambda \log \frac{1}{r}, \qquad (3.18)$$

where r is the perpendicular distance from P to the wire. Notice that the potential does not vanish at infinity, but rather at $r = 1$.

Hence, the potential of an infinite wire decreases logarithmically as the point of observation recedes from the wire, a property that will extend to infinitely extended bodies of any cross-sectional shape. Such potentials are called *logarithmic potentials* for obvious reasons. It can be verified easily that equations 3.17 and 3.18 satisfy

$$\mathbf{g}(P) = \nabla U(P),$$

and

$$\nabla^2 U(P) = 0, \qquad r \neq 0.$$

The attraction of an infinite wire (equation 3.17) can be regarded in two ways. First, of course, it represents the Newtonian attraction of a wire of great length. It also can be regarded as a new kind of point source located at the intersection of the wire and the x, y plane. The attraction of the point source is proportional to the density of the wire λ and inversely proportional to the distance from the wire to the point of observation.

It is easily shown by integration of equation 3.18 that the Newtonian potential of an infinitely long, uniformly dense cylinder of radius a is given by

$$U(P) = 2\pi a^2 \gamma \rho \log \frac{1}{r}, \qquad (3.19)$$

where ρ is density and r is the perpendicular distance to the axis of the cylinder. Hence, the potential of an infinitely long, uniform cylinder is identical to the potential of an infinitely long wire located at the axis of the cylinder. Likewise, it follows from equation 3.17 that the gravitational attraction of an infinitely long cylinder is given by

$$\mathbf{g} = -\frac{2\pi a^2 \gamma \rho \,\hat{\mathbf{r}}}{r}, \qquad (3.20)$$

where $\hat{\mathbf{r}}$ is directed from the axis of the cylinder to P. Computer subroutine B.2 in Appendix B provides a Fortran subroutine to calculate the two components of gravitational attraction at external points of an infinitely extended cylinder.

3.3.2 General Two-Dimensional Distributions

The density of a two-dimensional source, by definition, does not vary in the direction parallel to its long axis, and ρ is a function only of the two dimensions perpendicular to the long axis of the body, that is, $\rho(x, y, z) = \rho(x, y)$. Starting with equation 3.5 and referring to Figure 3.6, we write

$$U(P) = \gamma \int_R \frac{\rho(Q)}{r} \, dv$$

$$= \gamma \int_S \rho(S) \left(\int_{-a}^{a} \frac{1}{r} \, dz' \right) dS,$$

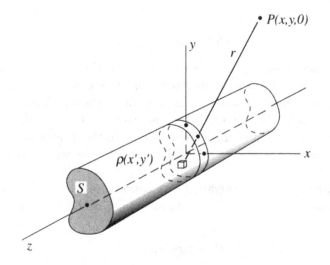

Fig. 3.6. Gravitational effects observed at point P due to infinitely extended body.

where S in this case represents the cross-sectional area of the two-dimensional source. As $a \rightarrow \infty$, the inner integral becomes the logarithmic potential of a wire with $\gamma\lambda = 1$, and the potential of the two-dimensional distribution is given by

$$U(P) = 2\gamma \int_S \rho(S) \log \frac{1}{r} \, dS. \qquad (3.21)$$

The gradient of equation 3.21 provides the gravitational attraction

$$\mathbf{g}(P) = -2\gamma \int_S \frac{\rho(S)}{r} \hat{\mathbf{r}} \, dS, \qquad (3.22)$$

which is perpendicular to the body. Because density is independent of the long dimension of the body, it is sometimes expressed as mass per cross-sectional area $\sigma(S)$, where σ/ρ has dimensions of length.

Equations 3.21 and 3.22 represent the Newtonian potential and attraction, respectively, of an infinitely long body, uniform in the direction parallel to the long dimension of the body. The attraction also can be considered as originating from a special kind of source: a two-dimensional wafer corresponding to the intersection of the body with the x, y plane (Figure 3.7). The attraction due to each element dS of the wafer is proportional to $\rho(S)$ and inversely proportional to distance.

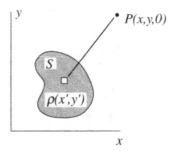

Fig. 3.7. Gravitational attraction of a two-dimensional body can be considered to originate from a special kind of source located in the x, y plane. Each element of the body has an attraction inversely proportional to distance.

Two-dimensional objects are generally easier to visualize than three-dimensional ones. Happily, certain geologic features, such as fault contacts and synclines, sometimes can be approximated by two-dimensional shapes thereby simplifying the interpretive process. In Chapter 9, we will describe the computation of the gravitational attraction of two-dimensional models with known cross section.

3.4 Gauss's Law for Gravity Fields

Consider a region R bounded by surface S. *Gauss's law states that the total mass in a region is proportional to the normal component of gravitational attraction integrated over the closed boundary of the region.* This can be seen by first applying the divergence theorem (Appendix A) to the normal component of gravity,

$$\int_S \mathbf{g} \cdot \hat{\mathbf{n}}\, dS = \int_R \nabla \cdot \mathbf{g}\, dv$$

$$= \int_R \nabla^2 U\, dv\,,$$

and then substituting Poisson's differential equation,

$$\int_S \mathbf{g} \cdot \hat{\mathbf{n}}\, dS = -4\pi\gamma \int_R \rho\, dv$$

$$= -4\pi\gamma M_T\,, \tag{3.23}$$

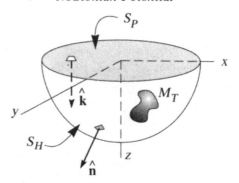

Fig. 3.8. An application of Gauss's law to find total excess mass. Gravity measurements are made on a horizontal surface S_P above all masses.

where M_T is the total mass. This relationship provides an important constraint in geophysical interpretations of gravity data, as we shall see in future chapters.

A well-known geophysical application of Gauss's law is the estimation of total excess mass below a surface on which the normal component of gravity is known (e.g., Hammer [109], LaFehr [152]). Suppose that the vertical component of gravity g_z is known over a horizontal surface S_P, as shown in Figure 3.8. All mass causing g_z is bounded in volume and located below S_P. The mass is enclosed by surface S, which is composed of S_P plus the $z > 0$ hemisphere S_H of radius a, as shown in Figure 3.8. The left side of equation 3.23 becomes

$$\int_S \mathbf{g} \cdot \hat{\mathbf{n}}\, dS = -\int_{S_P} g_z\, dS + \int_0^{2\pi} \int_{\frac{\pi}{2}}^{\pi} \frac{\partial U}{\partial r} r^2 \sin\theta\, d\theta\, d\phi. \qquad (3.24)$$

The potential of a volume distribution as viewed at a great distance is not dependent on the details of the distribution; that is, at large distances,

$$U(P) = \gamma \int_R \frac{\rho}{r}\, dv$$

$$\approx \frac{\gamma}{r} \int_R \rho\, dv$$

$$= \gamma \frac{M_T}{r},$$

where again M_T is the total mass. In other words, the potential of any bounded mass distribution appears as a point source when viewed sufficiently far away. Hence, as $a \to \infty$, $r^2 \frac{\partial U}{\partial r}$ can be moved outside the last integral of equation 3.24, and

$$\int_S \mathbf{g} \cdot \hat{\mathbf{n}} \, dS = -\int_{S_P} g_z \, dS - 2\pi\gamma M_T \int_{\frac{\pi}{2}}^{\pi} \sin\theta \, d\theta$$

$$= -\int_{S_P} g_z \, dS - 2\pi\gamma M_T \, .$$

Combining with equation 3.23 provides

$$\int_{S_P} g_z \, dS = 2\pi\gamma M_T \, , \qquad (3.25)$$

where S_P now includes the entire horizontal plane.

Hence, the vertical component of gravity integrated over an infinite plane is proportional to the total mass below the plane, so long as the mass is bounded in volume. In principle, equation 3.25 provides a way to estimate the total excess mass causing an anomaly in measured gravity if we can successfully isolate the field of the anomalous mass from all other gravitational sources. No assumptions are required about the shape of the source or how the density is distributed, so long as it is small with respect to the dimensions of the survey.

This may seem simple enough, but Gauss's law has many limitations in such applications. Gravity surveys are never available over infinite planes. The best that we can hope for is that the survey extends well beyond the localized sources of interest. Unfortunately, isolated sources never exist in nature, and it is often difficult to separate the gravitational anomaly caused by the masses of interest from anomalies caused by all other local and regional sources. We'll have more to say about this problem of "regional–residual" separation in a later chapter.

3.5 Green's Equivalent Layer

An argument was presented in Section 2.1.3 on the basis of Green's third identity that any given potential has an infinite variety of consistent

boundary conditions. Here we carry that point a little further and show
that a gravitational potential caused by a three-dimensional density distribution is identical to the potential caused by a surface density spread
over any of its equipotential surfaces (Ramsey [235]).

Let S_e be a closed equipotential surface resulting from a distribution
of mass with density ρ, and let R represent the region inside S_e. The
gravitational potential is observed at point P outside of S_e. Green's
second identity (Section 2.1.2) is given by

$$\int_R [U\nabla^2 V - V\nabla^2 U]\, dv = \int_{S_e} \left[U\frac{\partial V}{\partial n} - V\frac{\partial U}{\partial n} \right] dS\,,$$

where U and V are any functions with partial derivatives of first and
second order. Now let U be the potential of the mass and let $V = 1/r$,
where r represents the distance away from P. Because P is located
outside the region, the second identity reduces to

$$-\int_R \frac{\nabla^2 U}{r}\, dv = U_s \int_{S_e} \frac{\partial}{\partial n}\frac{1}{r}\, dS - \int_{S_e} \frac{1}{r}\frac{\partial U}{\partial n}\, dS\,,$$

where U_s is the constant potential of the equipotential surface. The first
integral on the right-hand side vanishes according to equation 2.2, and
substituting Poisson's equation into the integral on the left-hand side
provides

$$\gamma \int_R \frac{\rho}{r}\, dv = -\frac{1}{4\pi} \int_{S_e} \frac{1}{r}\frac{\partial U}{\partial n}\, dS\,. \tag{3.26}$$

The left-hand side of equation 3.26 is the potential of the density distribution observed at P. The right-hand side is the potential at P of a
surface distribution σ spread over S_e, where $\sigma = -\frac{1}{4\pi\gamma}\frac{\partial U}{\partial n}$. Hence, from
the perspective of point P, *the potential caused by a three-dimensional
density distribution is indistinguishable from a thin layer of mass spread
over any of its equipotential surfaces.* This relationship is called *Green's
equivalent layer.*

Furthermore, the total mass of the body is equivalent to the total
mass of the equivalent layer. This can be seen by integrating the surface

density over the entire surface and applying the divergence theorem (Appendix A), that is,

$$\int_{S_e} \sigma \, dS = -\frac{1}{4\pi\gamma} \int_S \frac{\partial U}{\partial n} \, dS$$

$$= -\frac{1}{4\pi\gamma} \int_R \nabla^2 U \, dv$$

$$= \int_R \rho \, dv.$$

Green's equivalent layer is of more than just academic interest. It shows that a potential can be caused by an infinite variety of sources, thus demonstrating the nonuniqueness of causative mass distributions. In later chapters, we will discuss applications of equivalent layers to the interpretation of gravity and magnetic data. The fact that the equivalent layer may have no resemblance to the true source will be of no importance in those applications. These hypothetical sources simply prove to be handy tools in manipulating the potential field.

3.6 Problem Set

1. Starting with the equation for gravitational attraction outside a uniform sphere, derive the "infinite slab formula"

$$\mathbf{g} = 2\pi\gamma\rho t\hat{\mathbf{k}}, \qquad (3.27)$$

   where ρ and t are the density and thickness of the slab, respectively, and $\hat{\mathbf{k}}$ is a unit vector directed vertically down. (Hint: Use superposition of two spheres and let their radii $\to \infty$.)

2. A nonzero density distribution that produces no external field for a particular source geometry is called an *annihilator* (Parker [207]). The annihilator quantitatively describes the *nonuniqueness* of potential field data because any amount of the annihilator can be added to a possible solution without affecting the field of the source. Find a simple annihilator ρ for a spherical mass of radius a as viewed from outside the sphere. (Hint: Let ρ represent *density contrast* so that ρ can reach negative values.)

3. Let the radius and density of the earth be represented by a and ρ, respectively.

(a) Show that the initial rate of decrease in g in descending a mine shaft is equal to g/a if ρ is constant.

(b) Assume that the earth has a spherical core of radius b and density $\rho_c \neq \rho$. Show that

$$\rho_c = \rho \left[1 + \left(\frac{1-\lambda}{2+\lambda} \right) \frac{a^3}{b^3} \right] ,$$

where $\lambda g/a$ is the initial rate of decrease of g in descending the shaft.

4. Use subroutine B.1 (Appendix B) to write a program that calculates the vertical attraction of gravity along a horizontal profile directly over a buried sphere.

(a) Use the program to verify that the shape (but not the amplitude) of the profile is independent of the sphere's radius.

(b) The horizontal profile has two points at which the horizontal gradient is maximum. Derive an expression for the depth d to the center of the sphere in terms of the horizontal separation of these two maximum horizontal gradients.

(c) Use the program to verify the answer in part (b).

5. An alluvium-filled basin lies within an otherwise homogeneous plain. Surrounding crustal rocks have a density everywhere of ρ_c. The thickness of the basin (i.e., the depth to basement) is D. At the surface, the alluvium has a density of ρ_a. The density of the basin, however, varies with depth because of compaction of the alluvium. The density of the alluvium at the surface is ρ_a, but the density contrast $\Delta\rho$ between alluvium and surrounding rocks decreases exponentially with increasing depth according to the equation

$$\Delta\rho = \Delta\rho_0 \, e^{-\lambda d} ,$$

where d is depth below the surface and $\Delta\rho_0 = \rho_a - \rho_c$. The horizontal dimensions of the basin are much larger than its depth, so the shape of the basin can be approximated as an infinite slab.

(a) Consider the gravitational attraction g measured above the center of the basin. Show that g reaches a limiting value as D increases.

(b) What is this limit in terms of ρ_a, ρ_c, and λ?

(c) Based on these results, discuss the limitations in trying to determine D from g for deep basins.

4

Magnetic Potential

There is about the Earth a magnetic field. Its cause and origin are veiled in mystery as in the case of that other great natural phenomenon, gravitation.

(J. A. Fleming)

It is well to observe the force and virtue and consequence of discoveries, and these are to be seen nowhere more conspicuously than in printing, gunpowder, and the magnet.

(Francis Bacon)

Under certain conditions, a magnetic field is uniquely determined by a scalar potential, analogous to the relationship between a gravity field and its corresponding gravitational potential. Happily for us, these special conditions are approximately obtained in typical geophysical measurements of the magnetic field. Consequently, much of what has been developed in previous chapters concerning Newtonian potentials will apply directly to magnetic fields as well. This chapter is primarily a review of the principles of electricity and magnetism, for which additional information is easily found (e.g., Panofsky and Phillips [201]).

4.1 Magnetic Induction

The discussion of Newtonian potentials in Chapter 3 began by investigating the mutual attraction of two point masses. We begin Chapter 4 in the same spirit by considering the mutual magnetic attraction of two small loops of electric current, the magnetic analog of two point masses. Consider the two loops of current shown in Figure 4.1 with currents I_a and I_b, respectively. The force acting on a small element $\mathbf{dl}_a$ of loop a caused by electric current in element $\mathbf{dl}_b$ of the second loop is given by the *Lorentz force*

$$\mathbf{df}_a = C_{\mathrm{m}} I_a I_b \frac{\mathbf{dl}_a \times (\mathbf{dl}_b \times \hat{\mathbf{r}})}{r^2} \,. \tag{4.1}$$

The factor C_{m} is a proportionality constant, analogous to Newton's gravitational constant in equation 3.1; it is used to adjust units and will be discussed subsequently.

In discussing Newtonian potential, we considered one mass to be a test particle with unit magnitude. Likewise, we now let loop a act as a "test loop" and define a vector $\mathbf{B}$ such that

$$\mathbf{dB}_b = C_{\mathrm{m}} I_b \frac{\mathbf{dl}_b \times \hat{\mathbf{r}}}{r^2} \,, \tag{4.2}$$

and

$$\mathbf{df}_a = I_a\,\mathbf{dl}_a \times \mathbf{dB}_b \,. \tag{4.3}$$

This is simply a derivative form of the Lorentz equation that describes the force acting on a charge Q moving with velocity $\mathbf{v}$ through a magnetic field:

$$\mathbf{F} = Q(\mathbf{v} \times \mathbf{B}) \,.$$

Integration of equation 4.2 around the loop of wire yields the *Biot–Savart law*:

$$\mathbf{B} = C_{\mathrm{m}} I_b \oint \frac{\mathbf{dl}_b \times \hat{\mathbf{r}}}{r^2} \,. \tag{4.4}$$

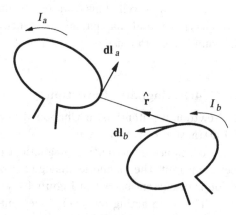

Fig. 4.1. Two loops of electric current I_a and I_b separated by a distance r. Unit vector $\hat{\mathbf{r}}$ is directed from an element $\mathbf{dl}_b$ of loop b to element $\mathbf{dl}_a$ of loop a.

Vector **B** in equation 4.4 is called the *magnetic induction, magnetic flux density,* or simply the *magnetic field* of a loop of current. Magnetic induction is defined as follows: Electric current induces a force on a moving charge; the force is the vector product of the magnetic induction field and the velocity of the charge. Note that, as in gravitational sources, $\hat{\mathbf{r}}$ is directed from the source (loop b in this case) to the point of observation.

Units

Two basic systems of units are commonly used in geophysical applications of magnetism. Most of the literature published prior to about 1980 employed the cgs system of units, also known as electromagnetic units (emu). More recently, the *Système Internationale* (International System, abbreviated SI) has become more common and, indeed, is required by most geophysical journals today. In the gravitational case (Chapter 3), the conversion between cgs and SI units was simply a matter of multiplying by an appropriate power of ten, but the magnetic case is more problematic, as will become particularly evident in Chapter 5.

For example, the proportionality constant C_m in equation 4.4 differs both in magnitude and dimensions between the two systems: In emu, $C_m = 1$ and is dimensionless, whereas in SI units, $C_m = \frac{\mu_0}{4\pi} = 10^{-7}$ henry·meter^{-1}, where μ_0 is called the *magnetic permeability of free space.* The difference in the proportionality constants in emu and SI units is a common source of frustration. We will attempt to avoid the problem in this book by employing the parameter C_m wherever possible without reference to any particular system of units. In most equations, letting $C_m = 1$ will make the equation compatible with emu, or letting $C_m = \frac{\mu_0}{4\pi}$ will transform the equation to SI units. However, we will have to abandon C_m in Chapter 5 because there even the mathematical derivation will depend on the system of units! Appendix D describes conversion factors for all magnetic units of importance to this text. For additional information, the interested reader is referred to books by Butler [47, pp. 15–18], Panofsky and Phillips [201, pp. 459–65], and the Society of Exploration Geophysicists [266]. Articles by Shive [255], Lowes [169], Payne [213], and Moskowitz [189] are also helpful.

In the emu system, magnetic induction **B** is reported in units of *gauss* (G), and current has units of *abamperes.* In SI units, **B** has units of weber·meter^{-2}, which is given the name *tesla* (T), and current is in units of amperes (1 ampere = 0.1 abampere). In geophysical studies, the gamma (emu) or the nanotesla (SI) is often used to express **B**, where

$$
\begin{aligned}
1 \text{ tesla} &= 10^4 \text{ gauss,}\\
1 \text{ nanotesla} &= 10^{-9} \text{ tesla}\\
&= 1 \text{ gamma}\\
&= 10^{-5} \text{ gauss.}
\end{aligned}
$$

4.2 Gauss's Law for Magnetic Fields

One of Maxwell's equations states that magnetic induction has no divergence, that is,

$$\nabla \cdot \mathbf{B} = 0, \tag{4.5}$$

and **B** is solenoidal. This statement holds for all points, even within magnetic media. Integration of equation 4.5 over a region R and application of the divergence theorem (Appendix A) provide a useful relationship:

$$
\int_R \nabla \cdot \mathbf{B} \, dv = \int_S \mathbf{B} \cdot \hat{\mathbf{n}} \, dS
$$

$$= 0 \tag{4.6}$$

for any R. Equation 4.6 shows that the normal component of all flux entering any region equals the normal component of flux leaving the region (Figure 4.2). This implies that no net sources (or sinks) exist anywhere in space; or put another way, magnetic monopoles do not exist, at least macroscopically.

Equation 4.6 is sometimes referred to as *Gauss's law* for magnetic fields and provides a useful constraint for many problems. For example, suppose that the field of a localized magnetic source is measured over a horizontal surface S_1 as in Figure 4.3. The net flux entering the region defined by S_1 and the hemisphere S_2 must be zero according to Gauss's law, that is,

$$
\int_{S_1} \mathbf{B} \cdot \hat{\mathbf{k}} \, dS_1 + \int_{S_2} \mathbf{B} \cdot \hat{\mathbf{n}} \, dS_2 = 0.
$$

As the limits of the survey are extended in the horizontal directions, surface S_2 moves arbitrarily far from the localized source, and it is easily shown that the integral over S_2 vanishes. Hence, a horizontal survey of the vertical component of **B** should average to zero if the lateral extent of the survey is large compared to the size of the magnetic sources. Put into geophysical terms, if regional-scale anomalies have been subtracted

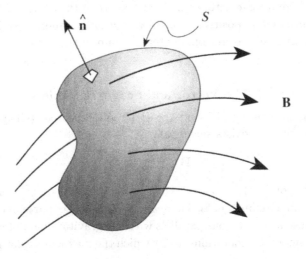

Fig. 4.2. The normal component of **B** integrates to zero over any closed surface. Hence, the net normal magnetic flux through any region is zero.

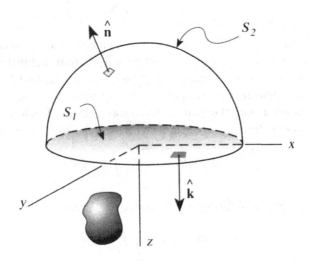

Fig. 4.3. A region R is bounded by planar surface S_1 and the hemispheric surface S_2. Magnetic sources are localized and lie beneath S_1.

properly from a large aeromagnetic survey, then the survey should contain about as many positive anomalies as negative anomalies. If it does not, then anomalies of regional extent remain in the data.

4.3 The Vector and Scalar Potentials

According to equation 4.5 and the Helmholtz theorem (Chapter 2), a vector potential $\mathbf{A}$ exists such that

$$\mathbf{B} = \nabla \times \mathbf{A}. \tag{4.7}$$

The vector potential for $\mathbf{B}$ always exists, but a scalar potential also exists under certain circumstances. The scalar potential is of particular interest here because it has obvious parallels with the gravitational potential. We now investigate the conditions under which the magnetic scalar potential is valid.

A second of Maxwell's equations states that the curl of $\mathbf{B}$ is equal to the vector sum of the various forms of charge moving in the region:

$$\nabla \times \mathbf{B} = 4\pi C_m \mathbf{I}_t$$

$$= 4\pi C_m \left(\mathbf{I}_m + \nabla \times \mathbf{M} + \frac{\partial \mathbf{D}}{\partial t} \right), \tag{4.8}$$

where $\mathbf{I}_t$ is the sum of all currents in the region, including macroscopic currents $\mathbf{I}_m$, currents related to magnetization $\mathbf{M}$ (to be defined shortly), and total displacement currents $\frac{\partial}{\partial t}\mathbf{D}$. The quantities $\mathbf{I}_m$ and $\mathbf{I}_t$ are *current densities* measured in units of current per unit area (e.g., A·m^{-2} in SI units). Now consider the relationship between $\mathbf{B}$ and $\mathbf{I}_t$ when averaged over a surface S. Integrating equation 4.8 over the surface gives

$$\int_S \nabla \times \mathbf{B} \cdot \hat{\mathbf{n}} \, dS = 4\pi C_m \int_S \mathbf{I}_t \cdot \hat{\mathbf{n}} \, dS,$$

and applying Stokes's theorem (Appendix A) provides

$$\oint \mathbf{B} \cdot \mathbf{dl} = 4\pi C_m \int_S \mathbf{I}_t \cdot \hat{\mathbf{n}} \, dS, \tag{4.9}$$

where $\mathbf{I}_t$ is the total of all electric currents, expressed as a current density, passing through surface S. Hence, the magnetic induction integrated around any closed loop is proportional to the normal component of all currents passing through the loop (Figure 4.4).

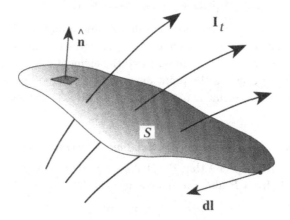

Fig. 4.4. The normal component of all currents I_t passing through a surface S is proportional to $\mathbf{B} \cdot \mathbf{dl}$ integrated around the loop that bounds S.

If no currents exist in the region of investigation, then

$$\nabla \times \mathbf{B} = 0,$$

and $\mathbf{B}$ is irrotational. According to the Helmholtz theorem, $\mathbf{B}$ then has a scalar potential V such that

$$\mathbf{B} = -\nabla V, \tag{4.10}$$

and $\mathbf{B}$ is a potential field in the absence of currents. The negative sign in equation 4.10 follows the convention of Kellogg [146], as discussed in Chapters 1 and 3.

In many geophysical situations, electrical currents are negligible in regions where the magnetic field is actually measured. Hence, equation 4.10 is often a suitable approximation outside of magnetic materials. This is a fortunate happenstance. For example, derivations of the magnetic fields caused by bodies of specified shape are generally more easily done by first deriving V and then applying equation 4.10. Moreover, under conditions where equation 4.10 holds, the potential of magnetic induction has obvious similarities to gravitational potential, and many of the statements made in earlier chapters apply immediately to magnetostatic theory. However, it should always be remembered that equation 4.10 is valid only outside of magnetic media and where line integrals encircle no currents.

4.4 Dipole Moment and Potential

The following pair of derivations result in some important expressions, namely the magnetic induction and scalar potential of a dipole. In Chapter 5, the dipole will be considered to be the elemental building block of magnetic sources, just as the point mass was considered in Chapter 3 to be the fundamental element of continuous density distributions.

4.4.1 First Derivation: Two Current Loops

It was shown in Chapter 1 that the change in potential caused by moving a test particle from P along a line element $\mathbf{dl}'$ is

$$dV(P) = -\mathbf{B} \cdot \mathbf{dl}'.$$

Now consider $\mathbf{B}$ to be the magnetic induction generated by a loop of current and substitute equation 4.4 into the previous equation:

$$dV(P) = -C_\mathrm{m} I \oint \frac{\mathbf{dl} \times \hat{\mathbf{r}}}{r^2} \cdot \mathbf{dl}'. \qquad (4.11)$$

The vector $\mathbf{dl}'$ is a constant, so placing it inside the integral is a legitimate maneuver. A vector identity (Appendix A) allows rearrangement of equation 4.11:

$$dV(P) = -C_\mathrm{m} I \oint \frac{\mathbf{dl} \times (-\mathbf{dl}') \cdot \hat{\mathbf{r}}}{r^2}. \qquad (4.12)$$

Figure 4.5 provides a geometric interpretation for the integrand of equation 4.12. Moving P along $\mathbf{dl}'$ has precisely the same effect on the potential at P as holding P stationary and moving the loop along $-\mathbf{dl}'$. Choosing the latter interpretation, we see that $\mathbf{dl} \times (-\mathbf{dl}')$ in equation 4.12 is a vector perpendicular to the shaded parallelogram in Figure 4.5 with magnitude equal to the area of the parallelogram. Therefore, the integrand of equation 4.12 is the solid angle of the shaded parallelogram as viewed from point P, and the integral is the elemental solid angle subtended at P by the entire ribbon:

$$dV(P) = +C_\mathrm{m} I\, d\Omega.$$

Clearly, $d\Omega$ is the increase in the solid angle caused by moving the loop along $-\mathbf{dl}'$, or moving P along $\mathbf{dl}'$. Because the solid angle of the loop is zero at infinity, the potential is given by

$$V(P) = C_\mathrm{m} I \frac{\hat{\mathbf{n}} \cdot \hat{\mathbf{r}}}{r^2} \Delta s,$$

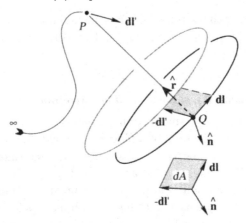

Fig. 4.5. Point P in the vicinity of a current loop. Moving P along $\mathbf{dl}'$ is equivalent to moving the loop along $-\mathbf{dl}'$.

where Δs is the area of the loop and $\hat{\mathbf{n}}$ is a unit normal vector (Figure 4.6).

Now let the current loop become small in diameter with respect to r and define

$$\mathbf{m} = I \,\hat{\mathbf{n}} \,\Delta s$$

as the *dipole moment*. Then

$$V(P) = C_{\mathrm{m}} \frac{\mathbf{m} \cdot \hat{\mathbf{r}}}{r^2}$$

$$= -C_{\mathrm{m}} \mathbf{m} \cdot \nabla_P \frac{1}{r}. \tag{4.13}$$

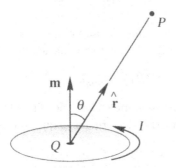

Fig. 4.6. Current loop observed at point P. Vector $\mathbf{m}$ has direction $\hat{\mathbf{n}}$ and magnitude equal to the current I times the area of the loop.

Equation 4.13 describes the potential of an elemental dipole. The dipole moment† has units of gauss·cm^3 in emu and ampere·m^2 in SI units, where 1 A·m^2 = 10^3 gauss·cm^3.

4.4.2 Second Derivation: Two Monopoles

Now that we've been through the previous tedious derivation, let's try an easier way to find the potential of a dipole. A second physical model for a dipole (as implied by its name) is two point masses (monopoles) of opposite sign in close proximity to each other (Figure 4.7). For now, let monopole 1 be at the origin and monopole 2 be at a distance $-\Delta z$ away on the z axis. The potential at P due to both monopoles is simply the sum of the potentials caused by each monopole, so

$$V(P) = V_1(P) + V_2(P).$$

The potential at P due to monopole 2 is simply the negative of the potential due to monopole 1 viewed a short distance away from P, namely, at $P + \Delta z$. Accordingly,

$$V(P) = -\left[V_1(P + \Delta z) - V_1(P)\right].$$

As Δz becomes small, this equation becomes the definition of the first derivative of $V_1(P)$:

$$V(P) = -\Delta z \frac{dV_1(P)}{dz}.$$

Because the gravitational potential of a point mass is given by $U = \gamma m/r$, it stands to reason that the magnetic potential at P due to a single monopole is simply

$$V_1(P) = C_m \frac{q}{r},$$

where q is *pole strength*, dimensionally equivalent to dipole moment per unit length. Rearranging the last two equations provides

$$V(P) = -C_m q \, \Delta z \frac{d}{dz} \frac{1}{r}.$$

In the general case, the monopoles are not aligned along any particular axis, and we should rewrite the previous equation as

$$V(P) = -C_m q \, \mathbf{ds} \cdot \nabla_P \frac{1}{r},$$

† Note that both mass and the magnitude of magnetic moment, $m = |\mathbf{m}|$, are represented traditionally and in this text by the letter m. In most cases, the meaning of m will be clear from context.

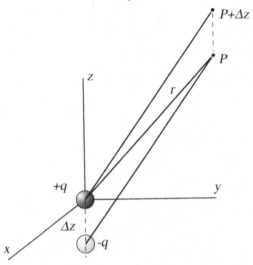

Fig. 4.7. Two monopoles of opposite sign, one at the origin and the other at $z = -\Delta z$ observed at point P.

where **ds** points from monopole 1 to monopole 2. Finally, we define the dipole moment as $\mathbf{m} = q\,\mathbf{ds}$, so that

$$V(P) = -C_{\mathrm{m}}\mathbf{m} \cdot \nabla_P \frac{1}{r},$$

which is identical to equation 4.13.

4.5 Dipole Field

Substitution of equation 4.13 into equation 4.10 provides the magnetic induction of a dipole at points other than the dipole itself:

$$\mathbf{B} = C_{\mathrm{m}}\frac{m}{r^3}\Big[3(\hat{\mathbf{m}} \cdot \hat{\mathbf{r}})\hat{\mathbf{r}} - \hat{\mathbf{m}}\Big], \qquad r \neq 0. \qquad (4.14)$$

Exercise 4.1 Derive equation 4.14 from equations 4.13 and 4.10. Hint: A vector identity for $\nabla(\mathbf{A} \cdot \mathbf{B})$ is helpful; see Appendix A.

Equation 4.14 describes the familiar vector field of a small bar magnet (Figure 4.8). The magnitude of **B** is proportional to the dipole moment and inversely proportional to the cube of the distance to the dipole. The direction of **B** depends on the directions of both $\hat{\mathbf{r}}$ and **m**. All flux lines of **B** emanate from the positive end of **m** and ultimately return to the negative end. Computer subroutine B.3 in Appendix B implements

Magnetic Potential

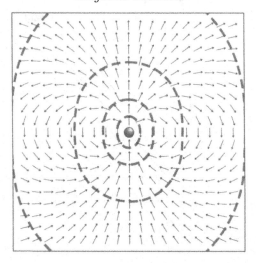

Fig. 4.8. Magnetic field of a dipole. Dipole **m** is oriented toward top of page. Vectors indicate the direction of **B** that would be observed at the center of each vector. Dashed contours indicate constant values of |**B**|, the value decreasing by a factor of 10 at each succeeding contour from the dipole; that is, if the closest contour to the dipole has the value |**B**| = 1, succeeding contours have the values 0.1, 0.01, and 0.001, respectively.

equation 4.14; it calculates the x, y, and z components of magnetic induction at any point (other than $r = 0$) due to a single dipole.

Exercise 4.2 Use equation 4.14 to prove that the field of a dipole satisfies $\nabla \cdot \mathbf{B} = 0$, even at $r = 0$. Hint: Use polar coordinates.

Figure 4.9 shows four limiting examples of the magnetic induction that would be measured on a horizontal surface above single dipoles: the vertical component of **B** due to a vertical dipole, vertical component of **B** due to a horizontal dipole, horizontal component of **B** due to a vertical dipole, and horizontal component of **B** due to a horizontal dipole. Figure 4.10 shows the same examples in profile form. Note how the "broadness" of the contours and profiles in Figures 4.9 and 4.10 depends on the depth z of the dipole. In particular, the horizontal distance between zero-crossings is $z\sqrt{2}$ for the horizontal component of **B** over a horizontal dipole, and $2z\sqrt{2}$ for the vertical component of a vertical dipole. We will exploit this general property of magnetic (and gravity) fields in a later chapter in order to estimate depth of sources from the broadness of magnetic and gravity anomalies. Also notice that

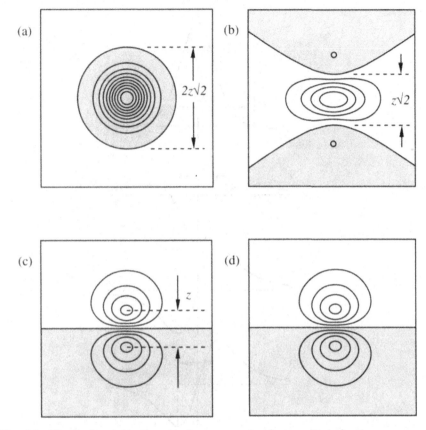

Fig. 4.9. The horizontal and vertical components of magnetic induction measured on a horizontal surface above horizontal and vertical dipoles. Contour interval is arbitrary. Gray regions indicate positive values of magnetic induction. (a) Vertical component of **B** due to vertical dipole; (b) horizontal component of **B** due to horizontal dipole; (c) horizontal component of **B** due to vertical dipole; and (d) vertical component of **B** due to horizontal dipole.

the horizontal component of **B** due to a vertical dipole equals the vertical component of **B** due to a horizontal dipole.

The symmetry of dipolar magnetic fields is apparent in Figures 4.9 and 4.10, and it is worthwhile to note the various components and magnitude of **B** in cylindrical coordinates. Equation 4.14 can be written as

$$\mathbf{B} = C_\mathrm{m} \frac{m}{r^3} \left[3 \cos \theta \, \hat{\mathbf{r}} - \hat{\mathbf{m}} \right],$$

Magnetic Potential

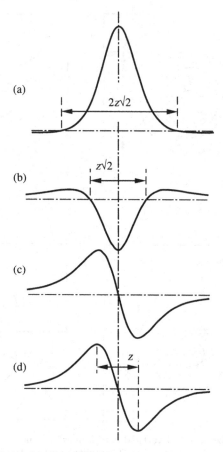

Fig. 4.10. The horizontal and vertical components of magnetic induction due to horizontal and vertical dipoles. Profiles are directly over dipole and parallel to the horizontal dipole. Labels a through d have the same meaning as described in the caption to Figure 4.9.

where θ is the angle between $\hat{\mathbf{m}}$ and $\hat{\mathbf{r}}$, as shown in Figure 4.6. Then

$$\mathbf{B} = -\nabla_P V,$$

$$B_r = -\frac{\partial}{\partial r} V,$$

$$= 2C_\mathrm{m} \frac{m \cos \theta}{r^3},$$

$$B_\theta = -\frac{1}{r} \frac{\partial}{\partial \theta} V$$

$$= C_{\mathrm{m}} \frac{m \sin \theta}{r^3},$$

$$|\mathbf{B}| = C_{\mathrm{m}} \frac{m}{r^3} \left[3 \cos^2 \theta + 1 \right]^{\frac{1}{2}}. \tag{4.15}$$

These equations show that the magnitude $|\mathbf{B}|$ of the dipole field along *any* ray extending from the dipole decreases at a rate inversely proportional to the cube of the distance to the dipole. The magnitude also depends on θ; for example, $|\mathbf{B}|$ is twice as great at a point along the axis of the dipole ($\theta = 0$ or $\theta = \pi$) as at an equivalent distance perpendicular to the dipole ($\theta = \pi/2$).

Although the derivations for the dipole moment are rather conceptual, many magnetic bodies exist in nature that are dipolar to first approximation. It's simply a matter of scale. To the nuclear physicist, the Bohr magneton, the fundamental magnetic moment caused by spinning and orbiting charged particles, is dipolar in nature. To an exploration geophysicist, a ground-based magnetic survey over a buried pluton may show a complex array of magnetic anomalies, but at the altitude of an aeromagnetic survey, the inhomogeneities of the pluton may average out so that the pluton appears similar to a dipole source. Indeed, the entire field of the earth appears nearly dipolar from the perspective of the other planets.

4.6 Problem Set

1. Magnetic induction is measured along a horizontal profile in the x direction directly above a single dipole located at a depth d below the profile. Derive expressions for the following horizontal distances in terms of d.

   (a) For a vertical dipole:

      i. Distance between zero-crossings of B_z.
      ii. Distance between maximum horizontal gradients of B_z.
      iii. Distance between maximum and minimum values of B_x.

   (b) For a horizontal dipole:

      i. Distance between zero-crossings of B_x for a horizontal dipole pointing in the x direction.
      ii. Distance between maximum horizontal gradients of B_x for a horizontal dipole pointing in the x direction.
      iii. Distance between maximum and minimum values of B_z for a horizontal dipole pointing in the x direction.

iv. Consider B_x measured along the x axis directly over a horizontal dipole oriented in the y direction. Show that B_x is of one sign along the entire profile.

2. Write a program that calculates the magnetic field of a dipole on a horizontal surface (Subroutine B.3 in Appendix B may be helpful). Use it to graphically demonstrate the foregoing relationships.

3. Let U be the Newtonian potential at a point P due to a point mass located at Q, and let V be the magnetic potential at P due to a dipole also located at Q. The dipole has a moment $\hat{m}$. Show that $V = C\hat{m} \cdot \nabla U$, where C is a constant. What is the value of C?

4. Let x, y, and z be orthogonal axes with z directed downward. A single magnetic dipole with moment $\mathbf{m}$ is located at $(0, 0, d)$ and directed at an angle I below the horizontal plane. Orient the coordinate system so that $\mathbf{m}$ lies in the x, z plane and consider the magnetic induction $\mathbf{B}$ as viewed along the x axis.

(a) Show that

$$
\begin{bmatrix} B_x \\ B_y \\ B_z \end{bmatrix} = \beta \begin{bmatrix} 2\alpha^2 - 1 & -3\alpha \\ 0 & 0 \\ -3\alpha & 2 - \alpha^2 \end{bmatrix} \begin{bmatrix} \cos I \\ \sin I \end{bmatrix},
$$

where $\alpha = x/d$ (dimensionless) and where

$$
\beta = \frac{C_m m}{d^3 (\alpha^2 + 1)^{\frac{5}{2}}}.
$$

(b) Sketch the horizontal and vertical components of $\mathbf{B}$ along the x axis for $I = 60°$.

5

Magnetization

Magnetes Geheimnis, erkläre mir das!
Kein grösser Geheimnis als Lieb und Hass.
[The mystery of magnetism, explain that to me!
no greater mystery, except love and hate.]
(Johann Wolfgang von Goethe)

We know that the magnet loves the lodestone,
but we do not know whether the lodestone
also loves the magnet or is attracted to it
against its will.
(Arab physicist of twelfth century)

5.1 Distributions of Magnetization

The magnetic induction **B** of an elemental dipole was derived in Chapter 4 by examining the magnetic induction of a vanishingly small loop of electrical current. Accordingly, the bulk magnetic properties of a volume of material can be considered either in terms of the net magnetic effect of all the dipoles within the volume, or in terms of the net effect of all the elemental electrical currents. Using the former concept, we define a vector quantity called *magnetization* **M** as follows: The magnetization of a volume V is defined as the vector sum of all the individual dipole moments $\mathbf{m}_i$ divided by the volume, that is,

$$\mathbf{M} = \frac{1}{V} \sum_i \mathbf{m}_i .$$

Magnetization is reported in units of ampere·meter^{-1} in the *Système Internationale* (SI) and in units of gauss† in the electromagnetic system (emu), where 1 gauss $= 10^3$ A·m^{-1}.

In Chapter 3, we considered a volume of mass with density $\rho(x, y, z)$ to be composed of a great many small masses $dm = \rho(x, y, z)\, dv$, each small mass acting like a point source. This led to an integral equation for the potential of a volume density distribution,

$$U(P) = \gamma \int_R \frac{\rho(Q)}{r}\, dv. \tag{5.1}$$

We do likewise here for magnetic sources. A small element of magnetic material with magnetization $\mathbf{M}$ can be considered to act like a single dipole $\mathbf{M}\, dv = \mathbf{m}$. The potential as observed at point P is given by

$$V(P) = -C_{\mathrm{m}}\, \mathbf{M} \cdot \nabla_P \frac{1}{r}\, dv,$$

where r is distance from P to the dipole. As discussed in Chapter 4, the constant C_{m} is used to balance units and has a value that depends on the system in use. In the emu system $C_{\mathrm{m}} = 1$ and is dimensionless, whereas in SI units $C_{\mathrm{m}} = \frac{\mu_0}{4\pi} = 10^{-7}$ henry·meter^{-1}, where μ_0 is the permeability of free space. In general, magnetization $\mathbf{M}$ is a function of position, where both direction and magnitude can vary from point to point, that is, $\mathbf{M} = \mathbf{M}(Q)$, where Q is the position of dv. Integrating this equation over all of the elemental volumes provides the potential of a distribution of magnetization

$$V(P) = C_{\mathrm{m}} \int_R \mathbf{M}(Q) \cdot \nabla_Q \frac{1}{r}\, dv, \tag{5.2}$$

analogous to equation 5.1. Magnetic induction at P is given by

$$\mathbf{B}(P) = -\nabla_P V(P)$$

$$= -C_{\mathrm{m}} \nabla_P \int_R \mathbf{M}(Q) \cdot \nabla_Q \frac{1}{r}\, dv. \tag{5.3}$$

In these last equations, we have changed the subscript of the gradient operator from P to Q when the operator is inside the volume integral. This is to indicate that the gradient is to be taken with respect to the source coordinates rather than with respect to the observation point.

† Magnetization is often reported in the geophysical literature in units of emu·cm^{-3}, where 1 emu·cm^{-3} $= 1$ gauss.

Exercise 5.1 Show that $\nabla_P \frac{1}{r} = -\nabla_Q \frac{1}{r}$.

5.1.1 Alternative Models

Distributions of Currents

Magnetization is the net effect of all elemental currents within the magnetic media. It should seem reasonable that the circulating current of one dipole will just cancel the current of its neighboring dipole if the dipoles are parallel to each other and have identical magnitude. If all dipole moments within a volume of matter are aligned parallel to one another and are uniformly distributed throughout the volume, then the net effect of all elemental currents will vanish except at the surface of the material (Figure 5.1). At the surface, all elemental currents coalesce into a surface current density $\mathbf{I_s}$ given by

$$\mathbf{I_s} = \mathbf{M} \times \hat{\mathbf{n}},$$

where $\hat{\mathbf{n}}$ is the unit vector normal to the magnetic material. If the magnetization is not uniform within the volume, then a volume current will exist at points where the circulating elemental currents fail to cancel. The volume current density $\mathbf{I_v}$ is given by the curl of the magnetization,

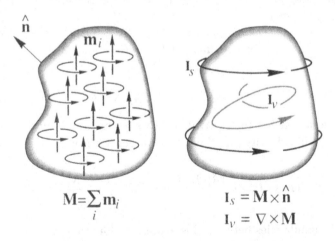

$$\mathbf{M} = \sum_i \mathbf{m}_i$$

$$\mathbf{I}_s = \mathbf{M} \times \hat{\mathbf{n}}$$
$$\mathbf{I}_v = \nabla \times \mathbf{M}$$

Fig. 5.1. Magnetization $\mathbf{M}$ of a volume is the vector sum of all dipole moments $\mathbf{m}_i$ divided by the volume. Magnetization can also be regarded in terms of the sum of all the elemental currents associated with the dipoles. The elemental currents coalesce into two components: a volume current $\mathbf{I_v}$ and a surface current $\mathbf{I_s}$.

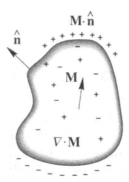

Fig. 5.2. A distribution of magnetization can be characterized as surface and volume distributions of magnetic charge.

as in Maxwell's equation 4.8; that is,

$$\mathbf{I_v} = \nabla \times \mathbf{M}.$$

In other words, a distribution of dipole moments is equivalent to a distribution of electrical currents, some on the surface of the volume and some within the volume. Both the dipole and current models provide the same magnetic induction outside or inside the material (Panofsky and Phillips [201]).

Distributions of Magnetic Charge

A vector identity from Appendix A,

$$\nabla \cdot (\phi \mathbf{A}) = \nabla \phi \cdot \mathbf{A} + \phi \nabla \cdot \mathbf{A},$$

allows equation 5.2 to be divided into a volume and surface integral. Applying this identity and then the divergence theorem to equation 5.2 yields

$$V(P) = C_{\mathrm{m}} \int_S \frac{\mathbf{M}(Q) \cdot \hat{\mathbf{n}}}{r} \, dS - C_{\mathrm{m}} \int_R \frac{\nabla \cdot \mathbf{M}(Q)}{r} \, dv. \qquad (5.4)$$

Note the similarities between this expression for magnetic potential and the expressions for gravitational potential of surface and volume densities (equations 3.5 and 3.10, respectively). Apparently the quantities $\nabla \cdot \mathbf{M}(Q)$ and $\mathbf{M}(Q) \cdot \hat{\mathbf{n}}$ are analogous to volume density and surface density, respectively. In fact, these quantities can be treated as volume and surface distributions of *magnetostatic charge* (Figure 5.2) given by

$$Q_v = -\nabla \cdot \mathbf{M}$$

$$Q_s = \mathbf{M} \cdot \hat{\mathbf{n}},$$

respectively, and substituting into equation 5.4 provides

$$V(P) = C_m \int_S \frac{Q_s}{r} \, dS + C_m \int_R \frac{Q_v}{r} \, dv.$$

Although magnetic charge does not physically exist, the conceptual model is a useful one as we will see in Chapter 9.

Hence, we have seen three ways to characterize the magnetization of a material: as a distribution of magnetic dipoles, as distributions of electrical current on the surface and within the material, and as surface and volume distributions of magnetostatic charge. All three concepts are used in geophysical applications and will reappear in future chapters.

5.2 Magnetic Field Intensity

Equation 4.8 in Chapter 4 shows that magnetic induction $\mathbf{B}$ originates from all currents, both microscopic and macroscopic. Now we consider a second kind of magnetic field which proves useful in the presence of magnetic materials. This new field is just that part of $\mathbf{B}$ arising from all currents other than atomic-level currents associated with magnetization. Assuming that dielectric properties of the region of interest are negligible (a good assumption in most geophysical applications), equation 4.8 can be written

$$\nabla \times \mathbf{B} = 4\pi C_m \left(\mathbf{I}_m + \nabla \times \mathbf{M} \right). \tag{5.5}$$

From here on, the derivation differs between the emu system and SI units. In each case, we begin with equation 5.5.

In the emu system,

$$\nabla \times (\mathbf{B} - 4\pi C_m \mathbf{M}) = 4\pi C_m \mathbf{I}_m,$$

$$\nabla \times \mathbf{H} = 4\pi C_m \mathbf{I}_m,$$

$$\mathbf{H} = \mathbf{B} - 4\pi \mathbf{M}, \tag{5.6}$$

whereas in SI units,

$$\nabla \times \left(\frac{\mathbf{B}}{4\pi C_{\mathrm{m}}} - \mathbf{M} \right) = \mathbf{I}_{\mathrm{m}},$$

$$\nabla \times \mathbf{H} = \mathbf{I}_{\mathrm{m}},$$

$$\mathbf{H} = \frac{\mathbf{B}}{\mu_0} - \mathbf{M}. \qquad (5.7)$$

Note that C_{m} has been replaced in equations 5.6 and 5.7 by its appropriate value for the particular system of units.

Vector $\mathbf{H}$ is a new quantity called the *magnetic field intensity*, and, as can be seen from the defining equations, is simply magnetic induction (except for a factor μ_0 in SI units) minus the effects of magnetization. Field intensity has units of oersteds (Oe) in the emu system and units of ampere·meter^{-1} in SI units. The conversion between the two systems is 1 Oe = $\frac{10^3}{4\pi}$ A·m^{-1}. Note that gauss and oersted units have equivalent magnitude and dimensions in the emu system; the first quantity is used for magnetic induction, the latter for magnetic field intensity. Moreover, $\mathbf{H}$ and $\mathbf{B}$ in the emu system are identical outside of magnetic materials. In SI units, however, $\mathbf{H}$ and $\mathbf{B}$ have the same direction outside of magnetic materials but differ in both magnitude and dimensions.

Equations 5.6 and 5.7 show that the magnetic field intensity is a hybrid vector function composed of two components with quite different physical meanings. Whereas magnetic induction $\mathbf{B}$ originates from all currents, both atomic and macroscopic, magnetic field intensity $\mathbf{H}$ arises only from true currents (again ignoring displacement currents). Magnetic field intensity has a remarkable property that can be seen from equations 4.9 and 5.7, namely, that

$$\oint \mathbf{H} \cdot \mathbf{dl} = \int_S \mathbf{I}_{\mathrm{m}} \cdot \hat{\mathbf{n}} \, dS \qquad (\mathrm{SI}),$$

that is, the line integral of field intensity around any closed loop is equal to the total macroscopic current crossing the surface bounded by the loop. In the absence of such currents,

$$\nabla \times \mathbf{H} = 0,$$

and a scalar potential exists such that

$$\mathbf{H} = -\nabla V'.$$

There is little to be gained from the distinction between **B** and **H** in most geophysical measurements of the magnetic field of the earth. Measurements of the earth's magnetic field, whether from aircraft, ship, or satellite, are made in environments very nearly free of magnetic material (e.g., Lowes [169]). Electrical currents can pose problems in some situations, such as problems associated with ionospheric currents at satellite altitudes, but this issue is not relevant to our choice of **B** versus **H**. Indeed, the geophysical literature dealing with the interpretation of magnetic anomalies uses both **B** and **H** interchangeably. Here we will attempt to use **B** whenever possible.

5.3 Magnetic Permeability and Susceptibility

Materials can acquire a component of magnetization in the presence of an external magnetic field. For low-amplitude magnetic fields, say on the order of the earth's magnetic field, this induced magnetization is proportional in magnitude and is parallel (or antiparallel) in direction to the external field, that is,

$$\mathbf{M} = \chi \mathbf{H}. \tag{5.8}$$

The proportionality constant χ is called the *magnetic susceptibility*. Equation 5.8 is the same in both the SI and emu systems. Susceptibility is dimensionless in both systems but differs in magnitude by 4π: Susceptibility in emu equals 4π times susceptibility in SI units.

A related quantity, the *magnetic permeability* μ, differs slightly between the two systems, and separate derivations are necessary. Starting with equations 5.6 and 5.7, respectively, the derivations are as follows.

In the emu system,

$$\mathbf{B} = \mathbf{H} + 4\pi\mathbf{M}$$

$$= \mathbf{H} + 4\pi\chi\mathbf{H}$$

$$= (1 + 4\pi\chi)\mathbf{H}$$

$$= \mu\mathbf{H},$$

$$\mu = 1 + 4\pi\chi, \tag{5.9}$$

whereas in SI units,

$$\mathbf{B} = \mu_0(\mathbf{H} + \mathbf{M})$$
$$= \mu_0(\mathbf{H} + \chi\mathbf{H})$$
$$= \mu_0(1 + \chi)\mathbf{H}$$
$$= \mu\mathbf{H},$$
$$\mu = \mu_0(1 + \chi). \tag{5.10}$$

Kinds of Magnetization

Although χ and μ are derived in a simplistic mathematical way, they are in fact complex products of the atomic and macroscopic properties of the magnetic material. The relationship between $\mathbf{M}$ and $\mathbf{H}$ is not necessarily linear as implied by equation 5.8; χ may vary with field intensity, may be negative, and may be represented more accurately in some materials as a tensor. This section provides a very cursory description of the magnetization of solid materials. Indeed, this subject is worthy of its own textbook, and the interested reader is referred to books by Chikazumi [57] and Morrish [187] for information on magnetic materials in general, and to the book by Butler [47] for applications related to paleomagnetic and geomagnetic problems specifically. The implications of rock magnetism for magnetic-anomaly studies have been reviewed concisely and comprehensively by Reynolds et al. [243].

There are many kinds of magnetization. *Diamagnetism*, for example, is an inherent property of all matter. In diamagnetism, an applied magnetic field disturbs the orbital motion of electrons in such a way as to induce a small magnetization in the opposite sense to the applied field. Consequently, diamagnetic susceptibility is negative. *Paramagnetism* is a property of those solids that have atomic magnetic moments. Application of a magnetic field causes the atomic moments to partially align parallel to the applied field thereby producing a net magnetization in the direction of the applied field. Thermal effects tend to oppose this alignment, and paramagnetism vanishes in the absence of applied fields because thermal effects act to randomly orient the atomic moments. All minerals are diamagnetic and some are paramagnetic, but in either case these magnetizations are insignificant contributors to the geomagnetic field.

There is, however, a class of magnetism of great importance to geomagnetic studies. Certain materials not only have atomic moments, but neighboring moments interact strongly with each other. This interaction is a result of a quantum mechanical effect called *exchange energy*, which is beyond the scope of this book. Suffice it to say that the exchange energy causes a *spontaneous magnetization* that is many times greater than paramagnetic or diamagnetic effects. Such materials are said to be *ferromagnetic*. There are various kinds of ferromagnetic materials too, depending on the way that the atomic moments align. These include ferromagnetism proper, in which atomic moments are aligned parallel to one another; *antiferromagnetism*, where atomic moments are aligned antiparallel and cancel one another; and *ferrimagnetism*, in which atomic moments are antiparallel but do not cancel.

At the scale of individual mineral grains, spontaneous magnetization of a ferromagnetic material can be very large. At the outcrop scale, however, the magnetic moments of individual ferromagnetic grains may be randomly oriented, and the net magnetization may be negligible. The magnetization of individual grains is affected, however, by the application of a magnetic field, similar to but far greater in magnitude than for paramagnetism. Hence, rocks containing ferromagnetic minerals will acquire a net magnetization, called *induced magnetization* and denoted by $\mathbf{M_i}$, in the direction of an applied field $\mathbf{H}$, where

$$\mathbf{M_i} = \chi \mathbf{H}.$$

Of course the earth's magnetic field produces the same response in such materials, and the material is magnetic in its natural state. In small fields, with magnitudes comparable to the earth's magnetic field, the relationship between induced magnetization and applied field is essentially linear, and the susceptibility χ is constant.

Induced magnetization falls to zero if the rock is placed in a field-free environment. However, ferromagnetic materials also have the ability to retain a magnetization even in the absence of external magnetic fields. This permanent magnetization is called *remanent magnetization*, which we denote here by $\mathbf{M_r}$. In crustal materials, remanent magnetization is a function not only of the atomic, crystallographic, and chemical make-up of the rocks, but also of their geologic, tectonic, and thermal history. In geophysical studies, it is customary to consider the total magnetization $\mathbf{M}$ of a rock as the vector sum of its induced and remanent magnetizations, that is,

$$\mathbf{M} = \mathbf{M_i} + \mathbf{M_r}$$
$$= \chi \mathbf{H} + \mathbf{M_r}.$$

The relative importance of remanent magnetization to induced magnetization is expressed by the *Koenigsberger ratio*

$$Q = \frac{|\mathbf{M}_r|}{|\mathbf{M}_i|}$$

$$= \frac{\mathbf{M}_r}{\chi \mathbf{H}}.$$

In subsequent discussions regarding magnetic fields of crustal materials, we will consider just two kinds of magnetization: induced and remanent. It is well to keep in mind, however, that both of these magnetizations arise from spontaneous magnetization, a complex property of the ferromagnetic minerals in the earth's crust.

The spontaneous magnetization is dependent on temperature. As a material is heated, the spacing between neighboring atomic moments increases until a point is reached where the spontaneous magnetization falls to zero. This temperature is called the *Curie temperature.* Hence, both induced and remanent magnetizations vanish at temperatures greater than the Curie temperature. Paramagnetic and diamagnetic effects persist at these temperatures, but from the perspective of magnetic-anomaly studies we may consider rocks above the Curie temperature to be nonmagnetic.

Magnetite (Fe_3O_4) and its solid solutions with ulvospinel (Fe_2TiO_4) are the most important magnetic minerals to geophysical studies of crustal rocks. Other minerals, such as hematite, pyrrhotite, and alloys of iron and nickel, are important in certain geologic situations, but the volume percentage, size, shape, and history of magnetite grains are of greatest importance in most magnetic surveys. Magnetite is a ferrimagnetic material with a Curie temperature of about 580°C.

Typical values of $|\mathbf{M}_r|$ and χ for representative rock types are provided by Lindsley, Andreasen, and Balsley [165] and Carmichael [54]. Generally speaking, mafic rocks are more magnetic than silicic rocks. Hence, basalts are usually more magnetic than rhyolites, and gabbros are more magnetic than granites. Also, extrusive rocks generally have a higher remanent magnetization and lower susceptibility than intrusive rocks with the same chemical composition. Sedimentary and metamorphic rocks often have low remanent magnetizations and susceptibilities. These statements and the compilations referenced are only statistical guidelines with many exceptions. Values of $\mathbf{M}_r$ and χ often vary by several orders of magnitude within the same outcrop, for example. Whenever feasible, interpretation of a magnetic survey should include

investigation of the magnetic properties of representative rock samples
from the area of study.

5.4 Poisson's Relation

Equations 3.1 and 4.13 show that the magnetic scalar potential of an
element of magnetic material and the gravitational attraction of an element of mass have some obvious similarities; for example, they both
have magnitudes that are inversely proportional to the squared distance
to their respective point sources. We can use this similarity to derive a
surprising relationship between gravity and magnetic fields.

Consider a body with uniform magnetization $\mathbf{M}$ and uniform density
ρ (Figure 5.3). The magnetic scalar potential is given by equation 5.2,

$$V(P) = C_{\mathrm{m}} \int_R \mathbf{M} \cdot \nabla_Q \frac{1}{r} \, dv$$

$$= -C_{\mathrm{m}} \mathbf{M} \cdot \nabla_P \int_R \frac{1}{r} \, dv . \tag{5.11}$$

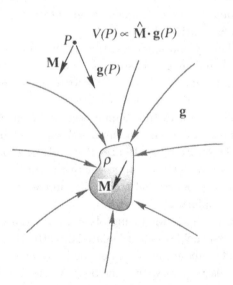

Fig. 5.3. Poisson's relation for a body with uniform magnetization and uniform
density. The magnetic potential at any point is proportional to the component
of gravity in the direction of magnetization.

The gravitational potential is written

$$U(P) = \gamma \int\limits_R \frac{\rho}{r}\, dv$$

$$= \gamma\rho \int\limits_R \frac{1}{r}\, dv,$$

so that

$$\int\limits_R \frac{1}{r}\, dv = \frac{U}{\gamma\rho}.$$

Substituting this last integral into equation 5.11 provides

$$V(P) = -\frac{C_\mathrm{m}}{\gamma\rho}\,\mathbf{M}\cdot\nabla_P U$$

$$= -\frac{C_\mathrm{m} M\, g_\mathrm{m}}{\gamma\rho}, \tag{5.12}$$

where g_m is the component of gravity in the direction of magnetization. Equation 5.12 is called *Poisson's relation*. It states that, if (a) the boundaries of a gravitational and magnetic source are the same and (b) the magnetization and density are uniform, then the magnetic potential is proportional to the component of gravitational attraction in the direction of magnetization (Figure 5.3).

What if the density and magnetization are not uniform? We can view both the gravity and magnetic sources as composed of elemental volumes. If the magnetization and density distributions are sufficiently well behaved, the density and magnetization within each elemental volume will approach a constant as the volume becomes arbitrarily small. Poisson's relation holds for each elemental volume, and by superposition must hold for the entire body. Hence, Poisson's relation is appropriate for any gravity and magnetic source where the intensity of magnetization is everywhere proportional to density and where the direction of magnetization is uniform.

Poisson's relation is an intriguing observation. With assumptions as stated, the magnetic field can be calculated directly from the gravity field without knowledge about the shape of the body or how magnetization and density are distributed within the body. Carried to its extreme, one might argue that magnetic surveys are unnecessary in geophysical investigations because they can be calculated directly from gravity surveys, or vice versa. In real geologic situations, of course, sources of gravity anomalies never have magnetization distributions in exact proportion to their density distributions. Nevertheless, Poisson's relation can be useful. First, it can be used to transform a magnetic anomaly into *pseudogravity*, the gravity anomaly that would be observed if the magnetization were replaced by a density distribution of exact proportions (Baranov [9]). We might wish to do this, not because we believe that such a mass actually exists, but because gravity anomalies have certain properties that simplify the determination of the shape and location of causative bodies. Thus, the pseudogravity transformation can be used to aid interpretation of magnetic data, a topic that will be discussed at some length in Chapter 12.

Second, Poisson's relation can be used to derive expressions for the magnetic induction of simple bodies when the expression for gravitational attraction is known. For example, the following sections use Poisson's relation to derive the magnetic induction of some simple bodies, such as spheres, cylinders, and slabs. We could do these derivations the hard way, by integrating equation 5.3. But we already know the gravitational attraction of these simple bodies because we derived them in Chapter 3. The magnetic expressions are more easily derived by simply applying Poisson's relation to the analogous gravitational expressions.

5.4.1 Example: A Sphere

From Chapter 3, the gravitational attraction of a solid sphere of uniform density is

$$\mathbf{g} = -\frac{4}{3}\pi a^3 \gamma \rho \frac{1}{r^2}\hat{\mathbf{r}}\,.$$

Substituting this into Poisson's relation yields the magnetic potential of a uniformly magnetized sphere (Figure 5.4), that is,

$$V = C_\mathrm{m}\frac{4}{3}\pi a^3 M \frac{1}{r^2}\,\hat{\mathbf{M}}\cdot\hat{\mathbf{r}}$$

$$= C_\mathrm{m}\frac{\mathbf{m}\cdot\hat{\mathbf{r}}}{r^2}\,,$$

where

$$\mathbf{m} = \frac{4}{3}\pi a^3 \mathbf{M}\,.$$

This is just the magnetic potential of a single dipole. Therefore, *the magnetic potential due to a uniformly magnetized sphere is identical to the magnetic potential of a dipole located at the center of the sphere with*

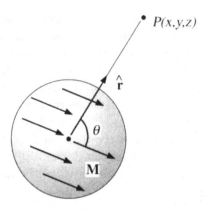

Fig. 5.4. Magnetic potential at point P due to a uniformly magnetized sphere.

dipole moment equal to the magnetization times the volume of the sphere. It follows that the magnetic field of a uniformly magnetized sphere is proportional to both its volume and its magnetization. Although the location of the center of the sphere can be determined directly from its magnetic field, the size of the sphere cannot be found without first knowing its magnetization. Subroutine B.3 in Appendix B calculates the three components of magnetic induction due to a uniformly magnetized sphere.

5.4.2 Example: Infinite Slab

As shown by equation 3.27, the gravitational attraction of an infinitely extended, uniformly dense layer is in the direction normal to the layer, is proportional to the thickness of the layer, and is independent of distance from the layer. For a horizontal layer,

$$\mathbf{g} = 2\pi\gamma\rho t\,\hat{\mathbf{k}},$$

where t is the thickness of the layer and $\hat{\mathbf{k}}$ is the unit vector directed toward and normal to the layer. If the magnetization is vertical, then Poisson's relation provides

$$V = -C_{\mathrm{m}}\frac{M}{\gamma\rho}\,g_z$$

$$= -2\pi C_{\mathrm{m}}\,M\,t,$$

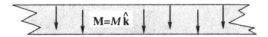

Fig. 5.5. Magnetic potential at point P of a vertically magnetized, infinitely extended slab.

Fig. 5.6. The magnetic field of a spherical cavity in a uniformly magnetized layer is equal to the field of a sphere with opposite magnetization.

and the magnetic potential of a uniformly magnetized slab is constant (Figure 5.5). Consequently, *the magnetic field of a uniformly magnetized slab is zero, and the slab cannot be detected through magnetic measurements alone.*

This remarkable fact can be used with the superposition principle to simplify certain problems. For example, the magnetic field caused by a spherical cavity within an infinitely extended layer with uniform magnetization $\mathbf{M}$ is identical to the field of an isolated sphere magnetized in the opposite direction, that is, with magnetization $-\mathbf{M}$ (Figure 5.6).

5.4.3 Example: Horizontal Cylinder

Equation 3.19 shows that the gravitational potential of an infinitely long cylinder with uniform density is given by

$$U(P) = 2\pi a^2 \rho \gamma \log \frac{1}{r},$$

where a is the radius of the cylinder and r is the perpendicular distance to the axis of the cylinder (Figure 5.7). Applying Poisson's relation to this expression provides the magnetic potential of a uniformly magnetized cylinder:

$$V = 2C_{\mathrm{m}}\pi a^2 \frac{\mathbf{M} \cdot \hat{\mathbf{r}}}{r}. \tag{5.13}$$

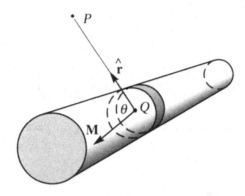

Fig. 5.7. Field at point P caused by a uniformly magnetized cylinder.

Note that, because $\hat{\mathbf{r}}$ is perpendicular to the cylinder, only the perpendicular components of $\mathbf{M}$ are significant to the cylinder's potential. In fact, the quantity $\pi a^2 \mathbf{M}$ has units of dipole moment per unit length, and equation 5.13 is equivalent to the potential of a line of dipoles located along the axis of the cylinder. Hence, *the potential and the magnetic field of a uniformly magnetized cylinder are identical to those of a line of dipoles*; that is, the potential of a line of dipoles is given by

$$V = 2C_{\mathrm{m}}\frac{\mathbf{m}' \cdot \hat{\mathbf{r}}}{r} \qquad (5.14)$$

where $\mathbf{m}'$ is dipole moment per unit length, and applying $\mathbf{B} = -\nabla V$ in cylindrical coordinates to equation 5.14 provides the magnetic field of a line of dipoles,

$$\mathbf{B} = \frac{2C_{\mathrm{m}}m'}{r^2}\left[2(\hat{\mathbf{m}}' \cdot \hat{\mathbf{r}})\hat{\mathbf{r}} - \hat{\mathbf{m}}'\right]. \qquad (5.15)$$

Exercise 5.2 What is the magnetic potential of an infinitely long cylinder magnetized in the direction parallel to its axis?

Note that the magnitude of the magnetic field of a line of dipoles is proportional to its dipole moment and inversely proportional to the square of its perpendicular distance from the observation point.

5.5 Two-Dimensional Distributions of Magnetization

The potential of a two-dimensional source can be derived from equation 5.2. The derivation is analogous to the derivation for gravitational potential of two-dimensional masses in Chapter 3. The body is first considered to be of finite length $2a$, as in Figure 3.6, and the volume integral in equation 5.2 is replaced with a surface integral over the cross-sectional area of the body and a line integral along its length:

$$V(P) = C_{\mathrm m} \int_R \mathbf{M}(Q) \cdot \nabla_Q \frac{1}{r} \, dv$$

$$= C_{\mathrm m} \int_S \mathbf{M}(Q) \cdot \left(\int_{-a}^{a} \nabla_Q \frac{1}{r} \, dz \right) dS ,$$

where S is the cross-sectional area of the body. As $a \to \infty$, the inner integral approaches the potential of an infinite line of dipoles of unit magnitude. Hence,

$$V(P) = 2C_{\mathrm m} \int_S \frac{\mathbf{M}(Q) \cdot \hat{\mathbf{r}}}{r} \, dS \qquad (5.16)$$

is the potential of a two-dimensional distribution of magnetic material, and applying $\mathbf{B} = -\nabla V$ provides

$$\mathbf{B}(P) = 2C_{\mathrm m} \int_S \frac{M(Q)}{r^2} \left[2(\hat{\mathbf{M}} \cdot \hat{\mathbf{r}})\hat{\mathbf{r}} - \hat{\mathbf{M}} \right] dS . \qquad (5.17)$$

As before, $\hat{\mathbf{r}}$ is understood to be normal to the long axis of the cylinder and r is the perpendicular distance. Notice that the component of magnetic induction parallel to the long axis of the body is zero, and the component of magnetization parallel to the long axis of the body contributes nothing to the magnetic field.

As discussed in Chapter 3 for the gravitational case, the magnetic field of an infinitely long body, uniform in the direction parallel to the long dimension of the body, can be thought of as originating from a two-dimensional source. The source is essentially a thin wafer corresponding to the intersection of the body and a normal plane containing the observation point (see Figure 3.7).

5.6 Annihilators

A nonzero distribution of magnetization (or density) that produces no external field for a particular source geometry is called an *annihilator* (Parker [207]). The annihilator can be added to its respective source geometry without affecting the magnetic (or gravity) field. We see that the magnetic annihilator for infinite slabs is just M = constant. A horizontal layer with variable magnetization M(x, y, z), therefore, is indistinguishable from a horizontal layer with the same magnetization plus any constant, that is, M(x, y, z) + M_0. Hence, an infinite variety of magnetizations can be conceived for the infinite slab, all producing precisely the same magnetic field. We will have more to say about annihilators in later chapters.

5.7 Problem Set

1. Use Poisson's relation to find the magnetic field of a uniformly magnetized spherical shell.

2. What nonzero distribution of magnetization within a sphere produces no magnetic field (i.e., what is the magnetic annihilator for a spherical source)?

3. Use Poisson's relation to prove the general statement that a uniformly magnetized slab has no magnetic attraction, regardless of the direction of magnetization.

4. Consider a line of dipoles aligned parallel to the y axis, with dipole moment per unit length $\mathbf{m} = (m_x, m_y, m_z)$, and intersecting the x, z plane at coordinates (x', z').

   (a) Let $f = 2C_m/(x'^2 + z'^2)^2$, $g = x'^2 - z'^2$, and $h = 2x'z'$. Show that the three components of magnetic induction observed at $(0, 0, 0)$ are given by

$$
\begin{bmatrix} B_x \\ B_y \\ B_z \end{bmatrix} = f \begin{bmatrix} g & 0 & h \\ 0 & 0 & 0 \\ h & 0 & -g \end{bmatrix} \begin{bmatrix} m_x \\ m_y \\ m_z \end{bmatrix}. \tag{5.18}
$$

   (b) Use this expression to show that the horizontal component of **B** due to a line of dipoles magnetized vertically equals the vertical component of **B** due to a line of dipoles magnetized horizontally.

   (c) Show that the vertical component of **B** due to a line of dipoles magnetized vertically equals the negative of the horizontal component of a line of dipoles magnetized horizontally.

5. A spherical cavity of radius a is buried with its center at depth a in a vertically and homogeneously magnetized layer. The layer has thickness t ($t > 2a$) and magnetization $\mathbf{M} = M\hat{\mathbf{k}}$.

   (a) Find an expression for **B** in terms of M at the point directly over the cavity.

(b) What is the dependence of **B** on a?

6. The crust of Planet X is permanently magnetized proportional to the field of an ancient centered dipole. The ancient field no longer exists. Consider the crust to be a shell with radii a_1 and a_2 and show that the field of the crust is zero at all external points. (Note: This problem is a demonstration of a more comprehensive theorem discussed in detail by Runcorn [248]: A shell magnetized proportional to the field of *any* internal source produces no external field. The magnetization is therefore an annihilator for the shell.)

6

Spherical Harmonic Analysis

But that to say in difficult problems the use of spherical harmonics is
laborious is not to slight the method, because any other accurate
treatment would be still more difficult.

(Sydney Chapman and Julius Bartels)

Gravity is the ballast of the soul, which keeps the mind steady.

(Thomas Fuller)

Physical quantities measured on or above the earth's surface are naturally suited to mathematical descriptions in spherical coordinates. The
most common such framework is *spherical harmonic analysis*. Spherical
harmonic analysis can be useful for any reasonably well-behaved, global
phenomenon and has been applied to a diverse range of data, from free
oscillations of the earth to global climate change. At the very least,
spherical harmonic analysis provides a way to synthesize from a scatter
of discrete measurements on a sphere an equation applicable to the entire
sphere. Such an equation then can be used to interpolate the behavior
of the phenomenon to regions of the sphere that have no measurements.

As the name implies, however, spherical harmonic analysis takes on
special meaning when applied to potential fields, because the building blocks of spherical harmonic analysis are a natural consequence of
Laplace's equation in spherical coordinates. In particular, the various
terms of a spherical harmonic expansion are sometimes related (with
caution) to specific physical phenomena. The most well-known example
is the separation of the dipole and nondipole components of the geomagnetic field: A spherical harmonic expansion based on discrete measurements of the geomagnetic field directly provides separate descriptions of
the dipole and nondipole fields, and these two fields often are attributed
to separate but linked processes in the earth's core.

This chapter outlines the basic principles of spherical harmonic analysis, relying heavily on the classic developments of Chapman and Bartels [56], Kellogg [146], Ramsey [235], and MacMillan [172]. Subsequent chapters will apply these principles to the gravity and magnetic fields of the earth.

6.1 Introduction

Before launching into spherical harmonic analysis, it will prove useful to start with a more familiar subject, Fourier analysis of periodic functions. A function $f(t)$, periodic over an interval T, can be synthesized by an infinite sum of weighted sinusoids, that is,

$$f(t) = \sum_{m=0}^{\infty} \left(a_m \cos \frac{2\pi mt}{T} + b_m \sin \frac{2\pi mt}{T} \right), \qquad (6.1)$$

where a_m and b_m are weighting coefficients which, as we shall see shortly, are determined directly from $f(t)$. Equation 6.1 is the well-known *Fourier series*. But why should sinusoids be used as the building blocks instead of some other function, say exponentials or logarithms? One reason is that $f(t)$ is periodic and so are sines and cosines. Another reason is that sines and cosines have a property called *orthogonality*, and this property simplifies the determination of the *best* coefficients a_m and b_m in equation 6.1. The property of orthogonality for sines and cosines is demonstrated by the following three integrals:

$$\int_0^T \cos \frac{2\pi mt}{T} \cos \frac{2\pi nt}{T} \, dt = \begin{cases} \frac{T}{2}, & \text{if } m = n \neq 0; \\ 0, & \text{if } m \neq n; \\ T, & \text{if } m = n = 0; \end{cases}$$

$$\int_0^T \sin \frac{2\pi mt}{T} \sin \frac{2\pi nt}{T} \, dt = \begin{cases} \frac{T}{2}, & \text{if } m = n; \\ 0, & \text{if } m \neq n; \end{cases}$$

$$\int_0^T \sin \frac{2\pi mt}{T} \cos \frac{2\pi nt}{T} \, dt = 0.$$

If equation 6.1 is multiplied by either $\sin \frac{2\pi nt}{T}$ or $\cos \frac{2\pi nt}{T}$ and integrated over the period T, all terms of the infinite sum are zero except one! The

Spherical Harmonic Analysis

infinite series then reduces to integral expressions for a_n and b_n,

$$a_n = \frac{2}{T} \int_0^T f(t) \cos \frac{2\pi n t}{T}\, dt, \qquad n = 1, 2, \ldots,$$

$$b_n = \frac{2}{T} \int_0^T f(t) \sin \frac{2\pi n t}{T}\, dt, \qquad n = 1, 2, \ldots,$$

$$a_0 = \frac{1}{T} \int_0^T f(t)\, dt .$$

Hence, the fact that sinusoids are orthogonal over the period of $f(t)$ provides an easy way to calculate a_n and b_n directly from $f(t)$.

Now let's wrap $f(t)$ around the circumference of a circle and let t be the angle from some fixed radius (Figure 6.1). The fundamental period T becomes 2π and equation 6.1 becomes

$$f(t) = \sum_{m=0}^{\infty} (a_m \cos mt + b_m \sin mt) .$$

Thus the function $f(t)$ defined by position on the circle is now represented by a sum of sinusoids about that circle.

Spherical harmonic analysis serves the same purpose for representing a function defined by position on a sphere. Let $f(\theta, \phi)$ be such a

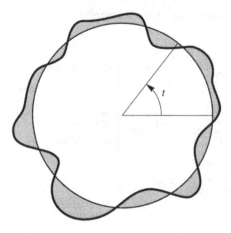

Fig. 6.1. Function $f(t)$ defined by position on a circle. The function has a fundamental period of 2π.

function, where θ is colatitude and ϕ is longitude (see Appendix A for a review of the spherical coordinate system), and consider just one circle of colatitude θ_0. Along this colatitude, $f(\theta_0, \phi)$ is a function of ϕ alone, has a fundamental period of 2π, and can be represented as before,

$$f(\theta_0, \phi) = \sum_{m=0}^{\infty} (a_m \cos m\phi + b_m \sin m\phi) \, .$$

A similar equation could be written for any colatitude on the sphere, each colatitude having its own set of coefficients. In other words, the coefficients are themselves functions of colatitude, and

$$f(\theta, \phi) = \sum_{m=0}^{\infty} (a_m(\theta) \cos m\phi + b_m(\theta) \sin m\phi) \, . \tag{6.2}$$

The question remains as to how best to formulate the coefficients $a_m(\theta)$ and $b_m(\theta)$, and that is the subject of the next two sections.

6.2 Zonal Harmonics

The dependence of $f(\theta, \phi)$ on θ is completely specified by the coefficients $a_m(\theta)$ and $b_m(\theta)$, but what form is most appropriate for these coefficients? We might expect that they could be approximated by a sum of weighted functions, just as the dependence on ϕ was approximated by a sum of weighted sinusoids. We also expect that the weighted functions should be orthogonal and periodic around any meridian.

The technique of least-squares analysis is a logical way to proceed. We can approximate the dependence of $f(\theta, \phi)$ on θ by a finite sum of weighted functions, requiring that the squared difference between $f(\theta, \phi)$ and the finite summation be a minimum when averaged over the surface of the sphere. (Fourier-series aficionados will recognize that a similar least-squares investigation of $f(t)$ over period T will discover the "best" weights for sines and cosines too.)

First let $f(\theta, \phi)$ be independent of ϕ so that equation 6.2 becomes

$$\begin{aligned} f(\theta, \phi) &= f(\theta) \\ &= a_0(\theta) \, , \end{aligned}$$

and approximate $f(\theta)$ by a weighted, finite sum of $k + 1$ orthogonal functions $P_n(\theta)$,

$$f_k(\theta) = C_0 P_0(\theta) + C_1 P_1(\theta) + \cdots + C_k P_k(\theta) \, . \tag{6.3}$$

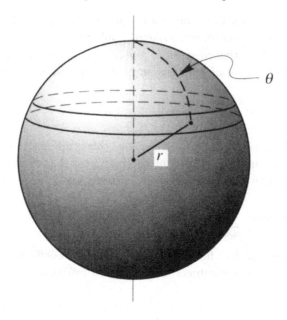

Fig. 6.2. Element of a spherical surface. The ribbon encircling the sphere has an area $2\pi r^2 \sin\theta \, d\theta$.

We wish to minimize the squared difference between $f(\theta)$ and $f_k(\theta)$ as averaged over the entire spherical surface (Figure 6.2). If the radius of the sphere is r, then the total area of the sphere is $4\pi r^2$, an element of area is $2\pi r^2 \sin\theta \, d\theta$, and the total squared error averaged over the spherical surface is given by

$$E = \frac{1}{4\pi r^2} \int_0^\pi \Big[f(\theta) - f_k(\theta) \Big]^2 2\pi r^2 \sin\theta \, d\theta$$

$$= \frac{1}{2} \int_0^\pi \Big[f(\theta) - f_k(\theta) \Big]^2 \sin\theta \, d\theta .$$

To simplify matters somewhat, we can make the substitutions $\mu = \cos\theta$ and $d\mu = -\sin\theta \, d\theta$, and the integral becomes

$$E = \frac{1}{2} \int_{-1}^{1} \Big[f(\mu) - f_k(\mu) \Big]^2 d\mu$$

$$= \frac{1}{2} \int_{-1}^{1} \left[f^2(\mu) - 2 f(\mu) f_k(\mu) + f_k^2(\mu) \right] d\mu . \tag{6.4}$$

Next we substitute the summation 6.3 into equation 6.4 and apply the orthogonality condition

$$\int_{-1}^{1} P_n(\mu) P_m(\mu) \, d\mu = \begin{cases} A_n, & \text{if } n = m; \\ 0, & \text{if } n \neq m, \end{cases}$$

where A_n is a constant. Then solving the least-squares condition

$$\frac{\partial E}{\partial C_l} = 0$$

for each C_l leads to

$$C_l = \frac{1}{A_l} \int_{-1}^{1} f(\mu) P_l(\mu) \, d\mu . \tag{6.5}$$

Exercise 6.1 Follow the previous instructions to derive equation 6.5 from equation 6.4.

Hence, by approximating $f(\theta)$ as a weighted sum of orthogonal functions, we find that the best weights, in a least-squares sense, are given by equation 6.5, an integral expression involving $f(\theta)$ itself. A remarkable property of orthogonal series is implicit in this result: The calculation of C_l is independent of k; that is, each C_l is the best possible coefficient regardless of how long the series is or how many terms are missing from the series.

At this point, any set of functions orthogonal over the interval $-1 \leq \mu \leq 1$ would do. One well-known set of functions is particularly appropriate: the *Legendre polynomials*, also known as *Legendre functions* or *zonal functions*. Legendre polynomials are given by *Rodrigues's formula*

$$P_n(\mu) = \frac{1}{n! \, 2^n} \frac{d^n}{d\mu^n} (\mu^2 - 1)^n , \tag{6.6}$$

where n is the degree of the polynomial. The first five Legendre polynomials are given in Table 6.1, and Figure 6.3 shows several of these graphically. One can prove (Ramsey [235]) that

$$\int_{-1}^{1} P_n(\mu) P_{n'}(\mu) \, d\mu = \begin{cases} 0, & \text{if } n \neq n'; \\ \frac{2}{2n+1}, & \text{if } n = n', \end{cases} \tag{6.7}$$

Spherical Harmonic Analysis

Table 6.1. *Legendre polynomials of degree 0 through 5.*

n	$P_n(\theta)$	$P_n(\mu)$
0	1	1
1	$\cos\theta$	μ
2	$\frac{1}{4}(3\cos 2\theta + 1)$	$\frac{1}{2}(3\mu^2 - 1)$
3	$\frac{1}{8}(5\cos 3\theta + 3\cos\theta)$	$\frac{1}{2}(5\mu^3 - 3\mu)$
4	$\frac{1}{64}(35\cos 4\theta + 20\cos 2\theta + 9)$	$\frac{1}{8}(35\mu^4 - 30\mu^2 + 3)$
5	$\frac{1}{128}(63\cos 5\theta + 35\cos 3\theta + 30\cos\theta)$	$\frac{1}{8}(63\mu^5 - 70\mu^3 + 15\mu)$

which is a demonstration of the orthogonality of Legendre functions over the appropriate interval. The reasons for selecting these particular orthogonal functions will become clear later in this chapter.

Hence, the function $f(\mu)$ (or $f(\theta)$) can be represented as an infinite sum of weighted Legendre functions,

$$f(\mu) = \sum_{n=0}^{\infty} C_n P_n(\mu), \qquad (6.8)$$

called a *zonal expansion*. It should be clear from the previous discussion how to determine the coefficients in this summation. The orthogonality of $P_n(\mu)$ makes this quite simple, at least conceptually; if C_j is needed, for example, we simply multiply both sides of the preceding equation by $P_j(\mu)$ and integrate both sides of the equation over the interval $-1 \le \mu \le 1$. The jth term of the expansion is the only nonzero term, and this single term provides

$$C_j = \frac{2j+1}{2} \int_{-1}^{1} f(\mu) P_j(\mu)\, d\mu. \qquad (6.9)$$

Hence, the orthogonality property of Legendre functions allows the determination of each coefficient directly from the function being represented, just as sines and cosines were successful in this regard for one-dimensional functions.

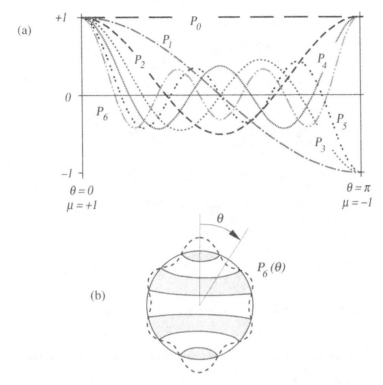

Fig. 6.3. A few low-degree Legendre functions. (a) Functions $P_0(\mu)$ through $P_6(\mu)$ are shown on the interval $-1 \leq \mu \leq 1$. (b) Function $P_6(\mu)$ is shown along the circumference of a circle; gray and white zones indicate areas where the function would be positive or negative, respectively, if wrapped around a sphere.

6.2.1 Example

An example is appropriate at this point to put all of the previous mathematics into perspective. Suppose $f(\theta, \phi)$ is a function of position on a sphere given by

$$f(\theta, \phi) = \begin{cases} +1, & \text{if } 0 \leq \theta < \frac{\pi}{2}; \\ -1, & \text{if } \frac{\pi}{2} < \theta \leq \pi, \end{cases}$$

as depicted in Figure 6.4. We can expect that such a discontinuous function would be difficult to represent with nicely behaved functions such as those in Figure 6.3, but let's give it a try.

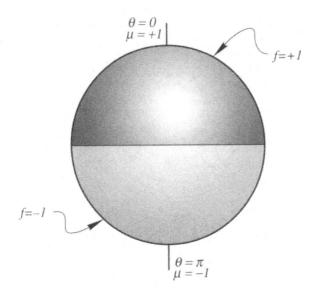

Fig. 6.4. Example of discontinuous $f(\theta, \phi)$ over surface of sphere.

First, we represent $f(\theta, \phi)$ as an infinite sum of Legendre functions.

$$f(\theta, \phi) = f(\theta)$$
$$= f(\mu)$$
$$= C_0 P_0(\mu) + C_1 P_1(\mu) + C_2 P_2(\mu) + \cdots.$$

The solution of equation 6.9 is simplified greatly by noting that $f(\mu)$ is an odd function over the interval $-1 < \mu < 1$, and the Legendre functions are even when n is even and odd when n is odd. Consequently, equation 6.9 becomes

$$C_n = \begin{cases} 0, & \text{if } n \text{ is even;} \\ (2n+1) \int\limits_0^1 P_n(\mu)\, d\mu, & \text{if } n \text{ is odd.} \end{cases}$$

In particular,

$$C_0 = 0,$$
$$C_1 = \frac{3}{2},$$
$$C_2 = 0,$$

$$C_3 = -\frac{7}{8},$$

$$C_4 = 0,$$

$$C_5 = \frac{66}{96},$$

$$\vdots$$

and

$$f(\theta,\phi) = \frac{3}{2}P_1(\mu) - \frac{7}{8}P_3(\mu) + \frac{66}{96}P_5(\mu) + \cdots$$

$$= \frac{3}{2}\mu - \frac{7}{8}\frac{1}{2}(5\mu^-3\mu) + \frac{66}{96}\frac{1}{8}(63\mu^5 - 70\mu^3 + 15\mu) + \cdots.$$

Figure 6.5 shows how well the discontinuous function of Figure 6.4 can be represented by these few low-degree Legendre polynomials. Of course, more terms in the summation would improve the approximation of $f_k(\theta)$ to $f(\theta)$.

Finally we should note some important properties of Legendre functions from Table 6.1 and Figure 6.3:

1. If n is odd, the last term of $P_n(\theta)$ is a multiple of $\cos\theta$; if n is even, the last term is a constant.
2. Each $P_n(\theta)$ has n zeroes between $\theta = 0°$ and $\theta = 180°$.
3. If n is even, $P_n(\theta)$ is symmetrical about $\theta = 90°$; if n is odd, $P_n(\theta)$ is antisymmetrical about $\theta = 90°$.

6.3 Surface Harmonics

Legendre polynomials work well in synthesizing a function that depends only on colatitude, but other orthogonal polynomials are more appropriate if the function varies with longitude as well. It was stated earlier that any series of functions that are orthogonal over the interval $0 \le \theta \le \pi$ (or $-1 \le \mu \le 1$) could represent the dependence of $f(\theta,\phi)$ on θ. The Legendre functions are actually a subset of another set of orthogonal functions that also represent the coefficients $a_m(\theta)$ and $b_m(\theta)$ over the interval $0 \le \theta \le \pi$ (or $-1 \le \mu \le 1$) in equation 6.2. These are called the *associated Legendre polynomials*, also known as *associated Legendre functions* or *spherical functions*. They are denoted by $P_{n,m}(\theta)$, where n is the *degree* and m is the *order* of the polynomial, and are given by

$$P_{n,m}(\theta) = \sin^m\theta\,\frac{\partial^m}{\partial(\cos\theta)^m}\,P_n(\cos\theta), \tag{6.10}$$

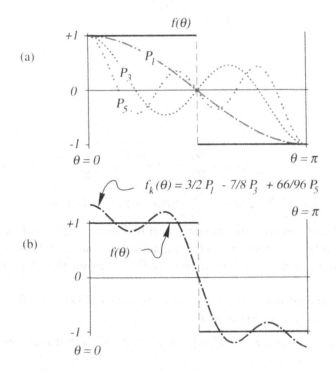

Fig. 6.5. Approximation of the discontinuous function shown in Figure 6.4 with the sum of three Legendre polynomials. (a) Unweighted Legendre functions of degree 1, 2, and 3; (b) the weighted sum of the three Legendre functions.

$$P_{n,m}(\mu) = (1 - \mu^2)^{\frac{m}{2}} \frac{\partial^m}{\partial \mu^m} P_n(\mu). \tag{6.11}$$

Notice that associated Legendre polynomials reduce to Legendre polynomials when $m = 0$. A few examples of the associated Legendre polynomials are as follows:

$$P_{1,1} = \sin\theta, \quad P_{2,1} = \tfrac{3}{2}\sin 2\theta, \quad P_{3,1} = \tfrac{3}{4}\sin\theta\,(5\cos 2\theta + 3),$$
$$P_{2,2} = 3\sin^2\theta, \quad P_{3,2} = \tfrac{15}{2}\sin\theta\sin 2\theta, \tag{6.12}$$
$$P_{3,3} = 15\sin^3\theta.$$

As promised, the associated Legendre polynomials are orthogonal over the interval $-1 \le \mu \le 1$ and with respect to degree n, that is,

$$\int_{-1}^{1} P_{n,m}(\mu) P_{n',m}(\mu)\, d\mu = \begin{cases} 0, & \text{if } n \ne n'; \\ \frac{2(n+m)!}{(2n+1)(n-m)!}, & \text{if } n = n'. \end{cases} \tag{6.13}$$

Exercise 6.2 Try equation 6.13 for $n = 4$ and $m = 1$. Try $n = 4$ and $m = 4$.

The results of Exercise 6.2 illustrate the large differences in the mean values (i.e., when integrated over the interval $-1 \leq \mu \leq 1$) of the squares of associated Legendre polynomials for any particular degree n. Later, we will normalize these functions in order to make their relative importance more alike in any given series.

In the last section, Legendre polynomials were shown to be a suitable building block for functions that are independent of longitude, that is,

$$f(\theta, \phi) = f(\theta)$$

$$= a_0(\theta)$$

$$= C_0 P_0(\theta) + C_1 P_1(\theta) + C_2 P_2(\theta) + \cdots .$$

The associated Legendre polynomials are more powerful in general because they also depend on order m, and this allows $f(\theta, \phi)$ to remain a function of ϕ in equation 6.2,

$$f(\theta, \phi) = \sum_{m=0}^{\infty} (a_m(\theta) \cos m\phi + b_m(\theta) \sin m\phi) .$$

In a later section of this chapter, we will see another important reason for switching to the associated Legendre functions.

Now we are in a position to rewrite equation 6.2, the original expansion of $f(\theta, \phi)$, using the associated Legendre polynomials. Similar to the derivation for the zonal harmonic expansion, we let

$$a_0(\theta) = C_0 P_{0,0}(\theta) + C_1 P_{1,0}(\theta) + C_2 P_{2,0}(\theta) + \cdots ,$$

$$a_1(\theta) = A_{1,1} P_{1,1}(\theta) + A_{2,1} P_{2,1}(\theta) + A_{3,1} P_{3,1}(\theta) + \cdots ,$$

$$b_1(\theta) = B_{1,1} P_{1,1}(\theta) + B_{2,1} P_{2,1}(\theta) + B_{3,1} P_{3,1}(\theta) + \cdots ,$$

$$a_2(\theta) = A_{2,2} P_{2,2}(\theta) + A_{3,2} P_{3,2}(\theta) + A_{4,2} P_{4,2}(\theta) + \cdots ,$$

$$b_2(\theta) = B_{2,2} P_{2,2}(\theta) + B_{3,2} P_{3,2}(\theta) + B_{4,2} P_{4,2}(\theta) + \cdots ,$$

$$\vdots$$

and substitute these equations into equation 6.2 to get

$$f(\theta, \phi) = C_0 P_{0,0}(\theta) + C_1 P_{1,0}(\theta) + C_2 P_{2,0}(\theta) + \cdots$$

$$+ \left[A_{1,1} P_{1,1}(\theta) + A_{2,1} P_{2,1}(\theta) + A_{3,1} P_{3,1}(\theta) + \cdots \right] \cos \phi$$

$$+ \left[B_{1,1} P_{1,1}(\theta) + B_{2,1} P_{2,1}(\theta) + B_{3,1} P_{3,1}(\theta) + \cdots \right] \sin \phi$$

$$+ \left[A_{2,2} P_{2,2}(\theta) + A_{3,2} P_{3,2}(\theta) + A_{4,2} P_{4,2}(\theta) + \cdots \right] \cos 2\phi$$

$$+ \left[B_{2,2} P_{2,2}(\theta) + B_{3,2} P_{3,2}(\theta) + B_{4,2} P_{4,2}(\theta) + \cdots \right] \sin 2\phi$$

$$+ \cdots .$$

Rearranging terms provides

$$f(\theta, \phi) = C_0 P_{0,0}(\theta)$$

$$+ \left[C_1 P_{1,0}(\theta) + A_{1,1} P_{1,1}(\theta) \cos \phi + B_{1,1} P_{1,1}(\theta) \sin \phi \right]$$

$$+ \left[C_2 P_{2,0}(\theta) + A_{2,1} P_{2,1}(\theta) \cos \phi + B_{2,1} P_{2,1}(\theta) \sin \phi \right.$$

$$\left. + A_{2,2} P_{2,2}(\theta) \cos 2\phi + B_{2,2} P_{2,2}(\theta) \sin 2\phi \right] + \cdots ,$$

which can be written

$$f(\theta, \phi) = \sum_{n=0}^{\infty} \left[C_n P_{n,0}(\theta) + \sum_{m=1}^{n} \left(A_{n,m} \cos m\phi + B_{n,m} \sin m\phi \right) P_{n,m}(\theta) \right].$$

$$(6.14)$$

Hence, $f(\theta, \phi)$ is represented by an infinite sum of functions, each function composed of associated Legendre polynomials, sines, and cosines.

For reasons that will become clear in Section 6.4, equation 6.14 is called a *spherical surface harmonic expansion*, and the functions $P_{n,m}(\theta) \cos m\phi$ and $P_{n,m}(\theta) \sin m\phi$ are called *spherical surface harmonics*. Notice that when $m = 0$, the spherical harmonic expansion reduces to a zonal harmonic expansion, as in equation 6.8. As we should expect, surface harmonics are orthogonal over the sphere; unless any two surface harmonics are identical, their product will average to zero over the surface of any sphere. For example,

$$\frac{1}{4\pi r^2} \int\limits_0^{2\pi} \int\limits_0^\pi P_{n,m}(\theta) \cos m\phi P_{n',m'}(\theta) \cos m'\phi \; r^2 \sin\theta \, d\theta \, d\phi$$

$$= \begin{cases} 0, & \text{if } n \neq n' \text{ or } m \neq m'; \\[2mm] \frac{(n+m)!}{2(2n+1)(n-m)!}, & \text{if } n = n' \text{ and } m = m' \neq 0; \\[2mm] \frac{1}{2n+1}, & \text{if } n = n' \text{ and } m = m' = 0, \end{cases} \qquad (6.15)$$

and similarly if $\cos m\phi$ is replaced with $\sin m\phi$ amd $\cos m'\phi$ is replaced with $\sin m'\phi$ in equation 6.15.

6.3.1 Normalized Functions

As illustrated in Exercise 6.3, the magnitude of an associated Legendre polynomial depends on its degree and order, so the magnitude of each coefficient must compensate accordingly. A spherical harmonic analysis would be more instructive if the magnitude of each coefficient reflected the relative significance of its respective term in the expansion.

This can be accomplished by normalizing the associated Legendre functions. Two normalizing schemes are in common usage. The *fully normalized functions*, commonly used in geodetic studies, are related to the unnormalized Legendre polynomials by

$$P_n^m(\theta) = \begin{cases} [2(2n+1)]^{\frac{1}{2}} \, P_{n,m}(\theta), & \text{if } m = 0; \\[2mm] \left[(2n+1)\frac{(n-m)!}{(n+m)!}\right]^{\frac{1}{2}} P_{n,m}(\theta), & \text{if } m > 0. \end{cases}$$

In geomagnetic studies, the *Schmidt functions* are more typical, and these are given by

$$P_n^m(\theta) = \begin{cases} P_{n,m}(\theta), & \text{if } m = 0; \\[2mm] \left[2\frac{(n-m)!}{(n+m)!}\right]^{\frac{1}{2}} P_{n,m}(\theta), & \text{if } m > 0. \end{cases}$$

Rewriting equation 6.14 but using, for example, the Schmidt functions yields

$$f(\theta,\phi) = \sum_{n=0}^\infty \left[A_n^0 P_n^0(\theta) + \sum_{m=1}^n (A_n^m \cos m\phi + B_n^m \sin m\phi) P_n^m(\theta) \right].$$

$$(6.16)$$

Table 6.2. *Surface harmonics of degree and order 0 through 3 expressed in terms of Schmidt functions.*

n	m	Normalized Surface Harmonic
0	0	1
1	0	$\cos\theta$
1	1	$\sin\theta \left\{ {\cos \atop \sin} \right\} \phi$
2	0	$\frac{1}{4}(3\cos 2\theta + 1) = \frac{3}{2}\cos^2\theta - \frac{1}{2}$
2	1	$\frac{\sqrt{3}}{2}\sin 2\theta \left\{ {\cos \atop \sin} \right\} \phi = \sqrt{3}\sin\theta\cos\phi \left\{ {\cos \atop \sin} \right\} \phi$
2	2	$\frac{\sqrt{3}}{2}\sin^2\theta \left\{ {\cos \atop \sin} \right\} 2\phi$
3	0	$\frac{1}{8}(5\cos 3\theta + 3\cos\theta) = \frac{5}{2}\cos^3\theta - \frac{3}{2}\cos\theta$
3	1	$\frac{\sqrt{6}}{8}\sin\theta(5\cos 2\theta + 3)\left\{ {\cos \atop \sin} \right\} \phi = \frac{\sqrt{6}}{4}\left(5\cos^2\theta - 1\right)\sin\theta \left\{ {\cos \atop \sin} \right\} \phi$
3	2	$\frac{\sqrt{15}}{4}\sin\theta\sin 2\theta \left\{ {\cos \atop \sin} \right\} 2\phi = \frac{\sqrt{15}}{2}\sin^2\theta\cos\theta \left\{ {\cos \atop \sin} \right\} 2\phi$
3	3	$\frac{\sqrt{10}}{4}\sin^3\theta \left\{ {\cos \atop \sin} \right\} 3\phi$

The magnitude of Schmidt surface harmonics, when squared and averaged over the sphere, are independent of their order, that is,

$$\frac{1}{4\pi r^2} \int_0^{2\pi} \int_0^{\pi} P_n^m(\theta) \left\{ {\cos m\phi \atop \sin m\phi} \right\} P_{n'}^{m'}(\theta) \left\{ {\cos m'\phi \atop \sin m'\phi} \right\} r^2 \sin\theta\, d\theta\, d\phi$$

$$= \begin{cases} 0\,, & \text{if } n \neq n' \text{ or } m \neq m'; \\ \frac{1}{2n+1}\,, & \text{if } n = n' \text{ and } m = m'. \end{cases} \tag{6.17}$$

Hence, the magnitudes of the coefficients A_n^m and B_n^m quickly indicate the relative "energy" of their respective terms in the series. Schmidt functions are commonly used in global representations of the geomagnetic field.

A few low-degree Schmidt functions are shown in Figure 6.6, and Table 6.2 shows several low-degree surface harmonics based on the Schmidt normalization. Subroutine B.4 in Appendix B provides a Fortran algorithm, modified from Press et al. [233], to calculate normalized associated Legendre functions.

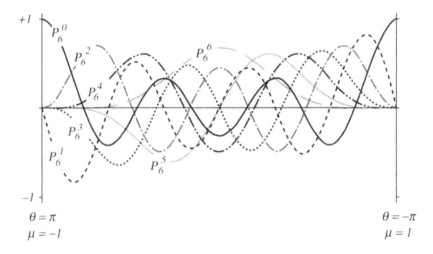

$\theta = \pi$ $\theta = -\pi$
$\mu = -1$ $\mu = 1$

Fig. 6.6. Normalized (Schmidt) surface harmonics of degree 6 and order 0 through 6.

Just as for zonal harmonic functions, the coefficients A_n^m and B_n^m can be found from measurements of $f(\theta, \phi)$ using the orthogonality property:

$$\left\{ \begin{matrix} A_n^m \\ B_n^m \end{matrix} \right\} = \frac{2n+1}{4\pi} \int\limits_0^{2\pi} \int\limits_0^{\pi} f(\theta, \phi) \, P_n^m(\theta) \left\{ \begin{matrix} \cos m\phi \\ \sin m\phi \end{matrix} \right\} \sin \theta \, d\theta \, d\phi . \quad (6.18)$$

In practice, this calculation can be done in two steps, by first numerically integrating the data over ϕ to find the coefficients $a_m(\theta)$ and $b_m(\theta)$, and then integrating over θ. The first step amounts to a Fourier series expansion. Equipped with A_n^m and B_n^m, $f(\theta, \phi)$ can be represented as the infinite sum of weighted surface harmonic functions by equation 6.16.

6.3.2 Tesseral and Sectoral Surface Harmonics

The normalized surface harmonic

$$P_n^m(\theta) \left\{ \begin{matrix} \cos m\phi \\ \sin m\phi \end{matrix} \right\}$$

vanishes along $(n-m)$ circles of latitude that correspond to the zeroes of $P_n^m(\theta)$. It also vanishes along $2m$ meridian lines from 0 to 2π due to the $\sin m\phi$ or $\cos m\phi$ term. The lines of latitude and meridian along which the normalized surface harmonics vanish divide the spherical surface into

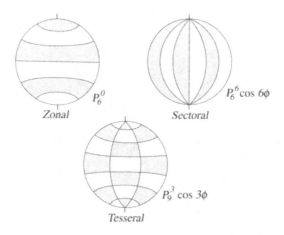

Fig. 6.7. Specific examples of zonal, sectoral, and tesseral surface harmonics.

patches of alternating sign. If $m = 0$, the surface harmonic only depends on latitude and is called a *zonal harmonic*. If $n - m = 0$, it depends only on longitude and is called a *sectoral harmonic* (like the "sectors" of an orange). If $m > 0$ and $n - m > 0$, the harmonic is termed a *tesseral harmonic*. Specific examples of each of these three types of normalized surface harmonics are shown in Figure 6.7.

As indicated by equations 6.14 and 6.16, any reasonably well-behaved function can be represented by an infinite sum of zonal, sectoral, and tesseral patterns, each weighted by an appropriate coefficient A_n^m or B_n^m, as shown by equation 6.16. This sum is just a three-dimensional analog of Fourier series, in which $f(t)$ also is represented by an infinite sum of patterns (sinusoids in the Fourier case) multiplied by appropriate coefficients.

6.4 Application to Laplace's Equation

The previous sections of this chapter showed how to synthesize a function $f(\theta, \phi)$ from measurements of the function on a sphere. The building blocks of this synthesis were Legendre polynomials and associated Legendre polynomials, selected for no particular reason except that they have the property of being orthogonal on a sphere. In this section, we investigate the special case where $f(\theta, \phi)$ is a potential field satisfying Laplace's equation. We will see that the same building blocks, Legendre polynomials and associated Legendre polynomials, are a natural consequence of and provide additional insights into the representation of harmonic functions on a sphere.

6.4.1 Homogeneous Functions and Euler's Equation

As we are well aware by now, potential fields outside of source regions satisfy Laplace's equation, which in cartesian coordinates is given by

$$\nabla^2 V = \frac{\partial^2 V}{\partial x^2} + \frac{\partial^2 V}{\partial y^2} + \frac{\partial^2 V}{\partial z^2}$$

$$= 0.$$

Every solution to Laplace's equation is a harmonic function if the first derivatives of the function are continuous and the second derivatives exist. Now let D or D' represent any spatial differential operation in cartesian coordinates, such as $\frac{\partial}{\partial x}$, $\frac{\partial^2}{\partial x \partial y}$, or ∇^2. Any two such operations are commutative, that is,

$$DD'f(x,y,z) = D'Df(x,y,z),$$

and in particular

$$D\nabla^2 U(x,y,z) = \nabla^2 DU(x,y,z).$$

Hence, *if $U(x,y,z)$ is a harmonic function, then so is any spatial derivative of $U(x,y,z)$.*

This theorem provides a means of generating a host of harmonic functions from known solutions to Laplace's equation. For example, because

$$\nabla^2 \frac{1}{r} = 0,$$

we know immediately that $\frac{\partial}{\partial z}\frac{1}{r}$, $\nabla\frac{1}{r}$, and $\mathbf{m} \cdot \nabla\frac{1}{r}$ (the magnetic potential of a dipole) are also harmonic. This is an important result for future chapters. If it can be shown, for example, that the scalar magnetic potential of the earth is harmonic, then any component of the magnetic field of the earth also must be harmonic.

A function V is said to be *homogeneous of degree n* if it satisfies *Euler's equation*,

$$x\frac{\partial V}{\partial x} + y\frac{\partial V}{\partial y} + z\frac{\partial V}{\partial z} = nV. \tag{6.19}$$

Homogeneous functions that also satisfy Laplace's equation are called *spherical solid harmonic functions* for reasons that will become apparent shortly. Specific examples are xyz (degree 3), xy (degree 2), x (degree 1), $\log \frac{r+z}{r-z}$ (degree 0), $\frac{1}{r}$ (degree -1), and $\frac{z}{r^3}$ (order -2).

Exercise 6.3 Give an example of a spherical harmonic function of degree 4.

It is easy to show that if V is homogeneous, then so is any derivative of V. Hence, all derivatives of $\frac{1}{r}$ are homogeneous, and in particular so is the potential of a magnetic dipole.

6.4.2 Point Source away from Origin

Consider a particle of mass located on the positive z axis at $z = a$ and observed at point $P(r, \theta, \phi)$ (Figure 6.8). The potential at point P is

$$U(r, \theta, \phi) = \frac{1}{R}$$

$$= \left[a^2 + r^2 - 2ar\mu \right]^{-\frac{1}{2}} .$$

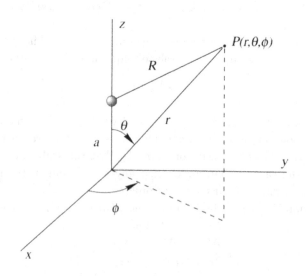

Fig. 6.8. Point mass on z axis observed at point P.

where $\mu = \cos\theta$.† First we consider the case where $r < a$. We factor out the parameter a,

$$\frac{1}{R} = \frac{1}{a}\left[1 + \frac{r^2}{a^2} - 2\frac{r}{a}\mu\right]^{-\frac{1}{2}}, \qquad (6.20)$$

expand the potential in a binomial series

$$\frac{1}{R} = \frac{1}{a}\left[1 + \left(-\frac{1}{2}\right)\left(\frac{r^2}{a^2} - 2\frac{r}{a}\mu\right) + \frac{(-\frac{1}{2})(-\frac{3}{2})}{2!}\left(\frac{r^2}{a^2} - 2\frac{r}{a}\mu\right)^2\right.$$

$$\left.+ \frac{(-\frac{1}{2})(-\frac{3}{2})(-\frac{5}{2})}{3!}\left(\frac{r^2}{a^2} - 2\frac{r}{a}\mu\right)^3 + \cdots\right]$$

$$= \frac{1}{a}\left[1 - \frac{1}{2}\left(\frac{r^2}{a^2} - 2\frac{r}{a}\mu\right) + \frac{3}{8}\left(\frac{r^4}{a^4} - 4\frac{r^3}{a^3}\mu + 4\frac{r^2}{a^2}\mu^2\right)\right.$$

$$\left.- \frac{15}{48}\left(\frac{r^6}{a^6} - 6\frac{r^5}{a^5}\mu + 12\frac{r^4}{a^4}\mu^2 - 8\frac{r^3}{a^3}\mu^3\right) + \cdots\right],$$

and rearrange terms into powers of r/a to get

$$\frac{1}{R} = \frac{1}{a}\left[1 + \left(\frac{r}{a}\right)\mu + \left(\frac{r}{a}\right)^2\left(-\frac{1}{2} + \frac{3}{2}\mu^2\right)\right.$$

$$\left.+ \left(\frac{r}{a}\right)^3\left(-\frac{3}{2}\mu + \frac{5}{2}\mu^3\right) + \cdots\right].$$

A quick comparison of this series with Table 6.1 shows that the factors containing μ are each a Legendre function, namely,

$$\frac{1}{R} = \frac{1}{a}\sum_{n=0}^{\infty}\left(\frac{r}{a}\right)^n P_n(\mu), \qquad r \le a.$$

Hence, the potential of a point mass displaced from the origin can be represented by an infinite sum of weighted Legendre functions.

---

† Note that Newton's gravitational constant has been dropped in the following discussion for convenience.

This series converges only if $r \leq a$. If on the other hand $r > a$, we simply factor out r rather than a in step 6.20. Then

$$\frac{1}{R} = \frac{1}{r}\left(1 + \frac{a^2}{r^2} - 2\frac{a}{r}\mu\right)^{-\frac{1}{2}}$$

$$= \sum_{n=0}^{\infty} \frac{1}{r}\left(\frac{a}{r}\right)^n P_n(\mu)$$

$$= \frac{1}{a}\sum_{n=0}^{\infty} \left(\frac{a}{r}\right)^{n+1} P_n(\mu), \quad r \geq a,$$

and to be complete, we write

$$\frac{1}{R} = \begin{cases} \frac{1}{a}\sum_{n=0}^{\infty} \left(\frac{r}{a}\right)^n P_n(\mu), & \text{if } r \leq a; \\ \frac{1}{a}\sum_{n=0}^{\infty} \left(\frac{a}{r}\right)^{n+1} P_n(\mu), & \text{if } r \geq a. \end{cases} \quad (6.21)$$

Note that if $a = 0$ (i.e., if the point mass is located at the origin), then only the $n = 0$ term is nonzero, and the expansion reduces to the equation $U = \frac{1}{r}$.

In the previous sections, we found that any function of latitude specified on a sphere could be approximated by a series of Legendre polynomials, and that the best coefficients in the series are provided by the orthogonality property of Legendre polynomials. More specifically, equation 6.21 is a special case of equation 6.8 and is an example of a zonal harmonic expansion. Now we see that the potential due to a point mass on the vertical axis is naturally approximated by a series of Legendre polynomials. It should be clear why Legendre polynomials were chosen in the previous sections to represent arbitrary functions on a sphere.

The first term of the expansion for $r > a$ is $\frac{1}{r}$, the potential of a monopole at the origin. Hence, this term of the expansion is called the *monopole term*. Similarly, the $n = 1$ term is $\frac{a\cos\theta}{r^2}$, which is the potential of a dipole located at the origin and pointing in the $\theta = 0$ direction; consequently, the second term of the expansion is referred to as the *dipole term*.

Exercise 6.4 What is the quadrupole term, and what is its physical meaning?

Therefore, equation 6.21 represents the potential of an off-centered monopole as a weighted sum of potentials caused by a series of masses (a monopole, dipole, and an infinite set of more complex sources) located at the origin. Each element of the expansion is harmonic and homogeneous and, therefore, a spherical solid harmonic.

6.4.3 General Spherical Surface Harmonic Functions

Now consider any function $V(x, y, z)$ that is homogeneous with degree n. Such functions have a fortunate property: When transformed to spherical coordinates, they can be factored into three functions where each function depends on only one of the variables r, θ, and ϕ. Consider, for example, the function

$$V(x, y, z) = x^i y^j z^k,$$

which is homogeneous with degree $i + j + k$. To transform to spherical coordinates, we make the substitutions

$$\begin{aligned} x &= r \cos \theta \cos \phi, \\ y &= r \cos \theta \sin \phi, \\ z &= r \sin \theta, \end{aligned} \tag{6.22}$$

and this provides

$$V(r, \theta, \phi) = r^{i+j+k}(\cos^{i+j} \theta \, \sin^k \theta)(\cos^i \phi \, \sin^j \phi),$$

which consists of three factors depending on r, θ, and ϕ, respectively. In general, a homogeneous function can be written in spherical coordinates as

$$V(r, \theta, \phi) = r^n S_n(\theta, \phi),$$

where $S_n(\theta, \phi)$ is independent of r.

Now suppose $V(r, \theta, \phi) = r^n S_n(\theta, \phi)$ is harmonic as well as homogeneous, that is, $V(r, \theta, \phi)$ is a spherical solid harmonic. Substituting $r^n S_n(\theta, \phi)$ into Laplace's equation in spherical coordinates (Appendix A) yields

$$r^{n-2} \left[n(n+1) S_n(\theta, \phi) + \frac{1}{\sin \theta} \frac{\partial}{\partial \theta} \left(\sin \theta \frac{\partial S_n}{\partial \theta} \right) \right.$$

$$\left. + \frac{1}{\sin^2 \theta} \frac{\partial^2 S_n(\theta, \phi)}{\partial \phi^2} \right] = 0,$$

$$r^{n-2}\left[n(n+1)S_n(\mu,\phi) + \frac{\partial}{\partial\mu}\left[(1-\mu^2)\frac{\partial S_n(\mu,\phi)}{\partial\mu}\right]\right.$$

$$\left. + \frac{1}{1-\mu^2}\frac{\partial^2 S_n(\mu,\phi)}{\partial\phi^2}\right] = 0\,,$$

and dropping the coefficient and rearranging terms provides

$$(1-\mu^2)\frac{\partial^2 S_n}{\partial\mu^2} - 2\mu\frac{\partial S_n}{\partial\mu} + n(n+1)S_n + \frac{1}{1-\mu^2}\frac{\partial^2 S_n}{\partial\phi^2} = 0\,. \qquad (6.23)$$

Equation 6.23 is called *Legendre's equation*; it is simply Laplace's equation in spherical coordinates applied to homogeneous, harmonic functions. Notice that this equation includes no dependence on r. Any function S_n that satisfies equation 6.23 is called a *spherical surface harmonic* of degree n because it indicates the θ and ϕ dependence of the spherical solid harmonic $r^n S_n(\theta,\phi)$ over any sphere with constant radius r.

The degree n of the harmonic affects equation 6.23 only in the $n(n+1)S_n$ factor. Because $r^n S_n(\theta,\phi)$ is harmonic and because $n(n+1)$ remains unchanged if we replace n by $-(n+1)$, it follows that

$$\frac{1}{r^{n+1}}S_n(\theta,\phi)$$

is also a harmonic function. This is the general result of the specific example $(V = \frac{1}{R})$ discussed in the previous section.

We need to prove now that the spherical surface harmonic $S_n(\theta,\phi)$ is really just a linear combination of associated Legendre functions, that is,

$$S_n(\theta,\phi) = \sum_{m=0}^{n}\left(A_n^m\cos m\phi + B_n^m\sin m\phi\right)P_n^m(\theta)\,. \qquad (6.24)$$

Since $V = r^n S_n(\theta,\phi)$ is homogeneous in x, y, and z, it is clear from equations 6.22 that $S_n(\theta,\phi)$ must involve $\sin\theta$ and $\cos\theta$ jointly to degree n (at most), and likewise for $\sin\phi$ and $\cos\phi$. It can be shown with arduous but straightforward trigonometry that

$$S_n(\theta,\phi) = \sum_{m=0}^{n}S_{n,m}(\theta)\cos(m\phi + \epsilon_n^m)\,, \qquad (6.25)$$

where coefficients $S_{n,m}$ are functions of θ alone. Rather than proving 6.25, we'll demonstrate its validity with an example. Let $V = xy/r^5$, or in spherical coordinates, $V = r^{-3} \cos^2 \theta \cos \phi \sin \phi$. The spherical surface harmonic in this example is

$$S_2(\theta, \phi) = \cos^2 \theta \cos \phi \sin \phi$$

$$= \left(\frac{1}{4} + \frac{1}{4} \cos 2\theta \right) \cos \left(2\phi - \frac{\pi}{2} \right),$$

which has the general form of equation 6.25 where

$$S_{2,0}(\theta) = 0,$$

$$S_{2,1}(\theta) = 0,$$

$$S_{2,2}(\theta) = \frac{1}{4} + \frac{1}{4} \cos 2\theta,$$

$$\epsilon_2^2 = -\frac{\pi}{2},$$

which was to be shown.

Substituting expression 6.25 into Legendre's equation 6.23 yields

$$\sum_{m=0}^{n} \left[(1 - \mu^2) \frac{d^2 S_{n,m}}{d\mu^2} - 2\mu \frac{dS_{n,m}}{d\mu} + n(n+1) S_{n,m} \right.$$

$$\left. - \frac{m^2}{1 - \mu^2} S_{n,m} \right] \cos(m\phi + \epsilon_n^m) = 0.$$

This expression is true for all ϕ, so

$$(1 - \mu^2) \frac{d^2 S_{n,m}}{d\mu^2} - 2\mu \frac{dS_{n,m}}{d\mu} + \left[n(n+1) - \frac{m^2}{1 - \mu^2} \right] S_{n,m} = 0. \quad (6.26)$$

The solution to equation 6.26 is $P_n^m(\theta)$ or some multiple of $P_n^m(\theta)$, such as $C_n^m P_n^m(\theta)$. To prove this statement, start with the identity

$$(\mu^2 - 1) \frac{d}{d\mu} (\mu^2 - 1)^n = 2n\mu(\mu^2 - 1)^n, \quad (6.27)$$

and perform the following steps for any desired integers n and m:

1. Differentiate both sides $n + 1$ times with respect to μ.
2. Note that this differentiation results in equation 6.26 with $m = 0$ if $S_{n,m}$ is replaced with Rodrigues's formula, equation 6.6,

$$P_n^0(\mu) = \frac{1}{n! \, 2^n} \frac{d^n}{d\mu^n} (\mu^2 - 1)^n. \quad (6.28)$$

3. Differentiate m times.

4. Note that the result is again equation 6.26, but $S_{n,m}$ has been replaced by

$$P_n^m(\mu) = (1 - \mu^2)^{\frac{m}{2}} \frac{d^m}{d\mu^m} P_n(\mu), \qquad (6.29)$$

which, from equation 6.11, is the associated Legendre polynomial of degree n and order m.

Hence,

$$S_{n,m}(\theta) = C_n^m P_n^m(\theta),$$

and rewriting equation 6.25 accordingly,

$$S_n(\theta, \phi) = \sum_{m=0}^{n} C_n^m P_n^m(\theta) \cos(m\phi + \epsilon_n^m).$$

This is the most general expression for a surface harmonic of degree n.

Exercise 6.5 Use the previous steps to prove that $S_{1,0}(\theta) = C_1^0 P_1^0(\theta)$. Prove that $S_{1,1}(\theta) = C_1^1 P_1^1(\theta)$.

Recall that

$$V(r, \theta, \phi) = \sum_{n=0}^{\infty} r^n S_n(\theta, \phi), \qquad (6.30)$$

or

$$V(r, \theta, \phi) = \sum_{n=0}^{\infty} r^{-(n+1)} S_n(\theta, \phi). \qquad (6.31)$$

Therefore, a harmonic function can be represented by an infinite sum of spherical solid harmonics (equation 6.25) where the dependence on θ and ϕ is contained within spherical surface harmonics, which in turn are represented by sums of weighted associated Legendre polynomials. It should be clear why, when we had to represent the θ dependence of $f(\theta, \phi)$ by an orthogonal series, we chose the associated Legendre polynomials.

Notice that if $S_n(\theta, \phi)$ can be found, perhaps from measurements of $V(r, \theta, \phi)$ on the surface of a sphere, equation 6.31 immediately provides $V(r, \theta, \phi)$ at any point outside the sphere. Hence, the potential field due to sources lying entirely within a sphere can be found at any point outside the sphere strictly from the behavior of the potential field on the sphere. Indeed, equations 6.30 and 6.31 together permit a determination

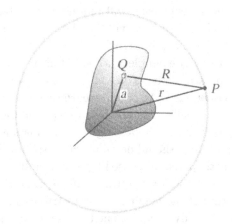

Fig. 6.9. Potential on a sphere due to a mass distribution within the sphere.

of the relative importance of sources inside and outside the sphere, a subject that is reserved for Chapter 8.

We might legitimately ask what all the foregoing mathematics has to do with potential fields caused by general distributions of density and magnetization. To provide a partial answer, consider the expression for gravitational potential observed at P and caused by a bounded density distribution (equation 3.5)

$$U(P) = \gamma \int_R \frac{\rho(Q)}{R} \, dv,$$

where Q is the location of the element of integration and R is the distance between P and Q (Figure 6.9). We can consider each volume element of such a mass to behave like a point source. Observed on the surface of a sphere surrounding the entire density distribution, each volumetric element of the mass would submit to a spherical harmonic analysis, similar to the example in Section 6.2.1 (but generally not so simple as a zonal expansion). Each term in the harmonic expansion represents the potential of an idealized mass (i.e., monopole, dipole, quadrupole, and so forth) located at the origin, and the coefficients in the expansion would reflect the importance of each of these origin-centered masses in building a model of the off-centered, volume element. The potential field caused by the entire density distribution also could be measured and then analyzed as a harmonic expansion, and by the superposition principle, the

resulting terms of the expansion will represent the integrated effects of
the entire mass. The coefficients will be related to the manner in which
mass is distributed about the origin.

Consider, as a trivial example, the potential of a uniform, spherical
mass measured at various points on a larger sphere symmetrically placed
around the spherical mass. A spherical harmonic analysis of these data
would find that all coefficients are zero except for the $n = 0$ term of
the expansion, immediately indicating that the mass is equivalent to a
centered monopole, as it should be. Similarly, a spherical harmonic analysis of the magnetic potential caused by a uniformly magnetized sphere
would reduce to just the $n = 1$ term of the expansion, immediately indicating the nature of the magnetization distribution, namely, that the
spherical magnetic body is equivalent to a centered dipole. Chapters 7
and 8 will apply some of these principles to global gravity and magnetic
fields, respectively.

6.5 Problem Set

1. Let $f(\theta, \phi) = A \sin 2\theta \cos 4\phi$ on a spherical surface. Find a representation for $f(\theta, \phi)$ in terms of normalized (Schmidt) associated Legendre polynomials.

2. Let $f(\theta, \phi)$ be defined on a sphere such that

   $$f(\theta, \phi) = \begin{cases} 1, & \text{if } 0 \leq \phi \leq \pi; \\ 0, & \text{if } \pi < \phi < 2\pi. \end{cases}$$

   Use Subroutine B.4 in Appendix B to calculate a ten-term approximation to $f(\theta, \phi)$ and plot the approximated values around the equator.

3. Are the following functions spherical solid harmonic functions? If so,
   find their degree.

   (a) $V = x^2 + y^2$.
   (b) $V = 3x^2 + xy$.
   (c) $V = x^2 y + xz$.
   (d) $V = \frac{z}{(x^2 + y^2 + z^2)^{3/2}}$.
   (e) $V = \frac{1}{x^2} + \frac{1}{y^2} + \frac{1}{z^2}$.

4. Show that if V is harmonic and homogeneous of degree n, then
   $\frac{\partial^a}{\partial x^a} \frac{\partial^b}{\partial y^b} \frac{\partial^c}{\partial z^c} V$ is harmonic and homogeneous of degree $n - a - b - c$.

5. Let $V = x^2 - y^2$. Find $S_n(\theta)$. Show that it satisfies equation 6.23 and
   that it can be represented as summation 6.25.

6. Consider the potential outside a uniformly magnetized sphere. Arrange the coordinate system so that the magnetization is in the direction $\theta = 0$.

   (a) Let the center of the sphere be at the origin and expand the potential in a zonal harmonic expansion.

   (b) Let the center of the sphere be displaced a distance a in the $\theta = 0$ direction and expand the potential in a zonal harmonic expansion.

   (c) Discuss what happens to the expansion if the dipole points in a direction other than $\theta = 0$.

7

Regional Gravity Fields

The reduction of gravity data is an excellent example of the enhancement of "signal" by the removal of predictable "noise."
(Robert W. Simpson and Robert C. Jachens)

La gravité est un mystère du corps inventé pour cacher les défauts de l'esprit. [Gravity is a mystery of the body invented to conceal the defects of the mind.]
(La Rochefoucauld)

7.1 Introduction

Later chapters will focus on techniques to characterize crustal and upper-mantle sources on the basis of their associated gravity and magnetic fields. A key initial step in such analyses is the proper separation of the field caused by the target source (the *residual* or *anomalous field*) from the field of the remainder of the universe (the *regional field*). This and the following chapter will attempt to characterize these two components for gravity and magnetic studies, respectively.

Our objective in this chapter is to reduce a gravity measurement made on or near the earth's surface to an anomaly value that reflects density variations in the crust and upper mantle. This involves a long series of operations with a well-established tradition. These operations account for the mass, shape, and spin of a "normal" earth, elevation of the measurements above sea level, tidal effects of the sun and moon, motion of the instrument, gravitational effects of terrain in the vicinity of the measurement, and effects of isostasy. Each of these operations will be addressed in this chapter. Readers wishing additional information are referred to texts by Heiskanen and Moritz [123] and Stacey [270] and to papers by Simpson and Jachens [259], Chovitz [58], and LaFehr [154].

128

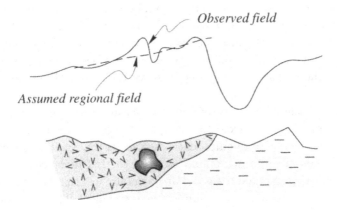

Observed field

Assumed regional field

Fig. 7.1. The definition of regional field depends on the scale of the problem.

Application of these various operations, if they are successful, will leave us with a gravity anomaly that reflects density variations in the crust and upper mantle. But even this product may not be quite what is needed. Our eventual objective may be the analysis of just one geologic element of the crust, such as a sedimentary basin or some plutonic complex, and the anomaly due to this one element ideally should be isolated from those of the surrounding geologic environment. This final aspect of regional–residual separation is often a subjective process and ultimately a matter of scale. A geophysicist attempting to estimate chromite potential within a buried ophiolite will probably consider the anomalies due to surrounding sources to be a nuisance, whereas another geophysicist studying the relationships between accreted terrains might regard as noise the anomaly due to the chromite (Figure 7.1).

7.2 The "Normal" Earth

Because the force of gravity varies from place to place about the earth, equipotential surfaces surrounding the earth are smooth but irregular. An equipotential surface of particular interest is the *geoid*, the equipotential surface described by sea level without the effects of ocean currents, weather, and tides. The geoid at any point on land can be thought of as the level of water in an imaginary canal connected at each end with an ocean. The shape of the geoid is influenced by underlying masses; it bulges above mass excesses (e.g., mountain ranges or buried high-density bodies) and is depressed over mass deficiencies (e.g., valleys or buried low-density bodies). Because the geoid is an equipotential surface, the force of gravity at any point on the geoidal surface must be perpendicular to the surface, thereby defining "vertical" and "level" at each point.

Because of the complexity of internal density variations, it is customary to reference the geoid to a simpler, smoother surface. By international agreement, that equipotential surface is the spheroidal surface that would bound a rotating, uniformly dense earth. Differences in height between this spheroid and the geoid are generally less than 50 m and reflect lateral variations from the uniform-density model. The shape of the reference spheroid was first investigated by measuring the arc lengths of degrees at various latitudes. It was recognized by the late 1600s that the spheroid is oblate (see Introduction). In fact, because of the competing forces of gravity and rotation, the spheroid very nearly has the shape of an ellipse of revolution and, consequently, is called the *reference ellipsoid*. It should be intuitive in any case that the spheroid is symmetric through its center and symmetric about the axis of rotation. Its shape is described by just two parameters, the equatorial radius a and polar radius c (Figure 7.2), and often is expressed in terms of the *flattening* parameter

$$f = \frac{a - c}{a}.$$

The earth is nearly spherical, of course, with flattening of only $1/298.257$, and this fact will permit several simplifying approximations in the following derivations.

The force of gravity on the earth is due both to the mass of the earth and to the centrifugal force caused by the earth's rotation. The total potential of the spheroid, therefore, is the sum of its self-gravitational potential U_g and its rotational potential U_r,

$$U = U_g + U_r, \tag{7.1}$$

where

$$U_r = \tfrac{1}{2}\omega^2 r^2 \cos^2 \lambda,$$

ω is angular velocity, and λ is latitude (Heiskanen and Moritz [123]).

Exercise 7.1 Show that U_r is the potential of a centrifugal force $\mathbf{f}_r$, in the sense that it satisfies $\mathbf{f}_r = \nabla U_r$. Is U_r harmonic?

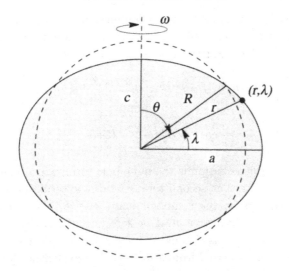

Fig. 7.2. Parameters involved in describing reference ellipsoid.

The gravitational potential U_g is harmonic outside the spheroid, and according to Section 2.1.1, is uniquely determined everywhere outside by its values on the surface. As we shall see shortly, U_g on the surface is determined completely by f, a, and the total mass of the earth. Hence, just these three parameters plus ω are sufficient to find the total potential U of the spheroid anywhere on or above its surface.

The self-gravitational potential is given by equation 6.31,

$$U_g = \sum_{n=0}^{\infty} \frac{1}{r^{n+1}} S_n(\theta, \phi)$$

$$= \frac{\gamma M}{r} \sum_{n=0}^{\infty} \left(\frac{a}{r}\right)^n \sum_{m=0}^{n} (\alpha_n^m \cos m\phi + \beta_n^m \sin m\phi) P_n^m(\theta), \quad (7.2)$$

where M is total mass, a is equatorial radius, ϕ is longitude, and θ is colatitude. Equation 6.31 was derived in Chapter 6 from Laplace's equation, $\nabla^2 V = 0$, in spherical coordinates with no particular physical meaning attached to V. In equation 7.2, however, the various terms in the expansion describe the gravitational potential in terms of an infinite set of idealized masses (monopole, dipole, and so forth) centered at the origin, the coefficients α_n^m and β_n^m describing the relative importance of each mass.

Symmetry of the spheroid greatly simplifies this equation. First, U_g has no dependence on ϕ, so all terms with $m \neq 0$ are zero. Therefore, with the help of Table 6.2, gravitational potential reduces to

$$U_g = \frac{\gamma M}{r} \left[\alpha_0^0 P_0^0(\theta) + \alpha_1^0 \frac{a}{r} P_1^0(\theta) + \alpha_2^0 \left(\frac{a}{r}\right)^2 P_2^0(\theta) + \cdots \right]$$

$$= \frac{\gamma M}{r} \left[\alpha_0^0 + \alpha_1^0 \frac{a}{r} \cos\theta + \alpha_2^0 \left(\frac{a}{r}\right)^2 \frac{1}{4}(3\cos 2\theta + 1) + \cdots \right]. \quad (7.3)$$

The first term of this equation is the monopole term, which must equal $\gamma M/r$. Hence, $\alpha_0^0 = 1$. The second term, the dipole term, must be zero because the origin is at the center of mass. Hence, $\alpha_1^0 = 0$, and all other coefficients of odd degree must be zero for the same reason. Consequently, the third term is the lowest term in the series that describes the departure of the spheroid from a sphere. The coefficient α_2^0 is generally expressed in terms of the *ellipticity coefficient* J_2, where $\alpha_2^0 = -J_2$. Its relationship to the flattening f of the spheroid is given approximately by

$$J_2 = \frac{2f - m}{3}$$

$$= 1.082626 \times 10^{-3}$$

(Stacey [270, p. 90]), where m is the ratio of the centrifugal force to the gravitational force at the equator, given by

$$m = \frac{\omega^2 a}{\gamma M/a^2}$$

$$= \frac{\omega^2 a^3}{\gamma M}$$

$$= 3.46775 \times 10^{-3}.$$

Dropping all higher terms in equation 7.3, changing from colatitude to latitude, and substituting U_g into equation 7.1 yields the total gravitational potential

$$U = \frac{\gamma M}{r} - \frac{\gamma M a^2 J_2}{2r^3}(3\sin^2\lambda - 1) + \frac{1}{2}\omega^2 r^2 \cos^2\lambda. \quad (7.4)$$

If the spheroid is approximately spherical, any normal to the spheroid will be very nearly parallel to r. Then total gravity, normal to the spheroid and directed inward, is given approximately by

$$g_0 = -\frac{\partial U}{\partial r}$$

$$= \frac{\gamma M}{r^2} - \frac{3}{2}\frac{\gamma M a^2 J_2}{r^4}(3\sin^2\lambda - 1) - \omega^2 r\cos^2\lambda, \qquad (7.5)$$

where g_0 is used here to denote the total gravity of the spheroid. Note that in previous chapters, the radial component of gravity was defined as $g_r = \frac{\partial U}{\partial r}$, so the force of gravity was negative in the direction of increasing distance from the spheroid. Equation 7.5 differs from this sign convention, and g_0 is always positive.

Equation 7.5 describes the total gravity of the ellipsoid anywhere on or outside the ellipsoid in a reference frame that moves with the spin of the earth. If we can express r in this equation in terms of a and λ, we can obtain a simplified view of how total gravity varies on the surface of the ellipsoid. The radius of an ellipsoid is given approximately by the relation

$$r = a(1 - f\sin^2\lambda). \qquad (7.6)$$

Exercise 7.2 Prove the previous statement. Hint: The defining equation for an ellipse, $x^2/a^2 + y^2/b^2 = 1$, is a good place to start, and the binomial expansion provides a useful approximation.

Because f is small, we can use this expression to expand $1/r^2$ in a binomial series,

$$\frac{1}{r^2} = \frac{1}{a^2}(1 + 2f\sin^2\lambda),$$

and substitute into the first term of equation 7.5. The last two terms of equation 7.5 are sufficiently small relative to the first term that the approximation $r = a$ will suffice. Making these substitutions leads to

$$g_0 = \frac{\gamma M}{a^2}(1 + 2f\sin^2\lambda) - \frac{3}{2}\frac{\gamma M}{a^2}J_2(3\sin^2\lambda - 1) - \omega^2 a(1 - \sin^2\lambda)$$

$$= \frac{\gamma M}{a^2}\left[\left(1 + \frac{3}{2}J_2 - m\right) + \left(2f - \frac{9}{2}J_2 + m\right)\sin^2\lambda\right]. \qquad (7.7)$$

At the equator, equation 7.7 becomes

$$g_e = \frac{\gamma M}{a^2}\left(1 + \frac{3}{2}J_2 - m\right),$$

and substituting this expression into equation 7.7 and rearranging terms
provides a simple relation describing the total gravitational attraction
of the spheroid,

$$g_0 = g_e(1 + f' \sin^2 \lambda), \qquad (7.8)$$

where

$$f' = \frac{2f - \frac{9}{2}J_2 + m}{1 + \frac{3}{2}J_2 - m}.$$

Equation 7.8 has the same form as equation 7.6, namely, that of an
ellipse. Hence, to first order, the total gravity of the spheroid varies with
latitude as the radius of a prolate ellipsoid. At the pole,

$$g_p = g_e(1 + f'),$$

so

$$f' = \frac{g_p - g_e}{g_e},$$

and the parameter f' in equation 7.8 is the gravitational analog of geometrical flattening.

The parameters g_e, g_p, and f' have values of 9.780327 m·sec^{-2},
9.832186 m·sec^{-2}, and 0.00530, respectively, and as we should have expected, the total gravity of the reference ellipsoid varies by only a small
amount over its surface, about 0.5 percent from equator to pole. As we
shall see, however, this small variation is nevertheless significant when
compared to gravitational attraction of geologic sources.

Theoretical Gravity

Carrying through the previous derivations to higher order, equation 7.8
can be cast more accurately as

$$g_0 = g_e(1 + \alpha \sin^2 \lambda + \beta \sin^2 2\lambda), \qquad (7.9)$$

where, as before, g_e is the equatorial attraction of the spheroid, and
α and β depend only on M, f, ω, and a. Equation 7.9 is a truncated
infinite series, but a closed-form expression for g_0 can be derived as well
(Heiskanen and Moritz [123, p. 70]),

$$g_0 = g_e \left(\frac{1 + k \sin^2 \lambda}{\sqrt{1 - e^2 \sin^2 \lambda}} \right), \qquad (7.10)$$

where k and e also depend only on M, f, ω, and a. This equation is
called the *Somigliana equation*.

Table 7.1. *Parameters of various geodetic reference systems, from Chovitz [58].*

System	a, km	f	J_2	γM, m^3·sec^{-2}	g_e, m·sec^{-2}
1924–30	6378.388	1/297.0	0.0010920	3.98633×10^{14}	9.780490
1967	6378.160	1/298.247	0.0010827	3.98603×10^{14}	9.780318
1980	6378.137	1/298.257	0.00108263	3.986005×10^{14}	9.780327

Hence, the gravitational attraction of the reference ellipsoid at any point (r, λ), whether expressed by equation 7.9 or 7.10, depends on only four observable quantities: γM, a, J_2 (or f), and ω. The quantity γM is considered one parameter here because the product of γM can be determined much more precisely than either γ or M separately. The equatorial radius a is found from arcs of triangulation, and rotation velocity ω is found from astronomical measurements. Prior to the first artificial satellites, γM and J_2 were based on surface gravity measurements. Now γM and J_2 are found from satellite observations and planetary probes (Chovitz [58]). Note that detailed knowledge of the earth's density is not required in order to specify the ellipsoid.

As knowledge of the defining parameters have evolved over recent years, so too has the reference ellipsoid. The ellipsoid is defined and refined by international agreement through the International Association of Geodesy (IAG) and its umbrella organization, the International Union of Geodesy and Geophysics (IUGG). Three international systems have been sanctioned in this way, and Table 7.1 shows the defining parameters for each of these systems. The first internationally accepted reference ellipsoid was established in 1930, and its associated parameters provided the 1930 International Gravity Formula,

$$g_0 = 9.78049(1 + 0.0052884 \sin^2 \lambda - 0.0000059 \sin^2 2\lambda),$$

where g_0 is in m·sec^{-2}. The advent of satellites provided a breakthrough in the accuracy of various geodetic parameters, and a new ellipsoid was adopted in 1967 called Geodetic Reference System 1967, thereby providing the 1967 International Gravity Formula,

$$g_0 = 9.78031846(1 + 0.0053024 \sin^2 \lambda - 0.0000058 \sin^2 2\lambda).$$

Most recently the IAG has adopted Geodetic Reference System 1980, which eventually led to the current reference field, World Geodetic

System 1984; in closed form it is given by

$$g_0 = 9.7803267714 \frac{1 + 0.00193185138639 \sin^2 \lambda}{\sqrt{1 - 0.00669437999013 \sin^2 \lambda}} . \tag{7.11}$$

The quantity g_0, expressed by equation 7.11 or its predecessors, is commonly referred to as *theoretical gravity* or *normal gravity*.

The Geoid

As discussed previously, the reference ellipsoid is the equipotential surface of a uniform earth, whereas the geoid is the actual equipotential surface at mean sea level. Differences in height between these two surfaces rarely exceed 100 m and generally fall below 50 m (Lerch et al. [163]). The shape of the geoid is dominated by broad undulations, with lateral dimension of continental scale but with no obvious correlation with the continents; they apparently are caused by widespread mantle convection (Hager [108]). Compared with these broad undulations, the response of the geoid to topography and density variations within the lithosphere are second-order effects, both low in amplitude and short in wavelength (Marsh et al. [175]; Milbert and Dewhurst [184]).

Gravity anomalies, to be discussed in the next sections, are referenced to the reference ellipsoid but involve various corrections relative to sea level (the geoid). This inconsistency is ignored in most crustal studies, and in the following discussion, we too will assume that g_0 represents theoretical gravity on the geoid. While this implicit assumption is acceptable for most geologic studies, the discrepancy between the reference ellipsoid and the geoid should be accounted for if the size of the study is on the order of the broad-scale undulations of the geoid.

7.3 Gravity Anomalies

The isolation of anomalies caused by local density variations from all other fields involves a series of corrections to observed gravity. They can be confusing to students because of the way they sometimes are described. For example, the free-air correction, to be discussed subsequently, is sometimes inaccurately described as "moving the observation point downward to sea level." It would be incorrect, however, to consider the observation point at sea level in subsequent calculations or graphical displays.

A better way to describe the series of corrections is to consider them each as contributors to observed gravity. The following sum shows the various components to observed gravity with the name of the corrections shown in parentheses:

observed gravity = attraction of the reference ellipsoid
+ effect of elevation above sea level (free-air)
+ effect of "normal" mass above sea level
(Bouguer and terrain)
+ time-dependent variations (tidal)
+ effect of moving platform (Eötvös)
+ effect of masses that support topographic loads
(isostatic)
+ effect of crust and upper mantle density
variations ("geology").

$$(7.12)$$

Our goal is to isolate the last quantity in this summation, the effect of crustal and upper mantle density variations, from all other terms. Unfortunately, this last quantity is a relatively minor part of observed gravity. The acceleration of gravity at the surface of the earth due to the whole earth is approximately 9.8 m·sec^{-2} (980 Gal), whereas anomalies caused by crustal density variations are typically less than 10^{-3} m·sec^{-2} (100 mGal), less than 0.01 percent of observed gravity. Portable gravity meters are quite capable of measuring gravity to within 10^{-7} m·sec^{-2} (0.01 mGal), or about one part in 10^8, but the various corrections to observed gravity involve assumptions that limit our ability to resolve the geologic component of observed gravity. Depending on a variety of factors, particularly the severity of the surrounding terrain, the actual resolution of the geologic component in field situations may range from 0.1 to 5 mGal.

We will use the simple crustal model shown in Figure 7.3 to help illustrate the various contributions to observed gravity. This cross section includes various examples of lateral variations in density: a topographic edifice, a low-density root that supports the topography in accordance with the principles of isostasy, and a dense body in the upper crust that extends both above and below sea level. Gravity is observed at the topographic surface along a west–east profile, and our goal is to isolate the anomaly caused by just the high-density body in the upper crust.

Exercise 7.3 How can we tell from Figure 7.3 that the profile is directed west–east and not north–south?

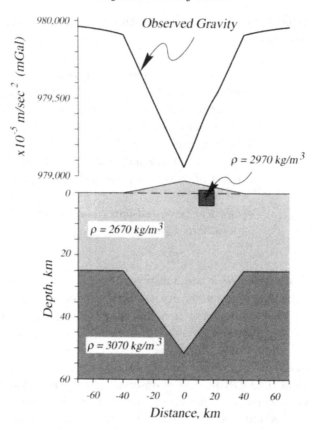

Fig. 7.3. Crustal cross section to describe various corrections to observed gravity. The crust and mantle are assumed to have densities of 2670 and 3070 kg·m^{-3}, respectively. The mountain range is isostatically compensated by a crustal root. A mass of rectangular cross section and density 2970 kg·m^{-3} represents a density variation due to upper-crustal geology. Vertical exaggeration 2.

The first correction described by equation 7.12 is easily accomplished with the results of the previous section. Equation 7.11 provides theoretical gravity, the normal gravitational attraction of a hypothetical earth containing no lateral density inhomogeneities. When this equation is evaluated and subtracted from gravity measurements, the remainder reflects departures of the earth's density from the homogeneous ellipsoid, in particular lateral density variations in the crust and mantle. The

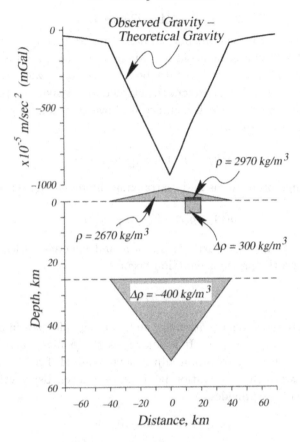

Fig. 7.4. Crustal cross section of Figure 7.3 after subtraction of theoretical gravity. The large negative anomaly is caused primarily by increasing distance between the gravity meter and the reference ellipsoid as the profile rises over the topographic edifice.

remainder also includes the effects of altitude, tides, and various other factors, and these will be discussed subsequently.

Figure 7.4 shows how the crustal cross section of Figure 7.3 is effectively changed by subtraction of theoretical gravity. The resulting gravity profile is dominated by a large negative anomaly caused primarily by the increasing altitude of the gravity meter as the profile goes over the topographic edifice. This contribution obviously is not related directly to crustal or mantle sources; it merely reflects changes in distance between the gravity meter and the center of the earth.

7.3.1 Free-Air Correction

Shipboard gravity measurements can be compared directly with the reference field g_0 because the geoid corresponds to sea level. Gravity measurements over land, however, must be adjusted for elevation above or below sea level. Let $g(r)$ represent the attraction of gravity on the geoid. The value of gravity a small distance h above the geoid is given by a Taylor's series expansion,

$$g(r + h) = g(r) + h\frac{\partial}{\partial r}g(r) + \cdots .$$

Dropping high-order terms and rearranging the remaining terms gives

$$g(r) = g(r + h) - h\frac{\partial}{\partial r}g(r).$$

If we assume that the earth is uniform and spherical, then $g(r) = -\gamma M/r^2$, and the previous equation becomes

$$g(r) = g(r + h) - \frac{2\,g(r)}{r}h.$$

The last term of this equation accounts for the difference in elevation between $g(r)$ and $g(r + h)$. It is known as the *free-air correction* g_{fa} because it is the only elevation adjustment required if no masses were to exist between the observation point and sea level. Using values of g and r at sea level provides

$$g_{\text{fa}} = -0.3086 \times 10^{-5}\,h, \tag{7.13}$$

where h is height above sea level. Equation 7.13 is the same in both SI units (g_{fa} in m·sec^{-2}, h in m) and cgs units (g_{fa} in Gal, h in cm) because g_{fa}/h has units of sec^{-2}. Application of the free-air correction provides the *free-air anomaly* given by

$$\Delta g_{\text{fa}} = g_{\text{obs}} - g_{\text{fa}} - g_0, \tag{7.14}$$

where g_{obs} is observed gravity. It should be clear that shipboard measurements minus g_0 are at once free-air anomalies.

Figure 7.5 shows the effect of the free-air correction on the hypothetical cross section of Figure 7.3. The large negative anomaly of the previous figure, which was caused by increasing elevation of the gravity meter over the topographic high, has been eliminated by the free-air correction. Over elevated areas of land, the free-air anomaly tends to rise to large values, which causes an often undesirable correlation between topography and gravity. This is apparent in Figure 7.5, where

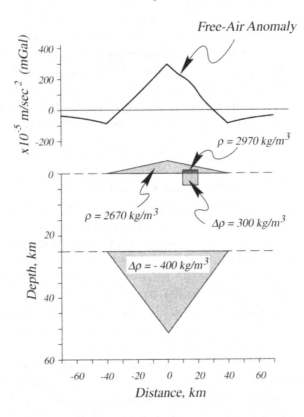

Fig. 7.5. Crustal cross section of Figure 7.3 after the free-air correction. Note that the observation points are not "moved to sea level" and that the free-air anomaly is strongly influenced by terrain.

although the free-air correction has accounted for the variation in elevation of the measurements, it has not accounted for the additional mass represented by the topographic edifice. Notice too that the crustal root in Figure 7.5, which isostatically supports the topography, also produces a long-wavelength, relatively low-amplitude, negative component in the free-air anomaly. Nevertheless, free-air anomalies are often used in geodesy for studies of the spheroid and geoid because they are very nearly equivalent to what would be observed if all the topographic masses were condensed onto the geoid.

Exercise 7.4 Sketch in profile form the free-air anomaly that would be observed across a vertical-sided iceberg.

The free-air correction is sometimes referred to as moving the observation point to sea level, but this description is misleading. More accurately, the free-air correction adjusts measured gravity for one factor not accounted for by the reference ellipsoid: the elevation of the gravity measurement above the reference ellipsoid. Although the free-air correction accounts for the elevation of the observation point, the observation point still remains fixed in space with respect to all causative masses (Figure 7.5).

7.3.2 Tidal Correction

Earth-tides caused by the sun and moon are of sufficient amplitude to be detected by gravity meters as time-varying gravity. The effect is both time- and latitude-dependent; it is greatest at low latitudes and has a strong periodic component with period on the order of 12 hours. The tidal effect never exceeds 3×10^{-6} m·sec^{-2} (0.3 mGal), a small quantity in comparison to other corrections to observed gravity. Nevertheless, tidal effects should be accounted for in high-precision surveys. Formulas exist to calculate the tidal effect at any time and at any place on the earth's surface (Longman [166]).

It may be appropriate in less precise surveys to assume that the tidal effect is linear over periods of several hours and to remove the tidal effect along with other temporal adjustments. For example, most gravity meters used in gravity surveys produce readings that drift slightly over the course of a day's fieldwork. This problem usually is treated by reoccupying certain observation points at various times during the day, assuming that drift has been linear between the repeated measurements, and subtracting the linear drift from all other readings. The tidal effect can be considered part of the instrumental drift.

7.3.3 Eötvös Correction

As discussed in Section 7.2, the attraction of the earth at a point fixed with respect to the earth is reduced by the centrifugal force related to the earth's rotation. It stands to reason that the angular velocity of an observer moving east is greater than for an observer remaining stationary with respect to the earth's surface, and consequently gravitational attraction will be slightly reduced for the moving observer. Likewise, gravitational attraction will be slightly increased for an observer moving in a westerly direction. This motion-related effect, called the *Eötvös effect*, must be accounted for in gravity measurements made on moving platforms, such as ships or aircraft. The *Eötvös correction* is given by

$$g_E = 7.503\,v\,\cos\lambda\,\sin\alpha + 0.004154\,v^2\,,$$

where v is in knots, α is heading with respect to true north, λ is latitude, and g_E is in mGal.

Exercise 7.5 Derive the Eötvös correction starting with equation 7.1.

The Eötvös correction can reach significant magnitudes in applications involving moving platforms. For example, the Eötvös correction is 5.4×10^{-5} m·sec^{-2} (5.4 mGal) for gravity measurements made on a ship at latitude 45°N heading easterly at 1 knot. Errors in heading or velocity, therefore, can produce errors in reduced gravity measurements that are similar in magnitude to anomalies caused by typical crustal sources. Indeed, the Eötvös correction is often the limiting factor in the precision of shipborne and airborne surveys.

7.3.4 Bouguer Correction

The free-air correction and theoretical gravity ignore mass that may exist between the level of observation and sea level. The *Bouguer correction* accounts for this additional mass. The *simple Bouguer correction* approximates all mass above sea level with a homogeneous, infinitely extended slab of thickness equal to the height of the observation point above sea level (Figure 7.6). The attraction of an infinite slab is described by equation 3.27,

$$g_{sb} = 2\pi\gamma\rho h\,,$$

where h is the thickness of the slab. Using a typical crustal density of 2670 kg·m^{-3}, the simple Bouguer correction becomes

$$g_{sb} = 0.1119 \times 10^{-5}\,h\,, \tag{7.15}$$

for both SI units (g_{sb} in m·sec^{-2}, h in m) and cgs units (g_{sb} in Gal, h in cm), where h is height above sea level. Therefore, ignoring tidal and Eötvös corrections, the *simple Bouguer anomaly* is given by

$$\Delta g_{sb} = g_{obs} - g_{fa} - g_{sb} - g_0\,. \tag{7.16}$$

For gravity measurements over water, the Bouguer correction amounts to replacing the water (density = 1000 kg·m^{-3}) with a slab of density 2670 kg·m^{-3} and thickness equal to bathymetric depth. Hence, the sign

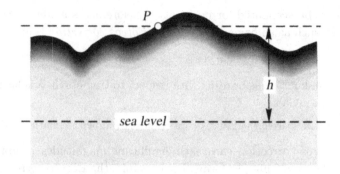

Fig. 7.6. Simple Bouguer, complete Bouguer, and terrain corrections to observed gravity.

of equation 7.15 is opposite, and the numerical coefficient is reduced for ocean measurements.

The Bouguer anomaly reflects "anomalous mass," masses with density above or below 2670 kg·m^{-3}. The choice of 2670 kg·m^{-3} as an average crustal density is appropriate for most geologic situations. In certain studies, such as over young volcanic terrain or sedimentary basins, another density may be more "normal." Figure 7.7 shows the effect of the simple Bouguer correction on our crustal cross section of Figure 7.5.

Exercise 7.6 Sketch the Bouguer anomaly, in profile form, across a vertical-sided iceberg.

Note that the anomaly now reflects the *density contrast* of the anomalous masses with respect to normal density, rather than their total densities.

The simple Bouguer anomaly ignores the *shape* of the topography (Figure 7.6). Mountains that rise above the observation level "pull up" on the gravity meter but are not accounted for in the slab approximation. Valleys that lie below the observation level form cavities within the slab approximation. In either case, a simple Bouguer correction tends to overcompensate measurements made near topographic features. The *terrain correction* g_t adjusts for this overcompensation and is an essential step in reducing measurements made in places of moderate to extreme topographic relief. The result is the *complete Bouguer anomaly*:

$$\Delta g_{cb} = g_{obs} - g_{fa} - g_{sb} - g_t - g_0 \,, \qquad (7.17)$$

where the sign of g_t is always negative. The terrain correction, which should include a term for the curvature of the earth (e.g., LaFehr [155]),

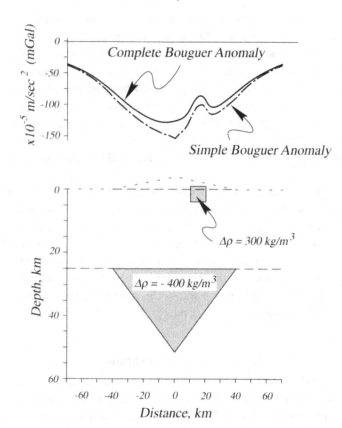

Fig. 7.7. Simple Bouguer and complete Bouguer anomalies over the crustal cross section of Figure 7.3. The light dashed line indicates the position of mea-surements with respect to causative masses. Note that the Bouguer anomaly includes a long-wavelength, negative component caused by the low-density mass (root) that isostatically compensates the topography.

is traditionally done by approximating the topography with a digital model and calculating the gravitational attraction of the model with techniques such as those to be discussed in Chapters 9 and 11. Terrain models, the bane of graduate students, are often developed by hand with templates and topographic contour maps (Spielman and Ponce [269]), but modern techniques of terrain correction, involving data bases of digital terrain (e.g., Plouff [230], Godson and Plouff [97]), greatly speed the procedure.

Bouguer anomalies typically show a strong inverse correlation with long-wavelength topography. The reason for this is apparent in Figure 7.7. Although the Bouguer correction has accounted for the direct effects of the topographic edifice, it has not accounted for the low-density root that isostatically supports the topography. Hence, the Bouguer anomaly in this figure is strongly negative because the gravitational effects of the compensating root remain in the anomaly. For similar reasons, Bouguer anomalies are negative over continental areas and positive over ocean basins because of the different crustal thicknesses between the two regimes.

The Bouguer correction is sometimes referred to as a stripping away of all material down to sea level. More accurately, it accounts for normal crust (i.e., density $= 2670$ kg·m^{-3}) above sea level, specifically that part of the "normal" earth not accounted for by theoretical gravity. The Bouguer correction, like the free-air correction, should not be thought of as physically translating the observation point to sea level, nor does it produce density discontinuities across the sea-level interface in masses that extend above and below sea level.

7.3.5 Isostatic Residual

Continents and ocean basins represent mass concentrations and deficiencies, respectively, with large lateral dimensions; it would seem that such profound masses should be reflected in low-order harmonic terms of the geoid. This is not the case, however, as can be seen from global-scale maps of the geoid (e.g., Lerch et al. [163]). Apparently, large mass concentrations and mass deficiencies are compensated at depth, so that total mass in each vertical section is laterally uniform to first order. It was recognized long ago (Pratt [231, 232], Airy [1]) that the extra mass of large topographic features is generally compensated at depth by mass deficiencies, whereas large topographic depressions are matched at depth by mass excesses. This phenomenon of compensation of topographic loads by deeper compensating masses is referred to as *isostatic compensation*.

Two models for isostatic compensation are shown in Figure 7.8. Pratt proposed that density varies laterally in the crust in order that every vertical crustal section have identical mass. Airy considered the compensating masses to be in the form of undulations of the crust–mantle interface; that is, below mountain ranges, low-density crustal roots extend into higher-density mantle, whereas below deep ocean

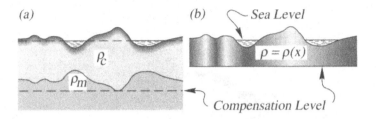

Fig. 7.8. Two early models for isostatic compensation of topographic loads. (a) Airy [1]; (b) Pratt [231].

basins, high-density mantle warps upward into the lower-density crust. The truth undoubtedly is more complicated than either of these models, and the mechanism varies from place to place depending on the geological setting. In any case, the anomalies caused by the compensating masses are generally long in wavelength and approximately negatively correlated with long-wavelength attributes of topography (Figure 7.7).

It is sometimes desirable to remove from gravity measurements the long-wavelength gravitational effects of compensating masses. For example, studies of crustal geology in mountainous terrain (e.g., Jachens and Griscom [136]) typically are concerned with lateral variations of crustal density rather than deep-crustal and upper-mantle sources that isostatically support the topography. Given a digital terrain model, it is a relatively straightforward procedure to (1) calculate the shape of the crust–mantle interface consistent with the Airy model for isostatic compensation and (2) calculate at each observation point the gravitational effect of the volume (Jachens and Roberts [138], Simpson, Jachens, and Blakely [260]). According to the Airy model of isostatic compensation, the total mass must be equal for all columns extending from the earth's surface to some depth of compensation. Consider the two columns in Figure 7.9. The mass in column 1 is proportional to $d_s\rho_c + (d_m - d_s)\rho_m$, where d_s is the depth of compensation at shorelines, d_m is the depth below sea level of the compensating root, ρ_c is crustal density, and ρ_m is mantle density. The mass in column 2 is proportional to $h\rho_t + d_m\rho_c$, where ρ_t is the average density of rocks that make up the terrain, and h is elevation of the observation point above sea level. Equating the two mass columns provides

$$d_m = h\frac{\rho_t}{\Delta\rho} + d_s\,, \tag{7.18}$$

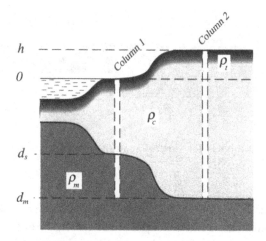

Fig. 7.9. Two crustal columns of equal total mass in an Airy model of isostatic compensation.

where $\Delta\rho = \rho_m - \rho_c$. Note that the depth to the root depends on the contrast across the crust–mantle interface, not on the absolute values of ρ_c and ρ_m.

Given a terrain model, equation 7.18 can be solved at each observation point, and the gravitational effect g_i of the root can be calculated using techniques to be discussed in Chapters 9 and 11. Subtraction of the isostatic regional anomaly g_i provides the *isostatic residual anomaly*,

$$\Delta g_i = g_{obs} - g_{fa} - g_{sb} - g_t - g_i - g_0 . \qquad (7.19)$$

The isostatic regional g_i is negative over continents and positive over oceans.

Exercise 7.7 Sketch the isostatic residual anomaly, in profile form, across a vertical-sided iceberg.

We see then, in comparing equation 7.19 with equation 7.12, that Δg_i is the gravitational effect of variations in crustal density, which was our initial goal. Figure 7.10 shows the isostatic residual anomaly over our simple crustal cross section. Now, within the limits imposed by our simplifying assumptions, we have isolated the gravitational effects of the small upper-crustal mass. Interested readers are referred to Simpson et al. [261] for additional discussion on the geologic implications of the isostatic residual anomaly.

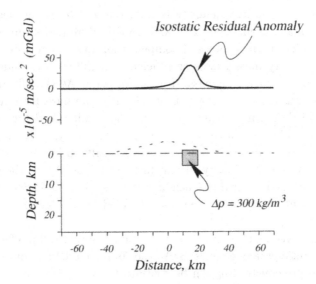

Fig. 7.10. Isostatic residual gravity profile over crustal cross section of Figure 7.3. The only remaining contribution to the profile originates from the crustal mass that is unrelated to isostatic compensation of topographic loads.

Decompensative Anomaly

The foregoing discussion has dealt with a correction to measured gravity for the gravitational effects of masses that isostatically support topographic loads. Of course, crustal masses unrelated to topography also seek isostatic equilibrium, and the effects of this compensation will not be accounted for in the isostatic correction. In particular, variations in crustal density, the very signal that we are trying to isolate, will be in isostatic equilibrium in most geologic situations, and the gravity anomaly due to density variations in the crust will be matched by long-wavelength anomalies of opposite sign caused by deeper compensating masses.

The *decompensative anomaly* (Zorin et al. [296]; Cordell, Zorin, and Keller [75]) attempts to account for the gravitational effects of masses that isostatically support variations in crustal density. In this view, the isostatic residual Δg_i (the anomaly adjusted, as in the previous section, for masses that isostatically support topography) is the sum of two parts,

$$\Delta g_i = \Delta g_c + \Delta g_d, \tag{7.20}$$

where Δg_c is the anomaly caused by variations in crustal density, and Δg_d is the anomaly due to deeper masses that support those crustal variations. The desired quantity is Δg_c, referred to as the decompensative anomaly. Zorin et al. [296] and Cordell et al. [75] assumed a very simple model for the process of compensation in order to solve for Δg_c from Δg_i: Every mass excess or deficiency in the crust is underlain at depth z by an identical mass of opposing sign. As the effects of the compensating masses are long in wavelength, the unrealistic nature of this model should not greatly affect the end result (Cordell et al. [75]). Hence, Δg_d, observed at the level of the gravity survey, is identical, apart from sign, to the anomaly due to crustal density variations observed at a height z above the gravity survey. In other words, Δg_d is the negative of Δg_c continued upward to a height z and, as discussed in Section 2.3.2, can be calculated from Δg_c without detailed knowledge of the sources of Δg_c or Δg_d.

We will investigate upward continuation in more detail in Chapter 12, but here anticipate one result: With the simple model for compensation described previously, Δg_c can be continued upward a distance z simply by multiplying the Fourier transform of Δg_c by the function $e^{-|k|z}$, where $|k|$ is inversely proportional to wavelength. Denoting Fourier transformation of a function $f(x,y)$ by the symbol $\mathcal{F}[f]$, equation 7.20 can be written as

$$\mathcal{F}[\Delta g_i] = \mathcal{F}[\Delta g_c] + \mathcal{F}[\Delta g_d]$$
$$= \mathcal{F}[\Delta g_c] - \mathcal{F}[\Delta g_c]\, e^{-|k|z} ,$$

and the desired quantity Δg_c is given in the Fourier domain by

$$\mathcal{F}[\Delta g_c] = \mathcal{F}[\Delta g_i]\,(1 - e^{-|k|z})^{-1}$$

(Cordell et al. [75]). Consequently, given an assumed value for z, the decompensative anomaly Δg_c can be found directly from the isostatic anomaly Δg_i by Fourier transforming Δg_i, multiplying by $(1-e^{-|k|z})^{-1}$, and inverse Fourier transforming the product.

7.4 An Example

Figure 7.11 compares the free-air, simple Bouguer, complete Bouguer, and isostatic residual anomalies over parts of the Klamath Mountains and Cascade Range in north-central California. The pre-Cenozoic sedimentary and volcanic rocks of the Klamath Mountains are generally more dense than the Tertiary and Quaternary volcanic rocks related to the Cascade arc. The Trinity ultramafic sheet, the largest ultramafic body in North America, lies in this part of the Klamath Mountains. Additional information regarding the geophysical setting of this area was provided by Griscom [105], LaFehr [153], and Blakely and Jachens [30].

Figure 7.11 illustrates the following:

1. Free-air anomalies (Figure 7.11(c)) are strongly correlated with terrain (Figure 7.11(b)). This correlation is particularly apparent over Mount Shasta, Medicine Lake Volcano, and the Klamath Mountains.

2. Simple and complete Bouguer anomalies (Figures 7.11(d) and 7.11(e)) over continental areas are strongly negative. This happens because the Bouguer correction has removed the effects of normal crust above sea level but has left the effects of deeper masses that isostatically support that crust.

3. Simple and complete Bouguer anomalies have regional-scale components that are approximately inversely correlated with very long-wavelength attributes of the terrain. In Figure 7.11(e), this regional component appears as a broad trend decreasing from west to east. The regional trend in Bouguer gravity is caused by an increase from west to east in the amount of deep, low-density material that supports the topography. In terms of the Airy model for isostatic compensation, the trend in Bouguer gravity is caused by a low-density root that thickens from west to east in order to support the continental edifice, although in actuality the support may be distributed in other ways; for example, through variable densities in the upper mantle.

4. The simple Bouguer map includes short-wavelength anomalies related to topography, whereas these effects are largely missing from the complete Bouguer map, which included a terrain correction. The large-amplitude negative anomaly directly over Mount Shasta on Figure 7.11(d) is the best example.

5. The isostatic residual anomaly (Figure 7.11(f)) most closely represents lateral variations in density of the middle and upper crust. It clearly shows, for example, the high-density Trinity ultramafic body and where it lies below pre-Cenozoic rocks of the Klamath Mountains (Griscom [105]). Low-density rocks of the Cascade Range are also apparent in the central and eastern parts of the map. The negative anomaly over and extending northeast from Mount Shasta is thought to be caused by low-density volcanic rocks (LaFehr [153]; Blakely and Jachens [30]).

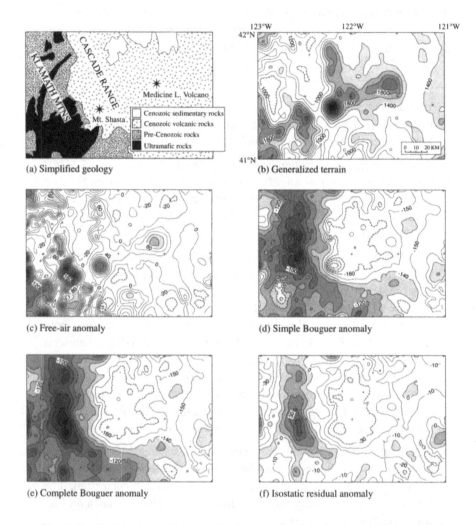

(a) Simplified geology

(b) Generalized terrain

(c) Free-air anomaly

(d) Simple Bouguer anomaly

(e) Complete Bouguer anomaly

(f) Isostatic residual anomaly

Fig. 7.11. Various corrections applied to gravity data from north-central California. (a) Simplified geology; (b) topography based on 5-minute averages, contour interval 200 m; (c) free-air anomaly, contour interval 20 mGal; (d) simple Bouguer anomaly, contour interval 10 mGal; (e) complete Bouguer anomaly, contour interval 10 mGal; (f) isostatic residual anomaly, contour interval 10 mGal. Gray shades indicate positive regions.

7.5 Problem Set

1. Imagine that you are chief science officer on a mission to investigate a newly discovered planet. Your spaceship is located in the plane of the planet's rotation at a distance r from the planet's center and a distance h from the planet's surface. Your spaceship is stationary with respect to the planet's rotation and is experiencing a gravitational attraction of G. Your sensors tell you that the planet is rotating with an angular velocity of ω, has the shape of an oblate spheroid with flattening f, and has dynamical properties much like Earth. Two landing parties are being sent to the planet: one to the equator and one to the north pole. The captain wants to know how much the force of gravity will differ between the two landing sites and is willing to accept some simplifying assumptions. What can you advise? Express your answer in terms of a, f, r, h, ω, and G.

2. Suppose that the earth begins to rotate more and more rapidly until the gravity experienced by an observer at the equator falls to zero. How long is the length of a day?

3. Show that to first order $f + f' = \frac{5}{2}m$. Note: This result is more important than it may seem; it shows that the geometrical shape of the ellipsoid can be determined from gravity measurements alone.

4. A sphere of density 3.17 g·cm^{-3} and radius 5 km is partially buried in a flat prairie so that the summit of the sphere is 2 km above the surrounding prairie. The prairie is 2 km above sea level. Assume crustal density is 2.67 g·cm^{-3} and use the Geodetic Reference System 1967 to sketch the following profiles (show appropriate shapes and amplitudes) as observed along a traverse directly over the center of the sphere:

   (a) observed gravity,
   (b) free-air anomaly,
   (c) Bouguer anomaly,
   (d) isostatic residual anomaly.

5. Explain why the following statement is wrong: "The isostatic residual anomaly is always zero over perfectly compensated topography."

6. The mass in Figure 7.10 is not isostatically compensated. What would the profile look like if it were?

8

The Geomagnetic Field

Magnus magnes ipse est globus terrestris.
[The whole earth is a magnet.]

(William Gilbert)

It has always been and still is [my] impression that
a magnetometer survey is just as much a means of
mapping geology as are the air photograph and the
surface geological traverse.

(Norman R. Paterson)

The previous chapter discussed the steps by which gravity measurements
are converted into gravity anomalies that reflect geological sources. The
present chapter treats magnetic anomalies in a similar vein. Whereas the
gravity field of the earth is largely time invariant, except for relatively
minor or long-term changes due to redistribution of mass (tides, moving magma, glacial rebound, erosion, mountain building, and so forth),
the geomagnetic field varies in both direction and intensity over time
scales ranging from milliseconds to millennia. It would seem that this
added complexity would make the reduction of magnetic measurements
significantly more difficult than that for gravity data, but in practice the
calculation of magnetic anomalies is relatively straightforward.

Our intent in this chapter is to characterize the global magnetic field
in order to isolate the magnetic field caused by crustal sources. This
agenda glosses over a large body of information that ordinarily would
be included in a chapter of this title, such topics as the origins of the
geomagnetic field (magnetohydrodynamic theories); the behavior of the
field in the geologic past (paleomagnetic studies); reversals of geomagnetic polarity; the magnetic properties of the sun, moon, meteorites,
and other planets; and the interaction of the earth's magnetic field with
solar phenomena. Other textbooks are dedicated to just these topics,

and readers are referred in particular to books by Merrill and McElhinny [183], Jacobs [139], Butler [47], Parkinson [211], and Chapman and Bartels [56] for more comprehensive discussions.

We begin this chapter with a general discussion of dividing potential fields measured on a sphere into parts originating from inside and outside the sphere. Later sections of this chapter will use spherical harmonic analysis to describe the internal geomagnetic field.

8.1 Parts of Internal and External Origin

Spherical harmonic analysis provides the means with which to determine from measurements of a potential field and its gradients on a sphere whether the sources of the field lie within or outside the sphere. Carl Friederich Gauss in 1838 was the first to describe the geomagnetic field in this way, and he came to the conclusion that the field observed at the earth's surface originates entirely from within the earth. We know now, with the benefit of satellites, space probes, and vastly more data, that he was only approximately correct; that is, a small part of the geomagnetic field originates from outside the earth. In this section, we investigate the general problem of separating a potential field into parts of external and internal origin, following the development of Chapman and Bartels [56]. We use magnetic induction in this derivation, but the discussion applies to any potential field.

Consider the magnetic induction $\mathbf{B}$ and its potential V, where $\mathbf{B} = -\nabla V$, and suppose that we have the ability to measure V or any component of $\mathbf{B}$ on a spherical surface with radius a. Assume further that in source-free regions V is harmonic and satisfies Laplace's equation,

$$\nabla^2 V = 0.$$

Specifically, V is harmonic on the surface of the sphere so long as sources of V do not extend across the surface. If no sources exist outside the sphere, then both V and $\frac{\partial V}{\partial r}$ must vanish as $r \to \infty$, and V can be represented by a spherical harmonic expansion similar to equation 6.31,

$$V^i = a \sum_{n=0}^{\infty} \left(\frac{a}{r}\right)^{n+1} \sum_{m=0}^{n} (A_n^{mi} \cos m\phi + B_n^{mi} \sin m\phi) P_n^m(\theta), \quad r \geq a,$$

$$(8.1)$$

where θ is colatitude, ϕ is longitude, and $P_n^m(\theta)$ is an associated Legendre polynomial of degree n and order m normalized according to the convention of Schmidt (Section 6.3.1). The superscript i identifies the potential and each harmonic coefficient as being due to internal sources.

On the other hand, if all sources lie outside the sphere, then V and $\frac{\partial V}{\partial r}$ must be finite within the sphere, and equation 6.30 is appropriate,

$$V^e = a \sum_{n=0}^{\infty} \left(\frac{r}{a}\right)^n \sum_{m=0}^{n} (A_n^{me} \cos m\phi + B_n^{me} \sin m\phi) P_n^m(\theta), \quad r \leq a, \quad (8.2)$$

where the superscript e denotes external sources. If sources exist both inside and outside the sphere, then the potential in source-free regions near the surface of the sphere is given by the sum of equations 8.1 and 8.2,

$$V = V^i + V^e$$

$$= a \sum_{n=0}^{\infty} \sum_{m=0}^{n} \left\{ \left[A_n^{mi} \left(\frac{a}{r}\right)^{n+1} + A_n^{me} \left(\frac{r}{a}\right)^n \right] \cos m\phi \right.$$

$$+ \left[B_n^{mi} \left(\frac{a}{r}\right)^{n+1} + B_n^{me} \left(\frac{r}{a}\right)^n \right] \sin m\phi \left. \right\} P_n^m(\theta)$$

$$= a \sum_{n=0}^{\infty} \sum_{m=0}^{n} \left\{ \left[C_n^m \left(\frac{r}{a}\right)^n + (1 - C_n^m) \left(\frac{a}{r}\right)^{n+1} \right] A_n^m \cos m\phi \right.$$

$$+ \left[S_n^m \left(\frac{r}{a}\right)^n + (1 - S_n^m) \left(\frac{a}{r}\right)^{n+1} \right] B_n^m \sin m\phi \left. \right\} P_n^m(\theta), \quad (8.3)$$

where

$$A_n^m = A_n^{mi} + A_n^{me},$$

$$B_n^m = B_n^{mi} + B_n^{me},$$

$$C_n^m = \frac{A_n^{me}}{A_n^m},$$

$$S_n^m = \frac{B_n^{me}}{B_n^m}.$$

The coefficients C_n^m and S_n^m in equation 8.3 are fractions ranging between 0 and 1; for any given degree n and order m, C_n^m and S_n^m indicate the relative importance of external sources to the total potential observed at the surface of the sphere.

Our objective now is to determine C_n^m and S_n^m from knowledge of the behavior of V just on the surface of the sphere. At $r = a$, equation 8.3 reduces to an expansion of spherical surface harmonics,

$$V = a \sum_{n=0}^{\infty} \sum_{m=0}^{n} (A_n^m \cos m\phi + B_n^m \sin m\phi) P_n^m(\theta)$$

$$= a \sum_{n=0}^{\infty} S_n(\theta, \phi). \tag{8.4}$$

Measurements of V will permit determination of A_n^m and B_n^m, as discussed in Chapter 6. Resolving C_n^m and S_n^m, however, requires information about how V changes in the direction normal to the surface of the sphere, and this is provided by the radial gradient of V (i.e., the radial component of $\mathbf{B}$). The radial gradient can be expressed as a spherical harmonic expansion, given at the surface by

$$\frac{\partial V}{\partial r} = \sum_{n=0}^{\infty} \sum_{m=0}^{n} (\alpha_n^m \cos m\phi + \beta_n^m \sin m\phi) P_n^m(\theta), \tag{8.5}$$

and measurements of $\frac{\partial V}{\partial r}$ would permit determination of α_n^m and β_n^m. But $\frac{\partial V}{\partial r}$ also can be derived from equation 8.3, so at $r = a$,

$$\frac{\partial V}{\partial r} = \sum_{n=0}^{\infty} \sum_{m=0}^{n} P_n^m(\theta) \Big\{ \Big[nC_n^m - (n+1)(1 - C_n^m) \Big] A_n^m \cos m\phi$$

$$+ \Big[nS_n^m - (n+1)(1 - S_n^m) \Big] B_n^m \sin m\phi \Big\}. \tag{8.6}$$

Equating the terms in equations 8.5 and 8.6 leads to

$$\alpha_n^m = [nC_n^m - (n+1)(1 - C_n^m)] A_n^m, \tag{8.7}$$

$$\beta_n^m = [nS_n^m - (n+1)(1 - S_n^m)] B_n^m. \tag{8.8}$$

The coefficients A_n^m and B_n^m can be obtained from a spherical harmonic analysis based on measurements of V using equation 8.4, and coefficients α_n^m and β_n^m can be derived from measurements of $\frac{\partial V}{\partial r}$ using equation 8.5. Having determined these coefficients, one can use equations 8.7 and 8.8 to provide C_n^m and S_n^m, the relative contribution from external sources to the potential at each harmonic. Hence, *knowledge of a potential and its radial gradient on a sphere determines the relative importance of sources internal and external to the sphere.*

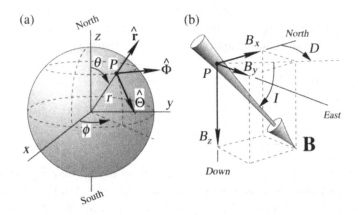

Fig. 8.1. (a) The spherical coordinate system. Point P is defined by coordinates r, θ, and ϕ, and a vector at point P is described in terms of three orthogonal unit vectors: $\hat{\mathbf{r}}$, $\hat{\boldsymbol{\Theta}}$, and $\hat{\boldsymbol{\Phi}}$. (b) The cartesian coordinate system at point P. The three components of vector $\mathbf{B}$ are shown: B_x is directed north ($B_x = -B_\theta$), B_y is east ($B_y = B_\phi$), and B_z is down ($B_z = -B_r$). Inclination I is the angle of $\mathbf{B}$ below horizontal, positive down; declination D is the azimuth of the horizontal projection of $\mathbf{B}$, positive east.

The potential is not measured directly in geomagnetic studies, so we must settle for some other way to determine the coefficients A_n^m and B_n^m in equations 8.7 and 8.8. Magnetometers can measure the three orthogonal components of magnetic induction. Assume for the moment that the earth is spherical. At the surface of the earth, we orient the cartesian coordinate system so that x is directed north, y is east, and z is down, as shown in Figure 8.1, and the relation $\mathbf{B} = -\nabla V$ leads to the expressions

$$B_x = -B_\theta$$

$$= \frac{1}{r}\frac{\partial V}{\partial \theta} \qquad \text{(north)}, \qquad (8.9)$$

$$B_y = B_\phi$$

$$= -\frac{1}{r\sin\theta}\frac{\partial V}{\partial \phi} \qquad \text{(east)}, \qquad (8.10)$$

$$B_z = -B_r$$

$$= \frac{\partial V}{\partial r} \qquad \text{(down)}. \qquad (8.11)$$

Equation 8.5 expresses the radial gradient of V as a surface harmonic expansion with coefficients α_n^m and β_n^m, so

$$B_z = \sum_{n=0}^{\infty} \sum_{m=0}^{n} (\alpha_n^m \cos m\phi + \beta_n^m \sin m\phi) \, P_n^m(\theta).$$

To find expressions for B_x and B_y in terms of the coefficients A_n^m and B_n^m, we substitute equations 8.3 into equations 8.9 and 8.10, respectively, and set $r = a$,

$$B_x = \sum_{n=0}^{\infty} \sum_{m=0}^{n} (A_n^m \cos m\phi + B_n^m \sin m\phi) \frac{\partial P_n^m(\theta)}{\partial \theta}, \tag{8.12}$$

$$B_y = \frac{1}{\sin \theta} \sum_{n=0}^{\infty} \sum_{m=0}^{n} (mA_n^m \sin m\phi - mB_n^m \cos m\phi) \, P_n^m(\theta). \tag{8.13}$$

Equation 8.13 is simply a surface harmonic expansion of the function $B_y \sin \theta$, so measurements of B_y provide all coefficients A_n^m and B_n^m needed in equations 8.7 and 8.8 at all degrees and orders except $m = 0$. Measurements of B_x can fill this remaining gap through application of equation 8.12. Equation 8.12 is not an expansion in terms of surface harmonics, but it can be shown that $\frac{\partial P_n^m(\theta)}{\partial \theta}$ can be expressed as such (Chapman and Bartels [53]). Measurements of B_z, as discussed earlier, provide the coefficients α_n^m and β_n^m. Hence, measurements of the southward, eastward, and downward components of **B** on the sphere are sufficient to assess the relative importance of external and internal sources to the geomagnetic field.

8.2 Description of the Geomagnetic Field

Equation 8.3 is written commonly in geomagnetic studies as

$$V = a \sum_{n=0}^{\infty} \left[\left(\frac{r}{a} \right)^n T_n^e + \left(\frac{a}{r} \right)^{n+1} T_n^i \right], \tag{8.14}$$

where

$$T_n^i = \sum_{m=0}^{n} \left(g_n^{mi} \cos m\phi + h_n^{mi} \sin m\phi \right) P_n^m(\theta)$$

$$T_n^e = \sum_{m=0}^{n} \left(g_n^{me} \cos m\phi + h_n^{me} \sin m\phi \right) P_n^m(\theta).$$

The new coefficients g_n^{mi}, g_n^{me}, h_n^{mi}, and h_n^{me} are called *Gauss coefficients*, have the same dimensions as magnetic induction, and generally are expressed in units of nanotesla (or gamma). As before, the superscripts e and i denote either external or internal terms. Gauss coefficients are related to the old coefficients (equation 8.3) according to

$$g_n^{mi} = (1 - C_n^m)A_n^m, \qquad h_n^{mi} = (1 - S_n^m)B_n^m,$$

$$g_n^{me} = C_n^m A_n^m, \qquad h_n^{me} = S_n^m B_n^m,$$

$$C_n^m = \frac{g_n^{me}}{g_n^{mi}+g_n^{me}}, \qquad S_n^m = \frac{h_n^{me}}{h_n^{mi}+h_n^{me}}.$$

Gauss made the first quantitative spherical harmonic analysis of the geomagnetic field in 1838. He determined the harmonic coefficients from measurements of B_x, B_y, and B_z at a total of only 84 points (spaced 30° apart in the ϕ direction along seven circles of latitude) and concluded that the external coefficients g_n^{me} and h_n^{me} are zero. Now we know that external sources contribute several tens of nT (and often much more) to the total magnetic field at the earth's surface, and that this contribution is highly variable in both time and space.

The magnetic field originating from inside the earth is approximately dipolar, as we shall see shortly, and would appear very much like Figure 4.8 if the earth were isolated in space. The earth, however, is continuously bathed by the *solar wind*, a stream of charged plasma emitted by the sun. The region of interaction between the solar wind and the internal magnetic field is called the *magnetosphere*, a region of considerably more magnetic complexity than depicted by Figure 4.8. It is this complex interaction between the earth's internal magnetic field and the solar wind, coupled with the earth's rotation, tidal forces, and thermal effects, that produces the external magnetic field. The *ionosphere*, which surrounds the earth at altitudes between roughly 50 and 1,500 km, is an important part of this interaction; the earth's rotation and tidal effects generate electrical currents in the ionosphere, which in turn produce magnetic fields that can reach magnitudes of up to 1,000 nT at the earth's surface.

The details of the origins of the external magnetic field are beyond the scope of this chapter, and the interested reader is referred to the text by Merrill and McElhinny [183] for a lucid and concise summary of this subject. The following discussion focuses on just the internal parts of the field. For convenience we will drop the superscripts i and e in subsequent spherical harmonic expansions with the understanding that g_n^m and h_n^m pertain to internal sources.

8.2.1 The Elements of the Geomagnetic Field

Vector quantities (e.g., **B**, **H**, and **M**) in geomagnetic studies must be described in a frame of reference fixed with respect to observation points on the earth's surface. This typically is done in one of two ways. The vector can be described in terms of three orthogonal components in a cartesian coordinate system, usually oriented so that x increases to the north, y is east, and z is down (Figure 8.1). For geomagnetic fields, these three components are often written in the literature as X, Y, and Z, respectively, and expressed in units of nT. Here we follow the convention established throughout this text and write the components as, for example, $B_x = X$, $B_y = Y$, and $B_z = Z$. The intensity of the horizontal component then is

$$H = \sqrt{B_x^2 + B_y^2}.$$

Alternatively, the vector can be described by its total intensity,

$$T = \sqrt{B_x^2 + B_y^2 + B_z^2}, \tag{8.15}$$

plus two angles, the *inclination* and *declination*. Inclination is the vertical angle between the vector and the horizontal plane, that is,

$$I = \arctan \frac{B_z}{\sqrt{B_x^2 + B_y^2}}.$$

By convention, inclination is positive when the vector is inclined below the horizontal plane and negative when above the horizontal plane. The vertical plane containing the vector is called the *magnetic meridian*, and *declination* is defined as the azimuth of the magnetic meridian, positive to the east and negative to the west, that is,

$$D = \arcsin \frac{B_y}{\sqrt{B_x^2 + B_y^2}}.$$

Contour maps describing these various elements of the geomagnetic field are called *isomagnetic maps*. *Isodynamic maps* indicate contours of equal field intensity, such as total intensity, vertical intensity, or horizontal intensity. *Isoclinal maps* show contours of equal inclination, and *isogonic maps* represent declination. Figure 8.2 shows various examples of these maps; similar charts are available from a variety of sources at more convenient scales (e.g., Peddie and Zunde [215]; Fabiano, Peddie, and Zunde [87]).

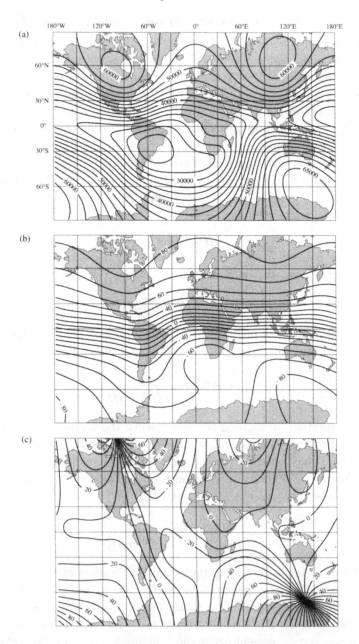

Fig. 8.2. The magnetic field of the earth based on IGRF 1990. (a) Isodynamic map showing total intensity, contour interval 2,500 nT; (b) isoclinic map
showing constant inclination, contour interval 10°; (c) isogonic map showing
constant declination, contour interval 10°.

The foregoing discussion has assumed that the earth is precisely spherical. Actual measurements of B_x, B_y, and B_z in studies of the main geomagnetic field, however, are generally oriented instead with respect to a spheroidal model (Langel [157]), and mathematical representations of the main field, such as those displayed on Figure 8.2, reflect this spheroidal coordinate system. Changing from a spherical to a spheroidal coordinate system has no effect on the B_y component and only a small effect on B_x and B_y, and the discrepancies are often ignored in studies of crustal magnetization.

8.2.2 The International Geomagnetic Reference Field

Like the reference ellipsoid and theoretical gravity (Section 7.2), the mathematical representation of the low-degree parts of the geomagnetic field is determined by international agreement. This mathematical description is called the International Geomagnetic Reference Field and is the purview of the International Association of Geomagnetism and Aeronomy (IAGA) and its umbrella organization, the International Union of Geodesy and Geophysics (IUGG). The IGRF consists of Gauss coefficients through degree and order 10 because, as we will discuss subsequently, these low-order terms are believed to represent in large part the field of the earth's core. Subtracting these low-order terms from measured magnetic fields provides in principle the magnetic field of the crust. As discussed in Section 8.3, however, a 10-degree harmonic expansion is not sufficient to isolate the crustal field.

The geomagnetic field changes with time, however, and so must its mathematical description. Because international agreement is not easily achieved on a day-to-day basis, IAGA adopts new IGRF models at five-year intervals which are intended to represent the geomagnetic field for the following five-year period, called an *epoch*. To provide this predictive ability, each Gauss coefficient for any particular IGRF model has a derivative term that predicts the field into the immediate future assuming each coefficient changes linearly with time.

Unfortunately the changes in the geomagnetic field are not entirely predictable, and differences between the predictive IGRF and the true geomagnetic field begin to grow over the course of each epoch. This divergence is treated in the short term by establishing a new IGRF model every five years. In the long term, it is possible to improve old IGRF models with the benefit of accumulating data. Therefore, IAGA periodically adopts models for past epochs, called Definitive Geomagnetic Reference Field (DGRF) models. DGRF models are not established until it becomes unlikely that the data sets will be significantly improved. DGRF models, therefore, become the official record of how the geomagnetic field has behaved in past epochs. Nine DGRF models are in effect at the present: DGRF 1945 through DGRF 1985, each representing the subsequent five-year epoch. These nine models, along with IGRF 1990 and its predictive terms, give a complete description of the geomagnetic field from 1945 through 1995. Shortly after 1995, IAGA will adopt a new DGRF 1990, to replace IGRF 1990, and establish a new IGRF 1995 complete with new predictive terms.

Table 8.1 shows the IGRF 1990 model and its predictive terms. A complete description of the IGRF has been published in various places (IAGA [134], Langel [157, 160]); these descriptions provide coefficients for all DGRF models from 1945 through 1985, list the coefficients and predictive terms for IGRF 1990, and include a concise discussion of the history of the IGRF. Figure 8.2 shows isodynamic, isoclinic, and isogonic maps based on the IGRF 1990 model.

8.2.3 The Dipole Field

Table 8.2 shows how the first few harmonic terms of the geomagnetic field have changed since the time of Gauss and illustrates some general characteristics of the earth's internal field. Notice that this table has no g_0^0 coefficient because, as discussed in Chapter 6, the zero-degree harmonic term varies as $1/r$ (for internal sources) and corresponds to the potential of a monopole; g_0^0 must vanish, therefore, because magnetic monopoles do not exist. Also notice the overwhelming dominance of the first-degree coefficients (g_1^0, g_1^1, and h_1^1). As discussed in Chapter 6 and in the following paragraphs, the first-degree harmonic describes the potential of a dipole centered at the center of the sphere, and the large amplitudes of these coefficients reflect the generally geocentric dipolar character of the main geomagnetic field. Finally, notice the systematic change of all coefficients (especially those of higher order) over time since 1835. Part of this change is due to improvements in field definition, but most of the change reflects real temporal variations of the geomagnetic field. We will return to this subject in Section 8.2.5.

We now investigate the $n = 1$ harmonic term of the geomagnetic field.

Table 8.1. *International Geomagnetic Reference Field for epoch 1990.*[a]

n	m	g	dg/dt	h	dh/dt	n	m	g	dg/dt	h	dh/dt
1	0	−29775	18.0			8	0	22	0.2		
1	1	−1851	10.6	5411	−16.1	8	1	5	−0.7	10	0.5
2	0	−2136	−12.9			8	2	−1	−0.2	−20	−0.2
2	1	3058	2.4	−2278	−15.8	8	3	−11	0.1	7	0.3
2	2	1693	0.0	−380	−13.8	8	4	−12	−1.1	−22	0.3
3	0	1315	3.3			8	5	4	0.0	12	0.4
3	1	−2240	−6.7	−287	4.4	8	6	4	−0.1	11	−0.5
3	2	1246	0.1	293	1.6	8	7	3	−0.5	−16	−0.3
3	3	807	−5.9	−348	−10.6	8	8	−6	−0.6	−11	0.6
4	0	939	0.5			9	0	4			
4	1	782	0.6	248	2.6	9	1	10	—	−21	—
4	2	324	−7.0	−240	1.8	9	2	1	—	15	—
4	3	−423	0.5	87	3.1	9	3	−12	—	10	—
4	4	142	−5.5	−299	−1.4	9	4	9	—	−6	—
5	0	−211	0.6			9	5	−4	—	−6	—
5	1	353	−0.1	47	−0.1	9	6	−1	—	9	—
5	2	244	−1.6	153	0.5	9	7	7	—	9	—
5	3	−111	−3.1	−154	0.4	9	8	2	—	−7	—
5	4	−166	−0.1	−69	1.7	9	9	−6	—	2	—
5	5	−37	2.3	98	0.4	10	0	−4			
6	0	61	1.3			10	1	−4	—	1	—
6	1	64	−0.2	−16	0.2	10	2	2	—	0	—
6	2	60	1.8	83	−1.3	10	3	−5	—	3	—
6	3	−178	1.3	68	0.0	10	4	−2	—	6	—
6	4	2	−0.2	−52	−0.9	10	5	4	—	−4	—
6	5	17	0.1	2	0.5	10	6	3	—	0	—
6	6	−96	1.2	27	1.2	10	7	1	—	−1	—
7	0	77	0.6			10	8	2	—	4	—
7	1	−64	−0.5	−81	0.6	10	9	3	—	0	—
7	2	4	−0.3	−27	0.2	10	10	0	—	−6	—
7	3	28	0.6	1	0.8						
7	4	1	1.6	20	−0.5						
7	5	6	0.2	16	−0.2						
7	6	10	0.2	−23	0.0						
7	7	0	0.3	−5	0.0						

[a] Note: Gauss coefficients are in units of nT; rate-of-change coefficients are predictions for the period 1990 to 1995 in nT per year.

Rewriting equation 8.14 for internal sources provides

$$V = a \sum_{n=1}^{\infty} \left(\frac{a}{r}\right)^{n+1} \sum_{m=0}^{n} \left(g_n^m \cos m\phi + h_n^m \sin m\phi\right) P_n^m(\theta), \qquad (8.16)$$

Table 8.2. *Comparison of various analyses of the geomagnetic field.*
Coefficients in units of nT.[a]

Source	Epoch	g_1^0	g_1^1	h_1^1	g_2^0	g_2^1	h_2^1	g_2^2	h_2^2
Gauss	1835	−32350	−3110	6250	510	2920	120	−20	1570
DF	1922	−30920	−2260	5920	−890	2990	−1240	1440	840
DGRF	1945	−30594	−2285	5810	−1244	2990	−1702	1578	477
DGRF	1965	−30334	−2119	5776	−1662	2997	−2016	1594	114
DGRF	1985	−29873	−1905	5500	−2072	3044	−2197	1687	−306
IGRF	1990	−29775	−1851	5411	−2136	3058	−2278	1693	−380

[a]Note: DF refers to Dyson and Furner [82].

and expanding the summations just through $n = 1$ yields

$$V^D = \frac{a^3}{r^2} \left[g_1^0 P_1^0(\theta) + (g_1^1 \cos\phi + h_1^1 \sin\phi) P_1^1(\theta) \right],$$

where superscript D refers to the dipole contribution. As discussed in Chapter 6, each coefficient of a harmonic expansion is the best possible coefficient, in a least-squares sense, regardless of the number of terms in the expansion. Thus, we know that the coefficients g_1^0, g_1^1, and h_1^1 in the previous equation still will do the best possible job in modeling the first-degree part of the field, even though all other terms of the expansion have been dropped. From Table 6.2, $P_1^0(\theta) = \cos\theta$ and $P_1^1(\theta) = \sin\theta$, and making these substitutions provides

$$V^D = \frac{a^3}{r^2} \left[g_1^0 \cos\theta + g_1^1 \cos\phi \sin\theta + h_1^1 \sin\phi \sin\theta \right]. \qquad (8.17)$$

Polar coordinates can be converted to cartesian coordinates using the usual relationships,

$$x = r\sin\theta\cos\phi,$$
$$y = r\sin\theta\sin\phi,$$
$$z = r\cos\theta,$$

where in this case x, y, and z are oriented as in Figure 8.1(a), with the origin at the center of the earth, the z axis aligned along the spin axis pointing north, and x positioned at Greenwich meridian. In cartesian coordinates, equation 8.17 becomes

$$V^D = \frac{a^3}{r^2}\left[g_1^0\frac{z}{r} + g_1^1\frac{x}{r} + h_1^1\frac{y}{r}\right]. \tag{8.18}$$

From Chapter 4, the potential of a dipole centered at the origin is given by

$$V = C_m \frac{\mathbf{m}\cdot\hat{\mathbf{r}}}{r^2}$$

$$= \frac{C_m}{r^2}\left[m_x\frac{x}{r} + m_y\frac{y}{r} + m_z\frac{z}{r}\right], \tag{8.19}$$

where $C_m = 1$ and is dimensionless in emu or $C_m = \frac{\mu_0}{4\pi} = 10^{-7}$ henry·meter^{-1} in SI units. Comparing equation 8.18 and 8.19, we see that *the $n = 1$ term of a spherical harmonic expansion describes the magnetic field of a dipole centered at the origin.* In particular, the first three nonzero Gauss coefficients are each proportional to one of the three orthogonal components of the dipole moment, that is, at the surface of the earth and in SI units,

$$m_x = \frac{4\pi}{\mu_0}a^3 g_1^1,$$

$$m_y = \frac{4\pi}{\mu_0}a^3 h_1^1,$$

$$m_z = \frac{4\pi}{\mu_0}a^3 g_1^0.$$

Using the coefficients from the 1990 International Geomagnetic Reference Field (IGRF) (Table 8.1) and letting $a = 6.371 \times 10^6$ m yields

$$m_x = -0.479 \times 10^{22},$$
$$m_y = 1.399 \times 10^{22},$$
$$m_z = -7.700 \times 10^{22},$$

each in units of A·m^2. The magnitude of the centered dipole, therefore, is

$$m = \sqrt{m_x^2 + m_y^2 + m_z^2}$$

$$= \frac{4\pi}{\mu_0}a^3\sqrt{(g_1^0)^2 + (g_1^1)^2 + (h_1^1)^2}$$

$$= 7.840 \times 10^{22} \quad (\text{A}\cdot\text{m}^2),$$

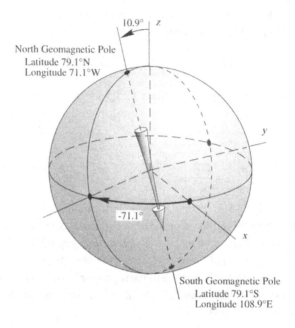

Fig. 8.3. The orientation of the dipole field as described by IGRF 1990.

and simple trigonometry then provides the orientation of the centered dipole:

$$\theta = \arccos \frac{m_z}{m}$$

$$= 169.1°,$$

$$\phi = \arccos \frac{m_x}{\sqrt{m_x^2 + m_y^2}}$$

$$= 108.9°,$$

where θ is colatitude, and ϕ is longitude with respect to Greenwich meridian. The field produced by this "best" geocentric dipole is called the *dipole field*.

As depicted in Figure 8.3, the extension of the positive end of the geocentric dipole intersects the earth's surface at coordinates $(\theta, \phi) = (169.1°, 108.9°)$, or at latitude 79.1°S and longitude 108.9°E; the negative end intersects the surface at $(\theta, \phi) = (10.9°, 288.9°)$, or at latitude 79.1°N and longitude 71.1°W in the Kane Basin between Ellesmere

Island and Greenland. These two points are called the south and north *geomagnetic poles*, respectively. The *magnetic poles*, on the other hand, are the points where the total field is normal to the surface of the earth, that is, the south magnetic pole is that point where the inclination is $-90°$ and the north magnetic pole is located where the inclination reaches $+90°$. If the geomagnetic field were perfectly dipolar (i.e., if $g_n^m = h_n^m = 0$ for all $n > 1$), the distinction between geomagnetic poles and magnetic poles would be unnecessary. Table 8.1 shows that the field departs significantly from that of a geocentric dipole, however, and the north magnetic pole, as defined by the IGRF 1990 field model, is located off the south coast of Ellef Ringnes Island in the Queen Elizabeth Islands, roughly 700 km from the geomagnetic pole. The departure of the geomagnetic field from that of a geocentric dipole will be the subject of the next section.

Exercise 8.1 Use Table 8.2 to describe the centered dipole of the earth in 1965. How many degrees did the dipole "rotate" between 1965 and 1980?

The first three nonzero coefficients of the spherical harmonic expansion define the single magnetic dipole, centered at the origin, that best fits the observed magnetic field. A better approximation results if the dipole is not constrained to be at the origin. The overall "best-fit" dipole is located about 400 km up the positive z axis from the earth's center, and the field that it produces is called the *eccentric dipole field*.

8.2.4 The Nondipole Field

Excluding the $n = 1$ harmonic from equation 8.16 eliminates the dipole term from the geomagnetic field. The remainder, given by

$$V^{\mathrm{N}} = a \sum_{n=2}^{\infty} \left(\frac{a}{r}\right)^{n+1} \sum_{m=0}^{n} (g_n^m \cos m\phi + h_n^m \sin m\phi) \, P_n^m(\theta), \qquad (8.20)$$

is called the *nondipole field*. To estimate the relative importance of the nondipole field, consider the vertical component B_r as measured at the north geographic pole ($r = a$, $\theta = 0°$). Differentiating equation 8.16 with respect to r yields

$$B_r = -\frac{\partial V}{\partial r}$$

$$= \sum_{n=1}^{\infty} (n+1) \left(\frac{a}{r}\right)^{n+2} \sum_{m=0}^{n} (g_n^m \cos m\phi + h_n^m \sin m\phi) \, P_n^m(\theta).$$

Now we let $r = a$ and $\theta = 0°$ and note from Table 6.2 that $P_n^m(0) = 1$ if $m = 0$ and $P_n^m(0) = 0$ otherwise. Thus the radial component of **B** at the north geographic pole is simply

$$B_r = \sum_{n=1}^{\infty} (n+1)g_n^0 \, .$$

Let B_r^{D} and B_r^{N} represent the radial components of the dipole and nondipole fields, respectively, at the north geographic pole. Then from Table 8.1,

$$B_r^{\mathrm{D}} = 2g_1^0$$

$$= -59,550 \qquad (\mathrm{nT}),$$

$$B_r^{\mathrm{N}} \approx \sum_{n=2}^{10} (n+1)g_n^0$$

$$\approx 3,518 \qquad (\mathrm{nT}),$$

and the percent provided by the nondipole field to the total vertical field at this particular location is

$$100 \times \frac{|B_r^{\mathrm{N}}|}{|B_r^{\mathrm{D}} + B_r^{\mathrm{N}}|} = 6.3 \qquad (\text{percent}).$$

Though this percentage is of only one component at only one location, it is characteristic of the main field as a whole: *The nondipole field comprises only about 10 percent of the main field, and thus the geomagnetic field is dipolar to a very good first approximation.* The spatial character of the nondipole field is one of large-scale anomalies (Figure 8.4), eight or ten in number, with spatial dimensions comparable to continental size (although no correlation is apparent), and with magnitudes on the order of 10^4 nT.

It is tempting but not advisable to attach physical significance to the nondipole and dipole geomagnetic fields. As discussed in Chapter 6, the various terms of a spherical harmonic expansion represent the potentials of highly idealized sources (monopoles, dipoles, quadrupoles, and so forth) situated precisely at the center of the sphere. These geocentric models are not unique, of course; the same potential could be modeled equally well by more complex, distributed sources, and Problem 2 at the end of this chapter provides one example. Moreover, it is well established on the basis of many lines of evidence that the low-degree terms of the

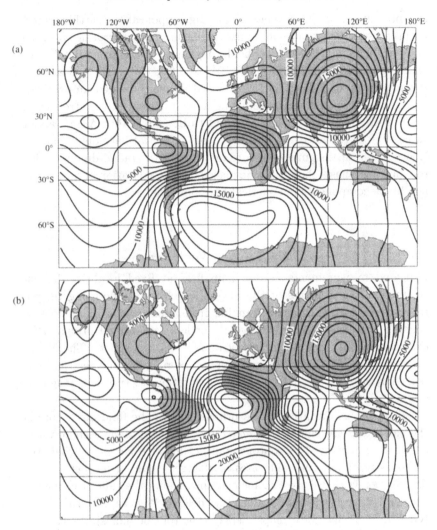

Fig. 8.4. The total intensity of the nondipole field at two different epochs. Contour interval 1000 nT. (a) Based on DGRF 1945. (b) Based on IGRF 1990.

geomagnetic field originate primarily from the fluid outer core, not from the precise center of the earth.

Furthermore, the origins of the dipole and nondipole fields are not independent of each other. Alldredge and Hurwitz [3] modeled the main field by placing a number of radial dipoles near the core–mantle interface, to represent the nondipole field, and a single dipole at the earth's center. By progressive iteration to minimize the mean squared error between the observed field and the modeled field, they found that eight radial dipoles (ranging in magnitude from 0.8×10^{22} to 3.6×10^{22} A $\cdot$ m^2) located at about 0.25 earth radii, plus the centered dipole, provided an excellent representation of the main geomagnetic field. However, the strength of the best-fit centered dipole was found to be 18.0×10^{22} A $\cdot$ m^2, about twice that of the dipole determined from spherical harmonic analysis. Hence, if physical significance is to be given to the radial-dipole model, too much emphasis should not be placed on the separation of dipole and nondipole components in spherical harmonic analyses.

8.2.5 Secular Variation

Repeated measurements of the main field at fixed localities demonstrate that the elements of the magnetic field are undergoing temporal changes over time scales ranging from milliseconds to millions of years. Short-period variations (yearly or less) are caused primarily by external sources, such as electric currents in the ionosphere. These temporal changes are manifested in various ways, ranging from very periodic behavior, such as the daily or *diurnal* variation, to transient magnetic storms.

Longer-period variations arise primarily from the fluid outer core of the earth and are called *geomagnetic secular variation*. Secular variation is often displayed by contour maps, called *isoporic maps*, where the contours represent constant rates of change, either in nT per year or degrees per year. Figure 8.5, for example, shows the rate of change of the total intensity of the geomagnetic field in 1990 based on the 1990 IGRF field model. Such maps display cells (or anomalies) of either increasing or decreasing field. The cells are continental in size, although no correlation with continents appears to exist, and persist in the same sense for decades or longer. At any given location, the nondipole field changes in amplitude at an average (rms) rate of about 50 nT per year with maximum rates of about 100 nT per year.

Figure 8.4 shows the nondipole field at two epochs separated in time by 45 years. Subtle differences in various anomalies are apparent, even over this relatively short time period. Many of the foci drifted noticeably westward over this period and have been doing so since at least the time of Gauss; for example, the positive anomaly over the Gulf of Guinea (off

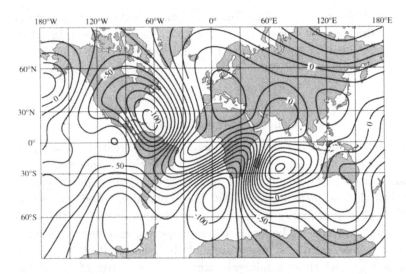

Fig. 8.5. Secular change in nT/yr based on IGRF 1990. Calculated by sub-tracting the total intensity in 1990 from that predicted in 1991.

the west coast of Africa), the positive anomaly south of Australia, the negative anomaly over the Mediterranean Sea, and the negative anomaly over Ecuador and Colombia. Indeed, the nondipole field in an average sense is drifting systematically westward at a rate of roughly 0.2° per year, although the rate varies with latitude (e.g., Bullard et al. [45], Yukutake [292]). On closer examination it has been shown that this west-ward drift is only an average characteristic of the nondipole field; some features drift westward more rapidly than 0.2° per year, while others remain stationary (e.g., Yukutake [293], Yukutake and Tachinaka [294]). The large positive anomaly over Mongolia in Figure 8.4, for example, has remained stationary over the 45-year period represented by this figure, and in fact for a much longer period of time. This dual behavior suggests that the nondipole field has a long-term stationary component in addi-tion to westward-drift (Yukutake [293], Yukutake and Tachinaka [294]), and no doubt even greater complexity (James [141, 142]).

These observations of secular variation are direct ones; i.e., as deter-mined from continuously operating observatories, repeat measurements, and surveys conducted from ships, aircraft, and satellites. Paleomagnetic studies are continuing to piece together the behavior of the geomagnetic field in the geologic past, including the long-term record of secular vari-ation, reversals of geomagnetic polarity, excursions of the geomagnetic poles, and true polar wander. This considerable body of information is beyond the scope of this text, and the interested reader is referred to books by Merrill and McElhinny [183] and Butler [47] for information on these subjects.

8.3 Crustal Magnetic Anomalies

The internal sources of the geomagnetic field are located primarily in two regions of the earth. The majority of the field is generated in the fluid outer core by way of complex magnetohydrodynamic processes and is called the *core field* or *main field*. The remainder, called the *crustal field*, originates primarily from a relatively thin outer shell of the earth where temperatures are below the Curie temperatures of important magnetic minerals, primarily magnetite and titanomagnetite (Chapter 5). The depth to which such minerals exist is still a matter of discussion, however. The mantle is generally considered to be nonmagnetic (e.g., Wasilewski, Thomas, and Mayhew [289]; Frost and Shive [92]), so according to this view, the depth extent of magnetic rocks is either the crust–mantle interface or the Curie-temperature isotherm, whichever is shallower. Some studies have concluded, on the other hand, that upper mantle rocks may have significant magnetizations, especially in oceanic regions (e.g., Arkani-Hamed [6]; Harrison and Carle [117]; Counil, Achache, and Galdeano [76]). In the following, we will loosely regard the crust–mantle interface as magnetic basement, with the understanding that rocks in the uppermost mantle in some geologic environments may also contribute to the geomagnetic field. Hence, the vast region between the Curie-temperature isotherm (or crust–mantle interface, whichever is shallower) and the core–mantle interface is generally considered to be nonmagnetic. The calculation of crustal magnetic anomalies then amounts to subtracting the core field from measurements of the total magnetic field. Paterson and Reeves [212, Figure 8] showed an excellent example of the enhancement of an airborne magnetic survey by this simple residual calculation.

The large difference in depth between the sources of the crustal and core field is reflected in spherical harmonic analyses. This depth information is perhaps best displayed by way of the power spectrum R_n, defined as the scalar product $\mathbf{B}_n \cdot \mathbf{B}_n$ averaged over the spherical surface; that is,

$$R_n = \frac{1}{4\pi a^2} \int_0^{2\pi} \int_0^{\pi} \mathbf{B}_n \cdot \mathbf{B}_n \, a^2 \sin\theta \, d\theta \, d\phi , \qquad (8.21)$$

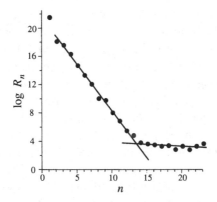

Fig. 8.6. Power spectrum of the geomagnetic field at the earth's surface based on 26,500 measurements from the Magsat satellite mission. Dots indicate calculated values of $\log R_n$; best-fit lines are shown for $2 \leq n \leq 12$ and $16 \leq n \leq 23$. Modified from Langel and Estes [160].

where

$$\mathbf{B}_n = -\nabla \left[a \left(\frac{a}{r} \right)^{n+1} \sum_{m=0}^{n} (g_n^m \cos m\phi + h_n^m \sin m\phi) \, P_n^m(\theta) \right]$$

evaluated at $r = a$. Using the orthogonality property of spherical surface harmonics, Lowes [167, 168] reduced equation 8.21 to

$$R_n = (n+1) \sum_{m=0}^{n} \left[(g_n^m)^2 + (h_n^m)^2 \right]^{\frac{1}{2}}. \tag{8.22}$$

It is clear from Table 8.1 that R_n decreases with increasing n at least through degree 10. Figure 8.6 shows R_n through degree 23, as calculated by Langel and Estes [160] from 26,500 low-orbit satellite measurements. The logarithm of R_n takes the form of two straight-line segments with a change in slope at about degree 14 (Figure 8.6), which is in general agreement with earlier studies (e.g., Cain, Davis, and Reagan [49]).

Within any range of n, the rate of decrease of R_n with increasing n is directly related to the depth of sources principally responsible for that part of the spectrum. To demonstrate this relationship, we first note from equation 8.21 that the power spectrum R_n is proportional to $(a/r)^{2n+4}$.

Exercise 8.2 Prove the previous statement; i.e., use equation 8.21 to show that R_n is proportional to $(a/r)^{2n+4}$.

Then to transform R_n based at the surface of the earth into a spectrum that would be determined at some new radius r, we simply have to multiply R_n by the factor $(a/r)^{2n+4}$, or add $(2n+4)\log(a/r)$ to $\log R_n$. This manipulation changes the slope of the logarithmic power spectrum by a constant $2\log(a/r)$. If $r > a$, R_n is transformed onto a larger sphere, the slope of $\log R_n$ is steepened, and the procedure is called *upward continuation*. If $r < a$, R_n is transformed to a smaller sphere (within the earth), the slope of $\log R_n$ is flattened, and the procedure is called *downward continuation*. Downward continuation is legitimate, however, only if all currents and other sources of magnetic fields are absent between radii a and r (e.g., Booker [34], Lowes [168]). We will have considerably more to say about upward and downward continuation in Chapter 12.

It is commonly assumed that the radius required to make $\log R_n$ as nearly constant as possible (i.e., to make the power spectrum "white") is the radius at which the important sources of the field are located (e.g., Lowes [168], Langel and Estes [160]). With this assumption, the principal sources are located at a radius given by the value of r that satisfies

$$S + 2\log \frac{a}{r} = 0\,,$$

where S is the slope of $\log R_n$. The line that best fits $\log R_n$ for $n < 14$ in Figure 8.6 has a slope of -1.309 (Langel and Estes [160]). Substituting this value into the previous equation provides $r = 3311$ km, which places the sources of this part of the spectrum at a radius about 174 km below the seismic core–mantle boundary. The spectrum at $n > 14$ indicates sources within the upper 100 km of the earth.

Hence, it is logical to interpret the steep part of the spectrum ($n < 14$) in Figure 8.6 to be caused by sources within the outer core, and the flatter part of the spectrum ($n > 14$) to be dominated by lithospheric sources (e.g., Bullard [44], Booker [34], Lowes [168], Cain et al. [49]; Langel and Estes [160]). It would seem, therefore, that a crustal magnetic map could be constructed from satellite data by subtracting a 13-degree spherical harmonic expansion derived from the same data (e.g., Mayhew [177]; Regan, Cain, and Davis [241]; Cain, Schmitz, and Math [50]; Langel, Phillip, and Horner [159]).

This assumption deserves consideration, however. Carle and Harrison [53] showed that residual fields calculated in this way may contain long-wavelength components, too long to be caused by near-surface sources. This happens in part because in practice we usually measure the total intensity of the geomagnetic field (discussed more fully in Section 8.3.1) rather than a single component of the field. The residual crustal anomaly in such cases is computed by subtracting the magnitude of a low-degree regional field from measurements of the geomagnetic intensity. The total intensity is the square root of the sum of the squares of three orthogonal components, as in equation 8.15. Although the potential is modeled in a spherical harmonic expansion as the sum of sinusoidal terms, in squaring the three components to form the intensity, each of those sinusoidal terms becomes a combination of both longer and shorter wavelength terms, as demonstrated by the following exercise.

Exercise 8.3 Consider a potential field given by $\mathbf{B} = -\nabla V$, where $V = a(a/r)^2 g_1^0 P_1^0(\theta)$. Show that each component of $\mathbf{B}$ (i.e., B_r and B_θ) has only one sinusoidal term of wavelength 2π, but that $|\mathbf{B}|$ has two terms, one with infinite wavelength and a second with wavelength π.

Hence, a value of n that represents the transition from dominantly core to dominantly crustal contributions to the geomagnetic field may not be appropriate for the total intensity of the field. A residual anomaly computed by subtracting the magnitude of a regional field truncated at n still will contain contributions from harmonics less than n.

Langel [156] agreed that the anomaly field includes long wavelengths, but that these can originate strictly from crustal sources by virtue of the way the anomaly field is calculated and do not necessarily imply contamination from sources in the core. Arkani-Hamed and Strangway [7] and Harrison, Carle, and Hayling [118] independently concluded that the crustal portion of the magnetic field dominates the total intensity at degrees 19 and greater; that is, to be sure that the residual anomaly represents only crustal sources, they recommended truncation of anomalies up through degree 18.

The important conclusion for our purposes here is that subtraction of a 10-degree IGRF model, such as that shown in Table 8.1, from a magnetic survey will be inadequate to eliminate the entire core field. Magnetic studies of continental or global scale should be evaluated carefully in the context of long-wavelength anomalies that might originate from the core and that may be confused with the crustal field. The long-wavelength shortcomings of the 10-degree IGRF are much less significant for local- or regional-scale magnetic studies applied to geologic problems. In such cases, additional regional fields can be removed subsequent to subtraction of the IGRF using a variety of techniques, such as simple curve fitting and digital filtering.

Chapter 7 focused on the reduction of measurements of the total gravitational attraction of the earth to gravity anomalies reflecting crustal density variations. This reduction procedure consists of a series of steps, including a subtraction of the gravitational effects of an average earth, the elevation of the measurement, and terrain. In crustal magnetic studies, the calculation of anomalies is often treated in a more cavalier way, usually consisting of only two steps: (1) an adjustment for daily variations and magnetic disturbances and (2) the subtraction of a suitable regional field, such as the IGRF model appropriate for the date of the survey. Some of the corrections routinely applied to gravity measurements generally are not attempted in magnetic studies simply because they are less tractable in the magnetic case. Terrain corrections in gravity studies, for example, although tedious are nevertheless straightforward because the density of the terrain is relatively uniform. Crustal magnetization, however, can vary by several orders of magnitude (and change sign) at essentially all spatial scales. This additional complexity requires less-straightforward techniques to correct for terrain effects (e.g., Clarke [61], Grauch [101]), and the effects of terrain are often left to the modeling and interpretation stage.

Exercise 8.4 Discuss the magnetic analog of the simple Bouguer correction.

8.3.1 Total-Field Anomalies

Total-field magnetometers are usually the instrument of choice for airborne and shipborne magnetic surveys. As the name implies, total-field magnetometers measure the magnitude of the total magnetic field without regard to its vector direction. The *total-field anomaly* is calculated from total-field measurements by subtracting the magnitude of a suitable regional field, usually the IGRF model appropriate for the date of the survey. If $\mathbf{T}$ represents the total field at any point, and $\mathbf{F}$ is the regional field at the same point, then the total-field anomaly is given by

$$\Delta T = |\mathbf{T}| - |\mathbf{F}| . \qquad (8.23)$$

Because ΔT will form the basis of future discussions on interpretation of magnetic data, it is important to establish under what conditions ΔT is harmonic. Let $\mathbf{\Delta F}$ represent the perturbation of $\mathbf{F}$ due to some anomalous magnetic source. Then the total field is given by

$$\mathbf{T} = \mathbf{F} + \mathbf{\Delta F}$$

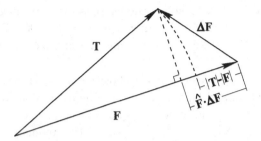

Fig. 8.7. Vector representation of total-field anomalies. Total field **T** is the vector sum of the regional field **F** and the anomalous field **ΔF**. Length $|\mathbf{T}| - |\mathbf{F}|$ represents the total-field anomaly, but length $\hat{\mathbf{F}} \cdot \mathbf{\Delta F}$ is a suitable approximation if $|\mathbf{F}| \gg |\mathbf{\Delta F}|$.

(Figure 8.7). Note that the total-field anomaly is not equivalent to the magnitude of the anomalous field because

$$\Delta T = |\mathbf{F} + \mathbf{\Delta F}| - |\mathbf{F}|$$
$$\neq |\mathbf{\Delta F}|\,.$$

Ideally we would like to know all three components of **ΔF**, or at least a single component of **ΔF**, in order to understand the source of the anomaly. Fortunately, under conditions that usually prevail in crustal magnetic studies, the total-field anomaly is a good approximation of one component of **ΔF** and moreover can be considered a harmonic function.

The first of these conditions is met if the anomalous field is small compared to the ambient field. If $|\mathbf{F}| \gg |\mathbf{\Delta F}|$, then

$$\begin{aligned}
\Delta T &= |\mathbf{F} + \mathbf{\Delta F}| - |\mathbf{F}| \\
&\approx (\mathbf{F} \cdot \mathbf{F} + 2\mathbf{F} \cdot \mathbf{\Delta F})^{\frac{1}{2}} - |\mathbf{F}| \\
&\approx (\mathbf{F} \cdot \mathbf{F})^{\frac{1}{2}} + \left(\tfrac{1}{2}\right)(2)(\mathbf{F} \cdot \mathbf{F})^{-\frac{1}{2}}(\mathbf{F} \cdot \mathbf{\Delta F}) - |\mathbf{F}| \\
&= \frac{\mathbf{F} \cdot \mathbf{\Delta F}}{|\mathbf{F}|}
\end{aligned}$$
$$\Delta T \approx \hat{\mathbf{F}} \cdot \mathbf{\Delta F}\,, \tag{8.24}$$

which is the projection of **ΔF** onto **F**. Figure 8.7 graphically illustrates equations 8.23 and 8.24. Hence, if the ambient field is much larger than the perturbing field, ΔT is approximately equal to one component of the field produced by the anomalous magnetic sources, namely the component in the direction of the regional field. Typical crustal anomalies measured in airborne and shipborne surveys range in magnitude from a few nT to several 1,000 nT, but rarely exceed 5,000 nT. Hence, the condition that $|\mathbf{F}| \gg |\mathbf{\Delta F}|$ is usually met in studies of crustal magnetization.

Another condition is also necessary, however, in order for ΔT to satisfy Laplace's equation and be harmonic over the dimensions of the survey. In general, the total-field anomaly is not harmonic because, as defined by equation 8.23, $\nabla^2 \Delta T \neq 0$. If, however, the anomaly field is small compared to the total field, then

$$\nabla^2 \Delta T = \nabla^2 (\hat{\mathbf{F}} \cdot \mathbf{\Delta F}).$$

If in addition the direction of the regional field is approximately constant over the dimensions of the survey, then $\hat{\mathbf{F}}$ is a constant and can be moved outside of the Laplacian, that is,

$$\nabla^2 \Delta T = \hat{\mathbf{F}} \cdot \nabla^2 \mathbf{\Delta F}.$$

As discussed in Section 6.4, specific components of a harmonic potential field are themselves harmonic. Hence, each component of $\mathbf{\Delta F}$ in the previous equation is harmonic, $\nabla^2 \Delta T = 0$, and ΔT itself is harmonic. The condition of invariant field direction depends on the scope of the study; it is a reasonable assumption for local and regional surveys, but not for studies of continental or larger scale. We will revisit this problem in Section 12.3.1.

In summary, the total-field anomaly at any point is approximately equal to the component of the anomalous field in the direction of the regional field if the anomaly field is small compared to the ambient field. In addition, the total-field anomaly is a potential and satisfies Laplace's equation if the direction of the ambient field is constant over the dimensions of the survey. Both conditions generally prevail in studies of local and regional scale, a conclusion with fortunate implications for discussions in future chapters.

8.4 Problem Set

1. Suppose that the northward-directed component of a force field $\mathbf{F}$ is known on the surface of a sphere with radius a. The field has a potential V that is harmonic in source-free regions. Sources may exist both inside and outside the sphere but not at the surface. Show from spherical harmonic analysis that the eastward-directed component of

F can be determined at any point on the sphere from knowledge of the northward-directed component.

2. Use a spherical harmonic expansion to show that the field of a single eccentric dipole is equivalent to the field produced by multiple sources located at the center of the sphere (geocentric dipole, quadrupole, and so forth).

3. On the surface of a planet with radius a the only nonzero Gauss coefficients are g_1^0 and g_3^3, and these have values 0.3 and 0.5, respectively.

    (a) Find the declination D, inclination I, and intensity $|\mathbf{B}|$ at a point with colatitude 90° and longitude 45°.

    (b) Repeat for a point with the same colatitude and longitude but at five planetary radii ($r = 5a$).

4. The magnetic field of Planet X is known on the basis of orbiting satellite measurements to have a power spectrum satisfying the equation $R_n = 0.5 \times 10^9 \cdot (0.3)^n$ (nT^2). The radius of the planet is 8000 km, and the satellite altitude was 500 km above the planet's surface.

    (a) Assume that the magnetic field is caused by a very thin, spherical, concentric, randomly magnetized layer. What is the radius of the layer?

    (b) Suppose instead that the field is actually caused by a thick layer of broadly circulating electrical currents. Discuss the errors involved in erroneously assuming a thin, randomly magnetized layer.

5. A vertically and uniformly magnetized sphere is buried at the magnetic equator. The radius of the sphere is 10 m, its center is 15 m below the profile, and its magnetization is 1 A·m^{-1}. Assume the surrounding rocks are nonmagnetic.

    (a) Calculate the total-field anomaly along a profile directly over the sphere.

    (b) Quantitatively describe the error in using the approximation expressed by equation 8.24.

6. The electrical current flowing in a spherical shell is adjusted so that the field inside the shell is uniform. Show that the magnetic field of the shell appears dipolar at points outside the sphere.

9

Forward Method

Physicists believe that known laws should suffice to explain the Earth's behavior, but the complexities of geology have defied simple explanation.

(J. Tuzo Wilson)

The purpose of models is not to fit the data but to sharpen the questions.

(Samuel Karlin)

9.1 Methods Compared

The magnetic or gravity survey is complete, the data are processed, and regional fields have been removed appropriately. Now comes the interesting challenge of interpretation. The problem is conceptually straightforward: Estimate one or more parameters of the source from observed gravity or magnetic fields, while incorporating all available geologic, geophysical, and other independent information.

The many techniques of interpretation can be divided into three categories (Figure 9.1). Each category has the same goal, to illuminate the spatial distribution of gravity or magnetic sources, but they approach the goal with quite different logical processes.

1. *Forward method:* An initial model for the source body is constructed based on geologic and geophysical intuition. The model's anomaly is calculated and compared with the observed anomaly, and model parameters are adjusted in order to improve the fit between the two anomalies. This three-step process of body adjustment, anomaly calculation, and anomaly comparison is repeated until calculated and observed anomalies are deemed sufficiently alike.

2. *Inverse method:* One or more body parameters are calculated automatically and directly from the observed anomaly. Simplifying assumptions are inevitable.

FORWARD METHOD

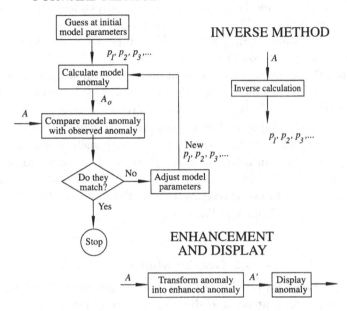

Fig. 9.1. Three categories of techniques to interpret potential field data. Measured anomaly is represented by A, calculated anomaly by A_0, and transformed measured anomaly by A'. Parameters $p_1, p_2, \ldots$ are attributes of the source, such as depth, thickness, density, or magnetization.

3. *Data enhancement and display:* No model parameters are calculated per se, but the anomaly is processed in some way in order to enhance certain characteristics of the source, thereby facilitating the overall interpretation.

The importance of employing all available independent information in the interpretive process cannot be overemphasized. Knowledge of the geologic and tectonic setting should be incorporated at each step of the process. Seismic reflection or refraction surveys, previous potential field studies, or other kinds of geophysical information may be available to guide the modeling. The interpretation in any case will be inherently nonunique, but incorporation of independent information may reduce the infinite set of mathematical solutions to a manageable array of models, still infinite in number but at least more geologically reasonable.

It may seem from the previous description that the inverse method is considerably simpler and more straightforward than the forward method.

This is not necessarily the case. Grossly simplified models are still required in the inverse method, and inclusion of independent information may be more difficult. The iterative process inherent in the forward method, on the other hand, facilitates the incorporation of independent information in the interpretive process.

This and the following chapters deal with these three approaches to interpretation of gravity and magnetic anomalies. The remainder of this chapter discusses various ways to calculate gravity and magnetic anomalies from relatively simple models, the essential ingredient of the forward method. The geophysical literature is replete with such techniques, far too many to include in a textbook of this scope. Instead we focus on a few of the "classic" techniques. Readers interested in additional discussions are referred to textbooks by Telford, Geldart, and Sheriff [279], Parasnis [203], and Grant and West [99].

9.2 Gravity Models

Equations 3.5 and 3.6 provide the gravitational potential U and gravitational attraction $\mathbf{g}$ at point P due to a volume of mass with density ρ, that is,

$$U(P) = \gamma \int_R \frac{\rho}{r}\, dv\,,$$

$$\mathbf{g}(P) = \nabla U$$

$$= -\gamma \int_R \rho \frac{\hat{\mathbf{r}}}{r^2}\, dv\,,$$

where r is the distance from P to an element of the body dv, and γ is the gravitational constant. In the following we use the tradition of directing the z axis vertically downward and arranging the x and y axes in a right-handed system (Figure 9.2).

Gravity meters measure the vertical attraction of gravity (i.e., in the direction of increasing z), here denoted by lowercase g. In cartesian coordinates, therefore,

$$g(x,y,z) = \frac{\partial U}{\partial z}$$

$$= -\gamma \int_{z'} \int_{y'} \int_{x'} \rho(x',y',z') \frac{(z-z')}{r^3}\, dx'\, dy'\, dz'\,, \qquad (9.1)$$

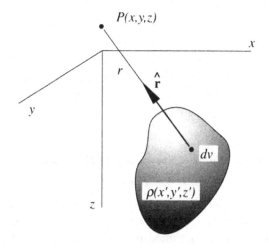

Fig. 9.2. A three-dimensional body with density $\rho(x', y', z')$ and arbitrary shape observed at point $P(x, y, z)$. Unit vector $\hat{\mathbf{r}}$ points from an element of the mass to P.

where

$$r = \sqrt{(x - x')^2 + (y - y')^2 + (z - z')^2} \,.$$

Notice that equation 9.1 has the general form

$$g(x, y, z) = \int\limits_{z'} \int\limits_{y'} \int\limits_{x'} \rho(x', y', z') \, \psi(x - x', y - y', z - z') \, dx' \, dy' \, dz' \,,$$

where

$$\psi(x, y, z) = -\gamma \frac{z}{(x^2 + y^2 + z^2)^{3/2}} \,.$$

As discussed in Section 2.3.2, $\psi(x, y, z)$ is called a Green's function. In equation 9.1, the Green's function is simply the gravitational attraction at (x, y, z) of a point mass located at (x', y', z'). More will be made of this in later chapters.

The forward method requires the repeated calculation of $g(x, y, z)$ using equation 9.1, simple enough in concept but not so simple in practice. The difficulty comes in trying to approximate complicated geologic situations by geometric shapes where the shapes are sufficiently simple to make the volume integral of equation 9.1 amenable to computers. Essentially, we must divide the hypothetical gravitational sources into N

simpler parts and convert equation 9.1 into something like

$$g_m = \sum_{n=1}^{N} \rho_n \psi_{mn} , \qquad (9.2)$$

where g_m is the vertical attraction at the mth observation point, ρ_n is the density of part n, and ψ_{mn} is the gravitational attraction at point m due to part n with unit density.

9.2.1 Three-Dimensional Examples

Rectangular Prisms

A collection of rectangular prisms provides a simple (but not particularly practical) way to approximate a volume of mass (Figure 9.3). If small enough, each prism can be assumed to have constant density. Then by the principle of superposition (Section 3.2), the gravitational anomaly of the body at any point could be approximated by summing the effects of all the prisms, as described by equation 9.2.

The gravitational attraction of a single rectangular prism is found by integration of equation 9.1 over the limits of the prism. For example, a rectangular prism with uniform density ρ and with dimensions described by the limits $x_1 \leq x \leq x_2$, $y_1 \leq y \leq y_2$, and $z_1 \leq z \leq z_2$ has a vertical

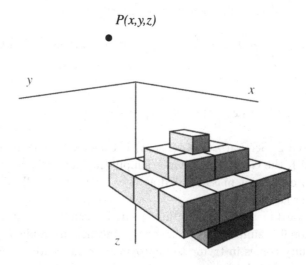

Fig. 9.3. Approximation of the three-dimensional mass by a collection of rectangular prisms.

attraction at the origin given by

$$g = \gamma\rho \int\limits_{z_1}^{z_2} \int\limits_{y_1}^{y_2} \int\limits_{x_1}^{x_2} \frac{z'}{[x'^2 + y'^2 + z'^2]^{\frac{3}{2}}} \, dx' \, dy' \, dz' \, .$$

Moving the observation point to the origin simplifies the integral, a common trick that we will use frequently. Plouff [229] provided a derivation of the preceding integral with the following result:

$$g = \gamma\rho \sum_{i=1}^{2} \sum_{j=1}^{2} \sum_{k=1}^{2} \mu_{ijk} \left[z_k \arctan \frac{x_i y_j}{z_k R_{ijk}} - x_i \log(R_{ijk} + y_j) \right.$$

$$\left. - y_j \log(R_{ijk} + x_i) \right] , \tag{9.3}$$

where

$$R_{ijk} = \sqrt{x_i^2 + y_j^2 + z_k^2},$$

$$\mu_{ijk} = (-1)^i (-1)^j (-1)^k \, .$$

Equation 9.3 can be used to calculate each ψ_{mn} in equation 9.2, and by summation thereby derive the gravitational attraction of bodies with arbitrary shape and variable density. Subroutine B.6 in Appendix B provides a Fortran subroutine to calculate equation 9.3 in order to provide the vertical attraction at a single point due to a single rectangular prism. The anomaly observed at any point and due to any body can be calculated, at least in principle, with repeated calls to this subroutine.

Stack of Laminas

Although conceptually straightforward, the previous approach would be cumbersome in practice. Geologic bodies are often difficult to model with rectangular blocks. Moreover, the computation does not take advantage of the fact that if the densities of neighboring prisms are identical, there is no need to include their mutual interface in the calculation. A more practical method was described by Talwani and Ewing [276]. Their technique approximates a body by a stack of infinitely thin laminas. The shape of each lamina is approximated by a polygon (Figure 9.4), the polygonal boundaries of individual laminas easily taken from topographic contour maps. Hence the method is often applied to the calculation of gravity anomalies over topographic or bathymetric features. It has led, for example, to a variety of computer algorithms to facilitate the correction for terrain effects in gravity measurements, as discussed in Chapter 7.

As before, we let the observation point be located at the origin and begin again with equation 9.1,

$$g(x, y, z) = \gamma\rho \int\limits_{z'} z' \, dz' \int\limits_{y'} \int\limits_{x'} \frac{dx' \, dy'}{(x'^2 + y'^2 + z'^2)^{3/2}}$$

$$= \gamma\rho \int\limits_{z'} z' \, G(z') \, dz' , \qquad (9.4)$$

where

$$G(z') = \int\limits_{y'} \int\limits_{x'} \frac{dx' \, dy'}{(x'^2 + y'^2 + z'^2)^{3/2}} .$$

The integrations over x' and y' represent a surface integration over a single horizontal lamina of the body. Eventually, we will consider the body to be a stack of laminas and replace the integration over z' with a summation, but first consider the surface integration over a single lamina.

As shown in Figure 9.4, this surface integration is equivalent to a two-step integration around the perimeter of the lamina. For example, consider integration of an integrand $f(x, y)$ over x and y. Referring to Figure 9.4, integration over x produces a new integrand $F(x, y)$ with x evaluated at limits l_1 and l_2:

$$\int \int f(x, y) \, dx \, dy = \int\limits_{y_1}^{y_2} F(x, y) \Big|_{x=l_1(y)}^{x=l_2(y)} dy$$

$$= \int\limits_{y_1}^{y_2} F(l_2(y), y) \, dy - \int\limits_{y_1}^{y_2} F(l_1(y), y) \, dy.$$

Notice that l_1 and l_2 are functions of y and represent two paths around the perimeter of the lamina. The two integrals taken together are equivalent to integration in a clockwise direction around the complete perimeter of the lamina, that is,

$$\int \int f(x, y) \, dx \, dy = \oint F(x, y) \, dy .$$

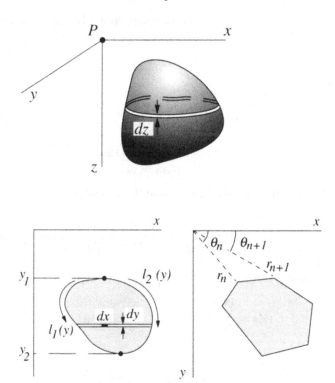

Fig. 9.4. Approximation of a three-dimensional body by a stack of laminas, each lamina approximated by a polygon.

Hence, the double integral of equation 9.4 becomes a line integral around the perimeter of the lamina,

$$G(z') = \oint \frac{x'\, dy'}{(y'^2 + z'^2)\sqrt{x'^2 + y'^2 + z'^2}},$$

and replacing the smooth integration around the perimeter with integration over M straight-line segments yields

$$G(z') = \sum_{m=1}^{M} \int_{y_m}^{y_{m+1}} \frac{x'\, dy'}{(y'^2 + z'^2)\sqrt{x'^2 + y'^2 + z'^2}}, \qquad (9.5)$$

where y_m and y_{m+1} are the y coordinates of the two endpoints of side m.

The variable x' can be eliminated in equation 9.5 by using the equation of a straight line

$$x' = \alpha_m y' + \beta_m \,,$$

where

$$\alpha_m = \frac{x_{m+1} - x_m}{y_{m+1} - y_m} \,,$$

$$\beta_m = \frac{x_m y_{m+1} - x_{m+1} y_m}{y_{m+1} - y_m} \,.$$

Substitution of α_m and β_m in equation 9.5 provides

$$G(z') = \sum_{m=1}^{M} \int_{y_m}^{y_{m+1}} \frac{(\alpha_m y' + \beta_m)\, dy'}{(y'^2 + z'^2)\sqrt{(\alpha_m^2 + 1)y'^2 + 2\alpha_m \beta_m y' + \beta_m^2 + z'^2}} \,,$$

which according to Grant and West [99] can be written

$$G(z') = \sum_{m=1}^{M} \{\arctan \Omega_{m+1} - \arctan \Omega_m\} \,, \qquad (9.6)$$

where

$$\Omega_m = \frac{z'(\beta_m y_m - \alpha_m z'^2)}{x_m[(1 + \alpha_m^2)z'^2 + \beta_m^2] - (\alpha_m^2 z'^2 + \beta_m^2)\sqrt{x_m^2 + y_m^2 + z'^2}} \,,$$

$$\Omega_{m+1} = \frac{z'(\beta_m y_{m+1} - \alpha_m z'^2)}{x_{m+1}[(1 + \alpha_m^2)z'^2 + \beta_m^2] - (\alpha_m^2 z'^2 + \beta_m^2)\sqrt{x_{m+1}^2 + y_{m+1}^2 + z'^2}} \,.$$

Although cumbersome in appearance, equation 9.6 can be programmed easily to provide $G(z')$ for any lamina; depth z' and the x, y coordinates of its M vertices are all that is required. Substituting equation 9.6 into equation 9.4 provides the vertical attraction of a stack of laminas. Integration over z' can be done with numerical quadrature techniques (e.g., Press et al. [233]).

Measured gravity anomalies over bodies of unknown shape can be modeled by trial-and-error adjustment of density and polygon vertices. If the anomalies are caused by known topographic or bathymetric features, the trial-and-error process is greatly simplified. This method is particularly powerful in such applications because the polygonal laminas can be constructed simply by digitizing contours on topographic or bathymetric maps.

In the method of Talwani and Ewing [276], the mass is approximated by a stack of infinitely thin laminas. Plouff [227, 229] took this representation one step farther. In effect, he used equation 9.4 to derive the gravitational attraction of a layer of finite thickness, with vertical sides and with top and bottom surfaces approximated by polygons. Analogous to the method of Talwani and Ewing [276], these polygonal layers can be stacked on top of one another in order to approximate three-dimensional bodies of arbitrary shape. Plouff [229] used the method to calculate gravitational effects of terrain, and the method has been implemented in various programs for removing the effects of terrain from gravity surveys (Plouff [230], Godson and Plouff [97]).

9.2.2 Two-Dimensional Examples

Geologic structures are often longer than they are wide. Fracture zones, faults, dikes, rift zones, and anticlines, for example, are often lineated in a particular horizontal direction, and the gravity or magnetic anomalies that they produce are similarly lineated (Figure 9.5). If anomalies are sufficiently "linear," it may be possible to consider the gravitational or magnetic sources as completely invariant in the direction parallel to the long direction. The y axis is directed parallel to the invariant direction leaving only the x and z dimensions to consider further; the body is said to be *two dimensional*. Density, for example, becomes

$$\rho(x, y, z) = \rho(x, z) .$$

Section 3.3 discusses some of the theory behind the potential of two-dimensional masses.

"Sufficiently linear" is, of course, a rather subjective criterion. Peters [219], in discussing a method for estimating depth to magnetic sources, considered a body to be two-dimensional when it produces closed anomaly contours roughly elliptical in shape, with long dimensions at least three times greater than their short dimensions. Grant and West [99], in discussing anomalies over ribbon-like sources, suggested that a shallowly buried ribbon must be at least 20 times longer than it is wide for the two-dimensional assumption to be legitimate. Problem 7 at the end of this chapter further investigates the two-dimensional assumption. As discussed subsequently, end corrections have been devised for some two-dimensional calculations to allow bodies that are not ideally two-dimensional (Rasmussen and Pedersen [237], Cady [48]).

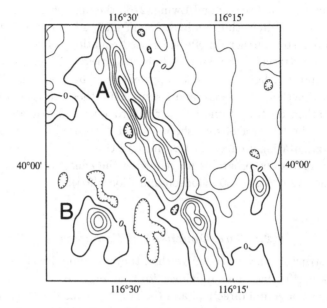

Fig. 9.5. Contour map showing examples of anomalies due to two- and three-dimensional sources. Contours represent the total-field anomaly of an area in north-central Nevada. A: two-dimensional anomaly caused by basaltic intrusions of the northern Nevada rift active during the middle Miocene (Zoback and Thompson [295]); B: three-dimensional anomaly probably caused by a Tertiary granitic intrusion (Grauch et al. [102]). Contour interval 100 nT. Data from Kucks and Hildenbrand [151].

Two-dimensional sources are easier to conceptualize and considerably easier to model than their three-dimensional counterparts, so there is an advantage in using them whenever the geologic situation permits. A bundle of parallel cylinders would constitute one simple kind of two-dimensional model. Then the anomaly could be approximated by equation 9.2, where in this case ψ_{mn} is the attraction at point m due to cylinder n with unit density. Subroutine B.2 in Appendix B provides a way to calculate ψ_{mn} for infinitely long cylinders.

A more useful way to approximate geologic situations is to replace the cross-sectional shape of two-dimensional bodies with simplified polygons. This method stems from an early paper by Hubbert [130], but Talwani, Worzel, and Landisman [278] first presented this method in a way suitable for adaptation to computer algorithms. This method and a similar magnetic method to be discussed subsequently are arguably the most widely used techniques in potential field interpretation today.

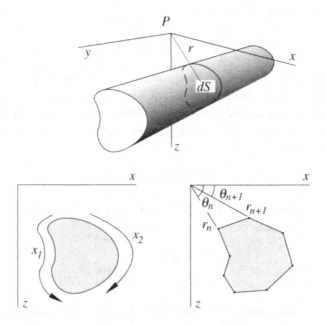

Fig. 9.6. Approximation of a two-dimensional body by an N-sided polygon.

The following derivation provides the same result as Talwani et al. [278] in a slightly different way but with many similarities to the derivation of the previous section. Equation 3.21 provided the gravitational potential of a two-dimensional body with volumetric density $\rho(x, z)$:

$$U = 2\gamma \int_S \rho(S) \log \frac{1}{r} \, dS \,,$$

where integration is over the cross-sectional surface S and where r is the perpendicular distance to an element of the body, given by

$$r = \sqrt{(x - x')^2 + (z - z')^2}$$

(Figure 9.6). To simplify matters, we now move the observation point to the origin and require the density to be constant. The vertical attraction of gravity is given by

$$g(P) = \frac{\partial U}{\partial z} = 2\gamma\rho \int \int \frac{z' \, dx' \, dz'}{x'^2 + z'^2} \,, \qquad (9.7)$$

and integration over x' yields

$$g = 2\gamma\rho \int \left[\arctan \frac{x_2'}{z'} - \arctan \frac{x_1'}{z'} \right] dz',$$

where x_1' and x_2' are both functions of z' and, as shown by Figure 9.6, represent separate paths around part of the perimeter of the cross-sectional surface. These two partial paths, when taken together and considering the change in sign, amount to a single clockwise integration around the perimeter, that is,

$$g = 2\gamma\rho \oint \arctan \frac{x'}{z'} dz'. \tag{9.8}$$

Now we replace the smooth perimeter with an N-sided polygon so equation 9.8 becomes

$$g = 2\gamma\rho \sum_{n=1}^{N} \int_{z_n}^{z_{n+1}} \arctan \frac{x'}{z'} dz', \tag{9.9}$$

where z_n and z_{n+1} are the z coordinates of the two endpoints of side n. Before continuing, we need an expression for x' in terms of z', and this is provided by the equation of a straight line:

$$x' = \alpha_n z' + \beta_n, \tag{9.10}$$

where

$$\alpha_n = \frac{x_{n+1} - x_n}{z_{n+1} - z_n},$$

$$\beta_n = x_n - \alpha_n z_n.$$

Substituting equation 9.10 into equation 9.9 provides

$$g = 2\gamma\rho \sum_{n=1}^{N} \int_{z_n}^{z_{n+1}} \arctan \left(\frac{\alpha_n z' + \beta_n}{z'} \right) dz'$$

$$= 2\gamma\rho \sum_{n=1}^{N} \left\{ \frac{\pi}{2}(z_{n+1} - z_n) + \left(z_n \arctan \frac{z_n}{x_n} - z_{n+1} \arctan \frac{z_{n+1}}{x_{n+1}} \right) \right.$$

$$+ \frac{\beta_n}{1 + \alpha_n^2} \left[\log \frac{\sqrt{x_{n+1}^2 + z_{n+1}^2}}{\sqrt{x_n^2 + z_n^2}} \right.$$

$$\left. \left. - \alpha_n \left(\arctan \frac{z_{n+1}}{x_{n+1}} - \arctan \frac{z_n}{x_n} \right) \right] \right\}.$$

The first two terms in parentheses of the summation add to zero around any closed polygon, so the previous equation simplifies to

$$g = 2\gamma\rho \sum_{n=1}^{N} \frac{\beta_n}{1+\alpha_n^2} \left[\log \frac{r_{n+1}}{r_n} - \alpha_n(\theta_{n+1} - \theta_n) \right] \qquad (9.11)$$

in such cases where r_n and θ_n are defined as shown in Figure 9.6.

Exercise 9.1 Some license was used in the preceding derivation. In particular, α_n and β_n are not defined if the ribbon is horizontal. What happens to equation 9.2.2 in this case and how might the problem be treated in a computer algorithm?

The gravitational attraction of the two-dimensional body, therefore, depends on the position of the N corners of the polygon. If we imagine N lines drawn from the observation point to each of the corners of the polygon, the gravitational attraction depends on the lengths of those lines and the angles that they make with the horizontal (Figure 9.6).

Exercise 9.2 What happens if the observation point falls on a corner of the polygon? How might this problem be treated in practice?

Equation 9.2.2 is implemented in Subroutine B.7 (Appendix B). This subroutine provides the vertical attraction at a single point due to a two-dimensional body with polygonal cross section. Similar subroutines form the core of various two-dimensional modeling systems available from commercial software companies and from public-domain sources (e.g., Saltus and Blakely [251]).

Strictly speaking, the preceding discussion applies to bodies that are infinitely extended along one horizontal axis. End corrections to these two-dimensional calculations have been developed (Rasmussen and Pedersen [237], Cady [48]) for anomalies that seem to be caused by two-dimensional bodies of limited extent. Such calculations are sometimes referred to as the "$2\frac{1}{2}$-dimensional" case.

9.3 Magnetic Models

Equation 5.3 describes the magnetic field of a volume of magnetic material,

$$\mathbf{B} = -C_m \nabla_P \int_R \mathbf{M} \cdot \nabla_Q \frac{1}{r} \, dv, \qquad (9.12)$$

where $\mathbf{M}$ is magnetization, and r is distance from the observation point P to element dv of the body. The value of the constant C_m depends on the system of units (see Chapter 4). Most magnetic surveys measure the total-field anomaly or a single component of **B**. As described in Section 8.3, the total-field anomaly is given approximately by

$$\Delta T = -C_{\mathrm{m}} \hat{\mathbf{F}} \cdot \nabla_P \int_R \mathbf{M} \cdot \nabla_Q \frac{1}{r} \, dv \,, \qquad (9.13)$$

where $\hat{\mathbf{F}}$ is a unit vector in the direction of the regional field.

Exercise 9.3 Write equation 9.13 in the form of equation 9.1. What is the physical meaning of $\psi(x, y, z)$ in this case?

Analogous to the discussion of three-dimensional gravity sources in Section 9.2, algorithms that implement equation 9.12 in order to calculate a component of **B** or the total-field anomaly, given a body's shape and distribution of magnetization, can be used for the forward method. As in the gravity case, the main difficulty arises in solving the volume integral. In practice, the body is approximated by collections of much simpler bodies, such as magnetic dipoles, rectangular prisms, or polygonal laminas.

9.3.1 A Choice of Models

As discussed in Chapter 4, magnetic material can be considered to be a collection of magnetic dipoles, magnetic charge, or circulating currents. These representations provide a variety of ways to model magnetic bodies (Figure 9.7), as discussed in the following sections.

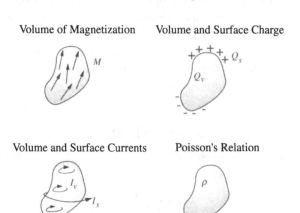

Fig. 9.7. Four ways to regard distributions of magnetization.

Volume of Magnetization

Equation 9.12 or 9.13 could be evaluated directly. In practice, however, this can be done analytically for only simple shapes and is not an appropriate strategy for trial-and-error forward modeling. Alternatively, a volume of magnetic material could be divided into N magnetic cells, analogous to equation 9.2. The three components of magnetic field become

$$\mathbf{B}_j = \sum_{i=1}^{N} M_i \, \mathbf{b}_{ij}, \tag{9.14}$$

where $\mathbf{B}_j$ is the magnetic field at the jth observation point, M_i is the magnitude of the magnetization of the ith cell, and $\mathbf{b}_{ij}$ is the magnetic field at the jth observation point due to the ith cell with unit magnetization,

$$\mathbf{b}_{ij} = -C_{\mathrm{m}} \nabla_P \int_i \hat{\mathbf{M}} \cdot \nabla_Q \frac{1}{r} \, dv. \tag{9.15}$$

If the cells are sufficiently small, they each can be considered to have uniform magnetization. In practice, cells would have to consist of simple shapes, such as rectangular prisms or magnetic dipoles, in order to easily compute equation 9.14 and so that the aggregate of all cells is easily visualized and adjusted.

Surface Charge

As discussed in Section 5.1, the volume integral of equation 9.12 can be converted into the sum of a volume and surface integral by first applying the vector identity $\nabla \cdot (\phi \mathbf{A}) = \nabla \phi \cdot \mathbf{A} + \phi \nabla \cdot \mathbf{A}$ and then applying the divergence theorem. The magnetic potential, for example, expands to

$$V = C_{\mathrm{m}} \int_R \mathbf{M} \cdot \nabla_Q \frac{1}{r} \, dv$$

$$= C_{\mathrm{m}} \int_S \frac{\mathbf{M} \cdot \hat{\mathbf{n}}}{r} \, dS - C_{\mathrm{m}} \int_R \frac{\nabla \cdot \mathbf{M}}{r} \, dv$$

$$= C_{\mathrm{m}} \int_S \frac{Q_{\mathrm{s}}}{r} \, dS + C_{\mathrm{m}} \int_R \frac{Q_{\mathrm{v}}}{r} \, dv. \tag{9.16}$$

The integrals in equation 9.16 have the same form as gravitational potential, where the scalar quantities Q_{s} and Q_{v} represent magnetic "charge" on the surface and within the interior of the body, respectively.

Exercise 9.4 Prove that the number of positive magnetic charges must equal the number of negative magnetic charges within and on the surface of any isolated magnetic volume.

If magnetization is uniform, then the second integral of equation 9.16 vanishes, and the potential is given by

$$V = C_{\mathrm{m}} \int_S \frac{\mathbf{M} \cdot \hat{\mathbf{n}}}{r} \, dS .\qquad (9.17)$$

If magnetization is uniform, therefore, the body can be completely represented by a distribution of magnetic charge on the body's surface. This representation has led to a number of powerful algorithms, several of which will be discussed in detail subsequently.

Surface Currents

The Biot–Savart law (equation 4.4) was used in Chapter 4 to show that the magnetic field of a small current loop appears at a distance like the field of a dipole. Hence a distribution of dipoles can be considered to be a volume of elemental currents. (This is not surprising; the ultimate sources of magnetization are circulating currents.)

As discussed in Section 5.1, circulating currents resolve into surface and volume currents,

$$\mathbf{I}_s = \mathbf{M} \times \hat{\mathbf{n}} ,$$
$$\mathbf{I}_v = \nabla \times \mathbf{M} .$$

If the magnetization is uniform, then the volume current vanishes, and a magnetic body can be replaced with an empty volume of identical shape carrying electrical currents on its surface (Figure 9.8).

This representation is rarely applied to forward modeling of magnetic anomalies but is useful in a variety of other geophysical applications. For example, a cylindrical rock sample uniformly magnetized along its axis is equivalent to a nonmagnetic cylinder with electrical currents on its surface. It is clear from Figure 9.9 that $\mathbf{I}_v = 0$ and $\mathbf{I}_s = \mathbf{M} \times \hat{\mathbf{n}}$. Hence, $\mathbf{I}_s$ is zero on the top and bottom of the cylinder and directed in a circular fashion around the cylindrical surface; that is, the current has the form of a simple solenoid, as shown in Figure 9.9. Laboratory magnetometers for measuring magnetization of cylindrical rock samples could be calibrated with a properly designed solenoid.

Exercise 9.5 Design a device to approximate cubical magnetic samples.

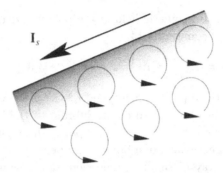

Fig. 9.8. The circulating currents responsible for the magnetization of a uniformly magnetized body are equivalent to currents on the body's surface.

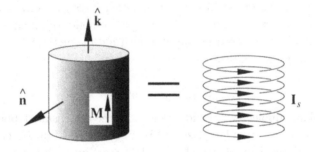

Fig. 9.9. A uniformly magnetized cylinder is equivalent to currents circulating on a nonmagnetic cylinder of identical size.

Poisson's Relation

Poisson's relation was discussed in Chapter 5. It is included here as a fourth way to derive forward-modeling algorithms. Any algorithm designed to calculate gravity anomalies in principle can be converted with Poisson's relation into an algorithm to calculate magnetic anomalies.

9.3.2 Three-Dimensional Examples

Dipoles

A three-dimensional magnetic body can be approximated by a collection of smaller elements, simple enough in form to possess analytical expressions for their magnetic field. The magnetic dipole is perhaps the simplest example. The magnetic field of a dipole was derived in Chapter 4 and is given by equation 4.14,

$$\mathbf{B} = C_{\mathrm{m}} \frac{m}{r^3} \left[3(\hat{\mathbf{m}} \cdot \hat{\mathbf{r}})\hat{\mathbf{r}} - \hat{\mathbf{m}} \right], \quad r \neq 0, \tag{9.18}$$

where $\mathbf{m} = m\,\hat{\mathbf{m}}$ is the dipole moment, and $\mathbf{r} = r\,\hat{\mathbf{r}}$ is the vector directed from the dipole to the observation point. Subroutine B.3 in Appendix B calculates the magnetic field of a single dipole. Although cumbersome in practice, the magnetic field of a body could be calculated by dividing the body into an array of small volume elements, assuming that each element appears at a distance as a dipole, using equation 9.18 to derive the field of each dipole, and finally summing the effects of all dipoles. The dipole moment m of each element is given by the product of its magnetization and its volume. In later chapters, we will discuss an inverse model that approximates the magnetic source by layers of dipoles. This type of approximation has been used to model the magnetization of the earth's crust based on satellite-altitude magnetic measurements (e.g., Mayhew [178, 179]).

Rectangular Prisms

Rather than dipoles, we could model a three-dimensional body with a collection of rectangular prisms. The magnetic field of a rectangular prism was derived by Bhattacharyya [13] starting with equation 9.13. Each prism is oriented parallel to the x, y, and z axes and has magnetization

$$\mathbf{M} = M(\hat{\mathbf{i}}\hat{M}_x + \hat{\mathbf{j}}\hat{M}_y + \hat{\mathbf{k}}\hat{M}_z),$$

and dimensions given by $x_1 \leq x \leq x_2$, $y_1 \leq y \leq y_2$, and $z_1 \leq z < \infty$. If the anomaly due to the prism is observed in a regional field directed parallel to $\hat{\mathbf{F}} = (\hat{F}_x, \hat{F}_y, \hat{F}_z)$, then the total-field anomaly observed at the origin is given by

$$\Delta T = C_{\mathrm{m}} M \left[\frac{\alpha_{23}}{2} \log\left(\frac{r - x'}{r + x'}\right) + \frac{\alpha_{13}}{2} \log\left(\frac{r - y'}{r + y'}\right) - \alpha_{12} \log(r + z_1) \right.$$

$$- \hat{M}_x \hat{F}_x \arctan\left(\frac{x'y'}{x'^2 + rz_1 + z_1^2}\right) - \hat{M}_y \hat{F}_y \arctan\left(\frac{x'y'}{r^2 + rz_1 - x'^2}\right)$$

$$\left. + \hat{M}_z \hat{F}_z \arctan\left(\frac{x'y'}{rz_1}\right) \right] \Bigg|_{x'=x_1}^{x'=x_2} \Bigg|_{y'=y_1}^{y'=y_2}, \tag{9.19}$$

where

$$\alpha_{12} = \hat{M}_x \hat{F}_y + \hat{M}_y \hat{F}_x,$$

$$\alpha_{13} = \hat{M}_x \hat{F}_z + \hat{M}_z \hat{F}_x,$$

$$\alpha_{23} = \hat{M}_y \hat{F}_z + \hat{M}_z \hat{F}_y,$$

$$r^2 = x'^2 + y'^2 + z_1^2.$$

Equation 9.19 provides the total-field anomaly of a prism with top at z_1 and bottom at infinity. If this equation is evaluated twice, once for $z_1 = z_t$ and $M = M_0$ and once for $z_1 = z_b$ and $M = -M_0$, then according to the superposition principle, the sum of the two calculations will provide the magnetic field of a prism with magnetization M_0, top at z_t, and bottom at z_b. Subroutine B.8 in Appendix B provides an algorithm to calculate equation 9.19.

By dividing the body into a collection of rectangular prisms, equation 9.19 could be used to iteratively model bodies of arbitrary shape. Rectangular prisms have also been used in the inverse method to directly derive vector **M** from the total-field anomaly (Vacquier [285]); more about this is given in Chapter 10.

Stack of Laminas

Talwani [275] used the volume integral of equation 9.12 to derive an algorithm for the calculation of magnetic fields due to bodies of arbitrary shape. His method and its derivation are analogous to the method of Talwani and Ewing [276] for calculating gravity anomalies over three-dimensional bodies, discussed earlier in this chapter. The magnetic body is approximated by a stack of laminas, and each lamina is approximated by a polygon. Hence the volume integral of equation 9.12 is solved analytically in the x and y directions and numerically in the z direction. The method is particularly amenable to anomalies caused by topographic or bathymetric features because contours can be digitized directly from topographic or bathymetric maps in order to represent individual laminas.

Plouff [228, 229] extended the method of Talwani [275] by replacing the infinitely thin laminas with layers of finite thickness, thereby providing more accurate representation of the modeled body. The theoretical expression for the magnetic field of an individual polygonal layer was derived (Plouff [228]) by integration with respect to depth of the field caused by one lamina.

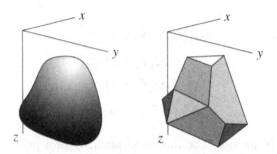

Fig. 9.10. A magnetic body of arbitrary shape modeled as a surface composed of polygonal facets.

In a later chapter, we will apply the methods of Talwani [275] and Plouff [229] to the inverse problem, that is, to derive the three components of magnetization directly from measured total-field anomalies. The inverse form of this method played a key role in our understanding of seamount magnetization, apparent polar wander, and seafloor spreading.

Polyhedrons

The previous section showed that, if magnetization is uniform, a magnetic body can be modeled by magnetic charge on the body's surface. Several workers (Bott [36], Barnett [11], Okabe [197], Hansen and Wang [114]) have exploited this simplification by developing methods that approximate the body's shape by a surface composed of flat polygonal facets (Figure 9.10). The following derivation follows the original work of Bott [36]. The technique of Hansen and Wang [114] will be discussed in Chapter 11.

Equation 9.17 provides the magnetic potential of a uniformly magnetized body; the magnetic field is given by

$$\mathbf{B} = -\nabla_P V$$

$$= C_{\mathrm{m}} \int_S \frac{\mathbf{M} \cdot \hat{\mathbf{n}}}{r^2} \, \hat{\mathbf{r}} \, dS. \qquad (9.20)$$

To simplify matters, the origin is placed at the observation point P. If the surface of the body is replaced by N polygonal facets, equation 9.20

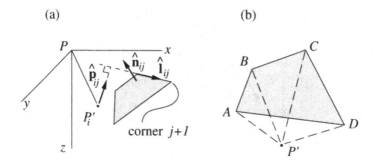

(a)

(b)

Area $ABCD$ = Area ABP' + Area BCP'
+ Area CDP' - Area ADP'

Fig. 9.11. (a) The ith facet of the polyhedron within an x, y, z coordinate system. Vector $\hat{\mathbf{l}}_{ij}$ is a unit vector along side j, $\hat{\mathbf{p}}_{ij}$ is a unit vector from P' perpendicular to side j, and PP_i' is normal to the face. (b) The area of a K-sided polygon is equal to the sum of the areas of K triangles.

becomes

$$\mathbf{B} = C_{\mathrm{m}} \sum_{i=1}^{N} (\mathbf{M} \cdot \hat{\mathbf{n}}_i) \int_{S_i} \frac{\hat{\mathbf{r}}}{r^2} \, dS \,, \qquad (9.21)$$

where S_i represents the surface of the ith facet and $\hat{\mathbf{n}}_i$ its outward normal. Now consider just the ith facet consisting of K_i corners (Figure 9.11). We always will regard the corners of any facet ordered in a clockwise sense as viewed from outside the body. Let P_i' be the intersection of the plane that contains the facet and a perpendicular line passing through P, and define $\hat{\mathbf{l}}_{ij}$ as the unit vector from corner j to corner $j+1$, and $\hat{\mathbf{p}}_{ij}$ as the unit vector from P_i' perpendicular to side j. Consider K_i triangles constructed such that each triangle has a vertex at point P_i' and an opposite side equal to one side of the facet. It should be clear from Figure 9.11 that the area of a K_i-sided facet equals the sum of the areas of K_i triangles weighted by $+1$ or -1 depending on whether $\hat{\mathbf{l}}_{ij}$ points left or right as viewed from P_i'. Hence the integral inside equation 9.21 can be replaced with the sum of integrals over K_i triangles,

$$\int_{S_i} \frac{\hat{\mathbf{r}}}{r^2} \, dS = \sum_{j=1}^{K_i} \mu_{ij} \Phi_{ij} \,, \qquad (9.22)$$

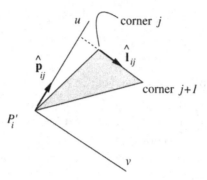

Fig. 9.12. Triangular facet within new u, v, w coordinate system. The facet is in the plane of the page.

where $\mu_{ij} = 1$ if $\hat{\mathbf{l}}_{ij} \times \hat{\mathbf{p}}_{ij}$ is parallel to $\hat{\mathbf{n}}_i$, $\mu_{ij} = -1$ if $\hat{\mathbf{l}}_{ij} \times \hat{\mathbf{p}}_{ij}$ is antiparallel to $\hat{\mathbf{n}}_i$, and $\mu_{ij} = 0$ if $\hat{\mathbf{l}}_{ij} \times \hat{\mathbf{p}}_{ij} = 0$, and where $\mathbf{\Phi}_{ij}$ is integration over the triangle including vertex P_i' and side j. At this point, it is convenient to move the origin to P_i and orient new axes u, v, and w parallel to $\hat{\mathbf{p}}_{ij}$, $\hat{\mathbf{l}}_{ij}$, and $\hat{\mathbf{n}}_i$, respectively (Figure 9.12). In this new coordinate system, the corners of the triangle are $(0, 0, d_i)$, (u_{ij}, v_{ij}, d_i), and $(u_{ij}, v_{i,j+1}, d_i)$, where d_i is the distance between P and P_i'. Then integration over the triangle becomes

$$\mathbf{\Phi}_{ij} = \int_{S_{ij}} \frac{\hat{\mathbf{r}}}{r^2} \, dS$$

$$= \int_0^{u_{ij}} \int_{v_a}^{v_b} \frac{u\hat{\mathbf{p}}_{ij} + v\hat{\mathbf{l}}_{ij} + d_i\hat{\mathbf{n}}_{ij}}{(u^2 + v^2 + d_i^2)^{3/2}} \, du \, dv$$

where $v_a = v_{ij}u/u_{ij}$ and $v_b = v_{i,j+1}u/u_{ij}$. Evaluation of this double integral yields

$$\mathbf{\Phi}_{ij} = \hat{\mathbf{p}}_{ij} \left[\frac{\alpha}{\sqrt{1+\alpha^2}} \log\left(\frac{r+\rho}{|d_i|}\right) - \frac{1}{2}\left(\log\frac{r+v}{r-v}\right) \right]$$

$$- \hat{\mathbf{l}}_{ij} \left[\frac{1}{\sqrt{1+\alpha^2}} \log\left(\frac{r+\rho}{|d_i|}\right) \right]$$

$$- \hat{\mathbf{n}}_i \left[\frac{d_i}{|d_i|} \arctan\left(\frac{\alpha(r-|d_i|)}{r+\alpha^2|d_i|}\right) \right] \Bigg|_{v=v_{ij}}^{v=v_{i,j+1}}, \qquad (9.23)$$

where

$$\alpha = v/u_{ij},$$

$$\rho = \sqrt{u_{ij}^2 + v^2},$$

$$r = \sqrt{u_{ij}^2 + v^2 + d_i^2}.$$

If P lies in the plane of the face, however, equation 9.23 becomes singular. This problem is treated by removing a small sector around P_i' from the triangle, with the result that

$$\Phi_{ij} = \hat{\mathbf{p}}_{ij} \left[\frac{\alpha}{\sqrt{1+\alpha^2}} \left(1 + \log \frac{r+\rho}{2\epsilon} - \frac{1}{2} \log \frac{r+v}{r-v} \right) \right]$$

$$- \hat{\mathbf{l}}_{ij} \left[\frac{1}{\sqrt{1+\alpha^2}} \left(1 + \log \frac{r+\rho}{2\epsilon} \right) \right] \Bigg|_{v=v_{ij}}^{v=v_{i,j+1}}, \qquad (9.24)$$

where ϵ is a small number. Equation 9.23 (or 9.24) is evaluated for each side of the facet, transformed back to the original coordinate system, and summed in equation 9.22. This result provides the attraction of facet i which then can be used in equation 9.21 for the entire polyhedron. Subroutine B.10 in Appendix B uses equations 9.23 and 9.24 to calculate the three components of magnetic field due to a polygonal sheet of magnetic charge. The magnetic anomaly of an arbitrary polyhedron can be found with repeated calls to this subroutine, one call for each face of the polyhedron.

9.3.3 Two-Dimensional Example

We again use the concept of surface magnetic charges, this time to develop an algorithm to model two-dimensional bodies. As in Section 9.2.2 of this chapter, we will replace the cross-sectional shape of the body with an N-sided polygon. If the body is uniformly magnetized, the magnetization can be replaced with magnetic "charge" on its surface (Figure 9.13). Hence, the problem reduces to the calculation of the magnetic attraction of N flat ribbons of charge, infinitely extended in the $+y$ and $-y$ directions. Although this two-dimensional forward method was first presented by Talwani and Heirtzler [277] nearly 20 years ago, it and its gravity counterpart (Talwani et al. [278]) probably remain the most widely used algorithms in potential field interpretation today.

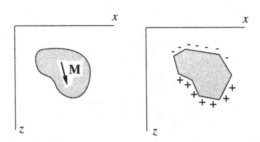

Fig. 9.13. Approximation of a two-dimensional body with infinitely extended ribbons of magnetic charge.

So far as I know, the following derivation differs from previously published discussions. It hinges on the following observation: The magnetic field of a uniformly magnetized body with volume R and surface S is given by equation 9.20,

$$\mathbf{B} = C_m \int_S \frac{\mathbf{M} \cdot \hat{\mathbf{n}}}{r^2} \hat{\mathbf{r}} \, dS \,,$$

and this equation has the same form as the gravity field of a hollow shell with identical shape, that is,

$$\mathbf{g} = -\gamma \int_S \frac{\sigma(S)}{r^2} \hat{\mathbf{r}} \, dS \,,$$

where $\sigma(S)$ is surface density in units of mass per unit area. If an expression for the gravitational field of shell S is known in terms of its surface density, the magnetic field of a uniformly magnetized body of volume R can be found by substituting $-\mathbf{M} \cdot \hat{\mathbf{n}}$ for $\sigma(S)$ and C_m for γ.

First consider a flat, horizontal ribbon with a surface density of σ, continuing infinitely far in the $+y$ and $-y$ directions, and extending from (x_1, z') to (x_2, z') (Figure 9.14(a)). One element dx of the ribbon penetrates the x, z plane at (x', z') and is equivalent to a wire infinitely extended parallel to the y axis and having mass per unit length $\lambda = \sigma \, dx$ (Figure 9.14(b)). As discussed in Chapter 3, the wire has a gravitational attraction observed at the origin given by

$$\mathbf{g} = -2\gamma\lambda\frac{\hat{\mathbf{r}}}{r}$$

$$= 2\gamma\lambda\frac{x'\hat{\mathbf{i}} + z'\hat{\mathbf{k}}}{x'^2 + z'^2} \,. \tag{9.25}$$

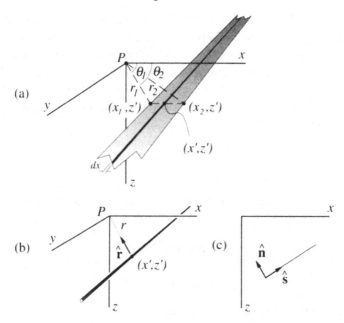

Fig. 9.14. (a) Horizontal ribbon of mass extending from (x_1, z') to (x_2, z') and infinitely extended parallel to the y axis. (b) Wire mass infinitely extended parallel to the y axis and piercing the x, z plane at (x', z'). (c) Horizontal ribbon in new coordinate system.

The gravitational attraction of the horizontal ribbon is found by letting $\lambda = \sigma\, dx$ in equation 9.25 and integrating over x,

$$g_x = 2\gamma\sigma \int_{x_1}^{x_2} \frac{x'}{x'^2 + z'^2}\, dx'$$

$$= 2\gamma\sigma \log \frac{r_2}{r_1},$$

$$g_z = 2\gamma\sigma z' \int_{x_1}^{x_2} \frac{dx'}{x'^2 + z'^2}$$

$$= 2\gamma\sigma(\theta_1 - \theta_2),$$

$$\mathbf{g} = 2\gamma\sigma \left[\hat{\mathbf{i}} \log \frac{r_2}{r_1} + \hat{\mathbf{k}}(\theta_1 - \theta_2) \right], \qquad (9.26)$$

where r_1 and r_2 are distances from P to edges 1 and 2, respectively, and θ_1 and θ_2 are angles between the x axis and those lines connecting edges

1 and 2, respectively (Figure 9.14(b)). Equation 9.26 is the gravitational attraction of a horizontal ribbon. To generalize to any ribbon, we rotate the ribbon an arbitrary amount and define two unit vectors $\hat{n}$ and $\hat{s}$ that remain normal and parallel to the ribbon, respectively (Figure 9.14(c)). Vector $\hat{s}$ is always directed parallel to the ribbon from edge 1 to edge 2; vector $\hat{n}$ is always normal to the ribbon and directed as in a right-handed system. Note that

$$\hat{n}_x = \hat{s}_z,$$
$$\hat{n}_z = -\hat{s}_x.$$

The components of gravitational attraction in the $\hat{s}$ and $\hat{n}$ directions are given by equation 9.26,

$$g_s = 2\gamma\sigma \log \frac{r_2}{r_1},$$

$$g_n = -2\gamma\sigma(\theta_1 - \theta_2),$$

and the x and z components are given by

$$g_x = \hat{i} \cdot \mathbf{g}$$
$$= \hat{s}_x g_s + \hat{n}_x g_n$$
$$= \hat{s}_x g_s + \hat{s}_z g_n$$
$$= 2\gamma\sigma \left[\hat{s}_x \log \frac{r_2}{r_1} - \hat{s}_z(\theta_1 - \theta_2) \right],$$

$$g_z = \hat{k} \cdot \mathbf{g}$$
$$= \hat{s}_z g_s + \hat{n}_z g_n$$
$$= \hat{s}_z g_s - \hat{s}_x g_n$$
$$= 2\gamma\sigma \left[\hat{s}_z \log \frac{r_2}{r_1} + \hat{s}_x(\theta_1 - \theta_2) \right].$$

Exercise 9.6 What happened to the y component of $\mathbf{g}$?

The previous equations provide the gravitational attraction of an infinitely extended ribbon of mass. To convert these equations to the magnetic case, we simply let $\gamma = C_m$ and $\sigma = -\mathbf{M} \cdot \hat{n}$,

$$B_x = -2C_m(\mathbf{M} \cdot \hat{n}) \left[\hat{s}_x \log \frac{r_2}{r_1} - \hat{s}_z(\theta_1 - \theta_2) \right], \qquad (9.27)$$

$$B_z = -2C_m(\mathbf{M} \cdot \hat{\mathbf{n}}) \left[\hat{s}_z \log \frac{r_2}{r_1} + \hat{s}_x(\theta_1 - \theta_2) \right] . \qquad (9.28)$$

Exercise 9.7 Describe B_x and B_z for both horizontal and vertical ribbons.

Equations 9.27 and 9.28 represent the magnetic attraction of a ribbon of magnetic charge; they can be used N times to calculate the magnetic attraction of an N-sided prism, that is

$$\mathbf{B} = \sum_{l=1}^{N} \left(\hat{\mathbf{i}} B_{lx} + \hat{\mathbf{k}} B_{lz} \right), \qquad (9.29)$$

where B_{lx} and B_{lz} are the x and z components of $\mathbf{B}$ due to side l. Finally, the total-field anomaly can be found by substituting equation 9.29 into equation 8.24,

$$\Delta T = \sum_{l=1}^{N} \left(\hat{F}_x B_{lx} + \hat{F}_z B_{lz} \right), \qquad (9.30)$$

where $\hat{F}_x$ and $\hat{F}_z$ are the x and z components of the ambient, unperturbed field.

The earlier comments concerning the advantages of two-dimensional forward modeling of gravity anomalies hold doubly for the magnetic case. Two-dimensional models are far easier to construct than three-dimensional models, and they generally should be used whenever the geologic situation permits. Subroutine B.15 in Appendix B performs the calculations described by equations 9.27 and 9.28. The magnetic attraction of an N-sided prism at a single point can be calculated by summing the results returned from N calls to this subroutine. Algorithms similar to Subroutine B.15 are the essential ingredients in a variety of two-dimensional forward-modeling programs (e.g., Saltus and Blakely [251]).

As discussed in Section 9.22, the question of whether an anomaly is sufficiently linear to permit the two-dimensional approximation is problematic. A useful variation on this two-dimensional method replaces the infinitely extended prisms with prisms of finite length (Shuey and Pasquale [256], Rasmussen and Pedersen [237], Cady [48]). This so-called $2\frac{1}{2}$-dimensional calculation comes in handy when the anomalies are somewhat linear but depart substantially from two-dimensionality. The anomaly in the south-central part of Figure 9.5 is an example of an anomaly that might suitably be treated with the $2\frac{1}{2}$-dimensional approach.

9.4 Problem Set

1. Find the vertical attraction of gravity and the vertical and horizontal components of magnetic induction for the tabular bodies shown in Figure 9.15. In each case, the body is 1 km thick, 1 km deep, and infinitely extended in both the y direction and in the direction indicated. The bodies have uniform density and magnetization of 2.7 g/cm^3 and 1 A/m, respectively. The directions of magnetization are shown by the arrows. The bodies are surrounded by nonmagnetic sedimentary deposits of density 1.5 g/cm^3.

   (a) Horizontal sill greatly extended in the $+x$ direction and magnetized in the $+x$ direction.

   (b) Horizontal sill greatly extended in the $+x$ direction and magnetized vertically downward.

   (c) Vertical dike extending to great depths and magnetized vertically upward.

2. For purposes of calibrating a magnetometer, you wish to generate a system of surface currents that will produce exactly the same magnetic field as a uniformly magnetized prism with rectangular cross section of 1 cm x 2 cm and a length of 5 cm.

   (a) Design an experimental arrangement for doing this that allows the simulated magnetization to be directed in any direction relative to the sides of the prism.

   (b) Discuss quantitatively the electrical currents needed to simulate a magnetization of 10 A/m parallel to the long dimension of the prism.

3. In the course of doing laboratory work in rock magnetism, it is sometimes necessary to measure the field strength of large magnets. This

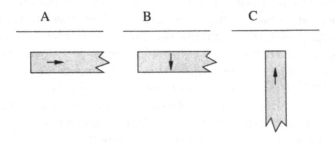

Fig. 9.15. Infinitely extended tabular bodies.

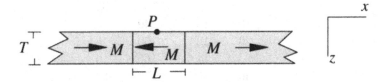

Fig. 9.16. Infinite slab containing two reversal boundaries.

may be done from first principles without recourse to secondary standards in the following way. Suspend a rectangular loop of wire from a balance such that the plane of the rectangle is perpendicular to flux lines between the pole pieces of the magnet, and such that one horizontal side of the loop lies completely in the uniform part of the magnetic field of the magnet and the other horizontal side lies outside the magnet gap. The weight of the loop is then related to the current in the loop and the strength of the magnet.

(a) What is the difference between the induction inside and outside the pole pieces of the magnet if the weight of the loop changes by 100 milligrams when the current in the loop increases from 0 to 100 milliamperes? The horizontal dimension of the loop is 1 cm.

(b) What do you think would limit the accuracy of magnetic induction measurements made in this way?

4. An infinitely extended slab of thickness T is horizontally magnetized with magnetization M in the x direction (Figure 9.16). The slab has two vertical reversal boundaries trending in the y direction and separated by a distance L.

(a) Solve in terms of L, T, and M for the values of B_x and B_z at point P midway between the two reversals and directly on the slab.

(b) What happens to B_x and B_z if L and T are both doubled?

(c) What happens to the values of B_x and B_z as $T \to \infty$?

5. The following vertical-field anomaly was measured over a narrow outcrop of magnetite in southern Alaska.

x (feet)	ΔB_z	x (feet)	ΔB_z
0	−14	130	95
10	−13	140	3
20	−12	150	−15
30	−11	160	−20
40	−10	170	−27
50	−8	180	−33
60	−4	190	−31
70	3	200	−21
80	11	210	−13
90	21	220	−11
100	37	230	−10
110	101	240	−10
120	180	250	−10

The heading of the traverse was 19°E over flat ground. Sensor height was about 4 ft above ground level. The outcrop is two dimensional and perpendicular to the traverse and is located between $x = 120$ ft and $x = 128$ ft. The direction of magnetization is believed to have an inclination of 75° and a declination of 24°E. The intensity of magnetization is about 500 A/m. Assume that the surrounding terrain is nonmagnetic.

(a) Write a computer program to calculate the magnetic anomaly over two-dimensional bodies. Hint: Subroutine B.15 in Appendix B could form the bulk of such a computer program.

(b) Use your program to estimate the size and shape of the magnetite body. To keep the problem simple, assume that the body has only four sides.

6. You are measuring the magnetic field with a proton-precession magnetometer above the geologic section shown in Figure 9.17. The section has a normal fault (erosion has removed the surface expression of the fault) with an east–west strike and a dip of 60°N. The rock units and their magnetic properties (in SI units) are shown in Figure 9.17. Happily, the regional field has zero declination, 60° inclination, and a magnitude of 0.05 mT. Assume that remanent magnetization is parallel to the regional field. Find the total-field anomaly in units of nT at point P, the surface trace of the fault. (Hint: Poisson's relation greatly simplifies this problem.)

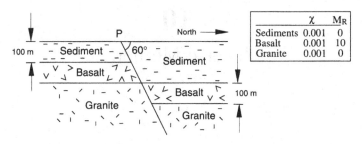

Fig. 9.17. Geologic cross section.

7. Consider a horizontal cylinder of length l, radius a, density ρ, and susceptibility χ. The long axis of the cylinder is oriented east-west and is at a depth d below ground level $(d > a)$. The inclination and declination of the earth's magnetic field are 60° and 0°, respectively. The gravity and total-field anomalies are measured at ground level along a line perpendicular to and directly above the midpoint of the cylinder.

(a) Discuss the errors involved in assuming that this finite-length cylinder has infinite length. How does the error change as l increases?

(b) Let $g_1(x)$ represent the gravity anomaly caused by the finite-length cylinder and $g_0(x)$ the anomaly of the infinite cylinder. At what length l does the maximum error along the gravity profile fall below 10 percent, where maximum error is defined as

$$E = \frac{\max |g_1(x) - g_0(x)|}{\max |g_0(x)|} ?$$

(c) Suppose that gravity and magnetic profiles are available over a finite-length cylinder but that the source of the anomalies is in fact unknown. What erroneous conclusions might be drawn about the source from a two-dimensional, forward-modeling study?

10
Inverse Method

The goal of inverse theory is to determine the parameters from the observations or, in the face of the inevitable limitations of actual measurement, to find out as much as possible about them.

(Robert L. Parker)

Errors using inadequate data are much less than those using no data at all.

(Charles Babbage)

Solid-earth geophysical studies generally aspire to learn something about the interior of the earth from measurements of physical quantities taken on or above the earth's surface. The previous chapter in particular discussed a general methodology, forward modeling, in which crustal and upper-mantle lithology and structure can be deduced from gravity or magnetic measurements. In that methodology, source parameters are estimated indirectly through trial-and-error calculations. The present chapter deals with another methodology, the *inverse method* (or *direct method*), in which source parameters are determined in a direct way from gravity or magnetic measurements.

The following discussion is sometimes heuristic in nature, relying on examples from the literature. Readers are referred to reviews by Parker [207] and Parker, Shure, and Hildenbrand [210] and to the textbook by Menke [182] for more rigorous discussions. We will revisit this subject in Chapter 11 where the inverse formulation will be transformed to the Fourier wavenumber domain.

10.1 Introduction

In Chapter 9, we became familiar with integral relations between potential fields and their causative sources. A few examples are listed

214

subsequently, where in each case R is the volume occupied by the causative source, P is the observation point located at (x, y, z) and always
outside of R, Q is the point of integration (x', y', z') within R, and $\mathbf{r}$ is
a vector directed from Q to P:

Vertical attraction of gravity

$$g(P) = -\gamma \int_R \rho(Q) \frac{z - z'}{r^3} \, dv. \tag{10.1}$$

Vertical magnetic field

$$B_z(P) = -C_\mathrm{m} \frac{\partial}{\partial z} \int_R \mathbf{M}(Q) \cdot \nabla_Q \frac{1}{r} \, dv$$

$$= C_\mathrm{m} \int_R \frac{\mathbf{M}(Q)}{r^4} \cdot \left[3\hat{\mathbf{r}}(z - z') - r\hat{\mathbf{k}} \right] dv. \tag{10.2}$$

Total-field anomaly (approximately)

$$\Delta T(P) = -C_\mathrm{m} \hat{\mathbf{F}} \cdot \nabla_P \int_R \mathbf{M}(Q) \cdot \nabla_Q \frac{1}{r} \, dv$$

$$= C_\mathrm{m} \int_R \frac{\mathbf{M}(Q)}{r^3} \cdot \left[3(\hat{\mathbf{F}} \cdot \hat{\mathbf{r}})\hat{\mathbf{r}} - \hat{\mathbf{F}} \right] dv. \tag{10.3}$$

In these equations, $\rho(Q)$ and $\mathbf{M}(Q)$ have the usual meaning of density
and magnetization, respectively. Unit vector $\hat{\mathbf{F}}$ is in the direction of
the unperturbed magnetic field, and unit vector $\hat{\mathbf{k}}$ is directed vertically
down. Factors γ and C_m are constants discussed in Chapters 3 and 4,
respectively.

Exercise 10.1 Fill in the missing steps leading to equations 10.2 and 10.3.

Equations 10.2 and 10.3 both have the general form

$$f(P) = \int_R \mathbf{s}(Q) \cdot \mathbf{G}(P, Q) \, dv. \tag{10.4}$$

We may be willing to specify the directional behavior of the magnetization in some situations. It is sometimes assumed, for example, that
magnetization is entirely induced by the ambient field. Then magnetization will be nearly unidirectional if susceptibility is isotropic and the

magnetic survey is not too large. In these cases, we can let $\mathbf{M}(Q) = M(Q)\hat{\mathbf{M}}$ in equations 10.2 and 10.3 and move the unit vector into the bracketed term. Then all three equations, 10.1 through 10.3, have the same general form

$$f(P) = \int_R s(Q)\psi(P,Q)\,dv,\qquad(10.5)$$

where $f(P)$ is the potential field at P, $s(Q)$ describes the physical quantity (density or magnetization) at Q, and $\psi(P,Q)$ is a function that depends on the geometric placement of observation point P and source point Q. Equation 10.5 is known as a *Fredholm equation* of the first kind (Morse and Feshbach [188]). As in Section 2.3.2, we call $\psi(P,Q)$ and $\mathbf{G}(P,Q)$ Green's functions. The remaining comments of this section refer to equation 10.5 but could be generalized readily to equation 10.4.

Equation 10.5 is a convenient vehicle with which to contrast the forward and inverse methods. The forward *calculation* is the calculation of $f(P)$ from known or assumed functions $s(Q)$ and $\psi(P,Q)$ and the volume R. For any given calculation of equation 10.5, $f(P)$ is completely determined with a complete knowledge of $s(Q)$, $\psi(P,Q)$, and R; that is, the forward calculation has a unique solution. The forward *method*, as discussed in Chapter 9, involves the repeated adjustment of $s(Q)$ and R, calculation of $f(P)$, and comparison with measured values of the field until the calculated field suitably "fits" the measured field. Although the forward calculation is unique in a mathematical sense, a model for magnetic or gravity sources developed by way of the forward method is, of course, not unique. The inverse method, on the other hand, inserts measurements of $f(P)$ directly into the left side of equation 10.5 and solves for some aspect of $s(Q)$ or R. Calculation of $s(Q)$ is known as the *linear inverse problem*, whereas calculation of some property of R is the *nonlinear inverse problem*.

First consider the linear problem. We could, for example, reformulate equation 10.5 as a matrix equation,

$$f_i = \sum_{j=1}^{N} s_j\psi_{ij},\qquad i = 1,2,\ldots,L,\qquad(10.6)$$

and if $L > N$ use least squares to find the N values of s_j. This is not as simple as it seems. The first fundamental difficulty is the problem of *nonuniqueness*. Even if we knew $f(P)$ precisely, we might not be able to determine a unique inverse solution for $s(Q)$. Uniqueness can be determined by asking whether there are any nontrivial solutions for $a(Q)$ in the equation

$$\int_R a(Q)\psi(P,Q)\,dv = 0.$$

If the answer is yes, then $s(Q)$ is nonunique. The class of all $a(Q)$ is known as the *annihilator* for that particular kernel $\psi(P,Q)$ and volume R (Parker [207]).

What can we do faced with limitations of nonuniqueness? There are two approaches: We could make simplifying assumptions about the source. We might, for example, be willing to assume that magnetization is uniform throughout the body or that the body is infinitely extended in one direction. Such assumptions may reduce the number of permissible solutions, but the solution may still be nonunique† and certainly will be much simpler than reality. Nevertheless, this is the tack generally taken in inverse studies. A second approach would be to attempt to find aspects about the source that are common to the entire infinite set of solutions. For example, we might be able to determine the maximum depth of burial of any realistic source. This leads to the theory of ideal bodies to be discussed subsequently. Regardless of which method is favored, independent geologic and geophysical information should be employed whenever possible to narrow the range of realistic solutions.

The second problem is *instability*. According to equation 10.5, the potential field at a single point depends on the entire source distribution. In fact, the potential field at a single point is a weighted average of all parts of the source, where the weighting function is $\psi(P,Q)$. In the language of linear systems analysis, $f(P)$ is a linear functional of $s(Q)$. For all $P \neq Q$, $\psi(P,Q)$ is a smoothly varying function. Hence, $f(P)$ is always "smoother" than $s(Q)$ as long as P is outside the body. Consequently, the inverse problem of deriving $s(Q)$ from equation 10.5 amounts to an "unsmoothing" of $f(P)$. Small changes to $f(P)$ cause large and unrealistic variations in $s(Q)$, and the solution is said to be *unstable*. Potential field inversion is notoriously unstable. There are ways to reduce the instability, as we will see in this and the next chapter, but only at the expense of giving up information about the source.

---

† Unique solutions may exist for a given anomaly if enough assumptions are stated. Smith [264] showed, for example, that a gravity anomaly has only one possible solution if the density of the body is uniform, the body is finite in extent, and every vertical line passes through the body at most one time.

Exercise 10.2 Demonstrate that calculation of potential fields from equation 10.5 is always a smoothing operation. Hint: Let $f(P)$ be the vertical attraction of gravity and let $s(Q) = \rho_0 \delta(P - Q)$. What is the physical meaning of this $s(Q)$?

The third problem is one of *construction*. Inverse methods are made tractable by modeling the source distribution with simple geometries. A particular method may assume, for example, that the causative body is composed of simple parts, such as dipoles, line sources, or thin sheets, and we will discuss a number of techniques in the next sections that are highly model dependent. Of course geology is never so simple, and our results almost certainly will be less than accurate.

10.2 Linear Inverse Problem

Some of the problems of the previous section are best demonstrated with some examples from the geophysical literature. From equation 10.5, magnetic or gravity fields are linearly dependent on magnetization or density, and estimation of magnetization or density from magnetic or gravity fields is a linear inverse problem.

Estimation of a single best density for a given mass is a simple example. Equation 9.4 describes the vertical attraction of gravity for a body of uniform density. If the shape of the body is known, the integral term in this equation could be calculated by one of the forward techniques of Chapter 9. The gravity anomaly measured at N discrete locations is then

$$g_i = \rho \psi_i \qquad i = 1, 2, \ldots, N,$$

and the constant ρ could be determined by simple linear regression. Plouff [229] and Ishihara [135] discuss this type of calculation in detail.

We could complicate this procedure by dividing the body into smaller compartments and use least-squares methods to solve for the density of each compartment. This approach is considered in the next section for the magnetic case.

10.2.1 Magnetization of a Layer

Early attempts to model the magnetization of ocean crust with inverse methods (Bott [37], Bott and Hutton [39], Emilia and Bodvarsson [84]) were motivated by an interest in seafloor spreading and how this dynamic process records geomagnetic field behavior. Magnetization was assumed

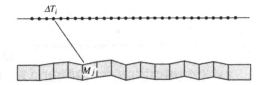

Fig. 10.1. Inverse model of the oceanic magnetic layer. Total-field anomaly ΔT_i is measured above a row of polygonal cells infinitely extended perpendicular to the page. Cells are uniformly magnetized with magnetization M_j.

to be two-dimensional (often a good assumption for anomalies related to seafloor spreading), so these early models consisted of a row of polygonal cells (Figure 10.1), arranged so that the tops of the cells correspond with the top of the magnetic layer. Then rewriting equation 10.6, the total-field anomaly at point i is given by

$$\Delta T_i = \sum_{j=1}^{N} M_j \psi_{ij}, \qquad i = 1, 2, \ldots, L, \qquad (10.7)$$

where M_j is the intensity of magnetization of cell j, and ψ_{ij} is the total-field anomaly at field point i due to cell j with unit magnetization. The left side of equation 10.7 is composed entirely of measured quantities. Matrix ψ_{ij} can be calculated with equation 9.30 (Subroutine B.15 would do most of this job). Hence, the only unknown quantities in equation 10.7 are the N values of magnetization, and if $N < L$, these can be calculated by least-squares methods. Claerbout [60] provided a good discussion of least-squares modeling.

This same sort of approach has been used to model magnetization of the earth's crust using satellite data (Mayhew [178, 179]; von Frese, Hinze, and Braile [287]; Ravat, Hinze, and von Frese [239]; Arkani-Hamed and Strangway [7]). Mayhew [178, 179], for example, constructed a model of magnetized crust with dipoles oriented parallel to the main field, and then used least-squares techniques to find the dipole moment of each dipole. Magnetizations were found by dividing the calculated dipole moments by the volumes that they represent.

It would be tempting to make the widths of the cells in Figure 10.1 as narrow as possible in order to learn about short-wavelength information in crustal magnetization. To do so in the case of seafloor spreading anomalies might answer questions concerning short-period behavior of

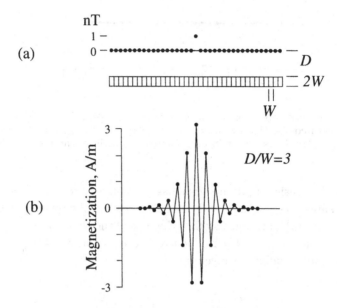

Fig. 10.2. Unstable solution for magnetization of magnetic layer. (a) Total-field anomaly is zero except at single point. Layer is composed of cells of width W and located at depth D. Thickness of layer is $2W$. (b) Resulting magnetization when $D/W = 3$. Modified from Bott and Hutton [38].

the earth's magnetic field (e.g., short polarity events and excursions of the main field) and about the process of seafloor spreading. Choosing cell width is a question of resolution; for a given depth to the magnetic layer, how much information can be squeezed out of the calculated magnetization? Bott and Hutton [38] showed that the resolving power of this inverse method has definite limits. As illustrated in Figure 10.2, they let $\Delta T_i = 0$ everywhere except at a single point, where $\Delta T_i = 1$ nT, and used equation 10.7 to solve for the magnetization. Clearly, a realistic solution for the M_j causing this special anomaly should be nearly zero for all j. Instead they found magnetization solutions that varied unrealistically (Figure 10.2), and the amount of the variation depended on the depth to the layer and the width of the cells (Figure 10.3). Bott and Hutton [38] concluded empirically that when cell width is made less than one or two times the depth to the layer, the inverse solution to equation 10.7 is unstable. In Chapter 11, we will be in a position to show this relationship in a more deductive way.

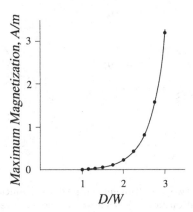

Fig. 10.3. Magnitude of instability as a function of resolution attempted. Curve represents various trials of the experiment described in Figure 10.2. Modified from Bott and Hutton [38].

The reason for the instability can be seen from simple algebraic considerations. Rewriting equation 10.7 in matrix notation provides

$$
\begin{bmatrix} \Delta T_1 \\ \Delta T_2 \\ \Delta T_3 \\ \vdots \\ \Delta T_L \end{bmatrix} = \begin{bmatrix} \psi_{11} & \psi_{12} & \cdots & \psi_{1N} \\ \psi_{21} & \psi_{22} & \cdots & \psi_{2N} \\ \psi_{31} & \psi_{32} & \cdots & \psi_{3N} \\ \vdots & \vdots & & \vdots \\ \psi_{L1} & \psi_{L2} & \cdots & \psi_{LN} \end{bmatrix} \begin{bmatrix} M_1 \\ M_2 \\ M_3 \\ \vdots \\ M_N \end{bmatrix} .
\qquad (10.8)
$$

Each column of matrix ψ_{ij} represents the total-field anomaly along a profile over a single cell, namely, cell j. If cell widths are small relative to depth, then the profile over single cell j will be very similar to profiles over cell $j+1$ or cell $j-1$. In other words, small cell width causes neighboring columns of matrix ψ_{ij} to be similar. In the parlance of matrix algebra, the matrix becomes *ill conditioned*.

To see how this situation might affect solutions for M_j, consider just two simultaneous equations,

$$z_1 = a_{11}x + a_{12}y,$$

$$z_2 = a_{21}x + a_{22}y,$$

representing some experiment, such as a simple case of equation 10.8; z_1 and z_2 are measured quantities, a_{11}, a_{12}, a_{21}, and a_{22} are calculated

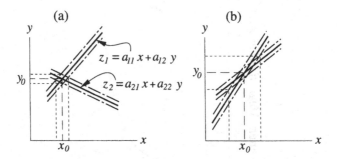

Fig. 10.4. Solution of two simultaneous equations. Equations are represented by solid lines and solution (x_0, y_0) is shown by their intersection. Errors in z_1 or z_2 cause lines to shift up or down, as shown by dashed-dotted lines. (a) If two solid lines make a large angle to each other, small errors in z_1 or z_2 will not affect the solution greatly. (b) If two lines are nearly parallel, small errors in z_1 or z_2 will greatly affect the solution.

quantities, and x and y are to be determined. As shown in (Figure 10.4), these two equations define lines in x, y space, and the solution (x_0, y_0) of the two equations is given by the intersection of the two lines. Errors in measurements of z_1 or z_2 cause parallel displacement of the lines. If the lines make a large angle with each other, as shown in Figure 10.4(a), slight displacements will not greatly affect the determination of (x_0, y_0). However, if the two lines are nearly parallel (Figure 10.4(b)), slight errors in z_1 or z_2 will cause significant errors in the determination of (x_0, y_0), and the solution is unstable.

The two lines will be nearly parallel if $a_{11}/a_{12} \approx a_{21}/a_{22}$. In terms of the magnetic layer (Figure 10.1), this kind of situation would occur if the field at point i due to cell j is similar to the field at point i due to cell $j + 1$ and if the field at point i due to cell j is similar to the field at point $i + 1$ due to cell j.

Equation 10.8 is simply an N-dimensional extension of these two simultaneous equations. Rows and columns of ψ_{ij} are smoothly varying functions. Hence, the forward calculation of ΔT_i from M_j is a smoothing operation, whereas the inverse calculation is an unsmoothing operation. Moreover, the deeper the layer is relative to cell width, the smoother is the matrix ψ_{ij}. If cell width is too small relative to the depth to the layer, the matrix ψ_{ij} becomes ill-conditioned, and small changes in ΔT_i will cause unrealistic values in the calculated M_j.

10.2.2 Determination of Magnetization Direction

Uniform magnetization

In some geologic situations, it may be appropriate to assume that a body is uniformly magnetized and to solve for the single vector that best describes that magnetization. A well-known application of this inverse method is the determination of seamount magnetization. In this case, measurements are made, usually at the sea surface, of the bathymetry and total-field anomaly of the seamount. The magnetic part of the seamount is assumed (1) to be uniformly magnetized, (2) to be bounded by an upper surface equal to the bathymetric surface, and (3) to have a known lower surface (usually flat and at the same depth as surrounding seafloor). Then the magnetic field can be inverted to determine the single vector that best describes its uniform magnetization. The magnetization thus determined is in effect a paleomagnetic sample, assuming that the magnetization is primarily remanent and was recorded at the time that the seamount formed. A collection of magnetization vectors from seamounts of different ages, therefore, provides a record of movement of the lithospheric plate on which the seamounts reside (e.g., Francheteau et al. [91], Harrison et al. [119]).

Equation 10.3 can be rewritten as

$$\Delta T(P) = -C_{\mathrm{m}} \hat{\mathbf{F}} \cdot \nabla_P \int_R \left[M_x(Q) \frac{\partial}{\partial x'} \frac{1}{r} + M_y(Q) \frac{\partial}{\partial y'} \frac{1}{r} \right.$$

$$\left. + M_z(Q) \frac{\partial}{\partial z'} \frac{1}{r} \right] dv \, .$$

If $\mathbf{M}$ is uniform, then

$$\Delta T(P) = M_x \left[-C_{\mathrm{m}} \hat{\mathbf{F}} \cdot \nabla_P \int_R \frac{\partial}{\partial x'} \frac{1}{r} \, dv \right]$$

$$+ M_y \left[-C_{\mathrm{m}} \hat{\mathbf{F}} \cdot \nabla_P \int_R \frac{\partial}{\partial y'} \frac{1}{r} \, dv \right]$$

$$+ M_z \left[-C_{\mathrm{m}} \hat{\mathbf{F}} \cdot \nabla_P \int_R \frac{\partial}{\partial z'} \frac{1}{r} \, dv \right]$$

$$= M_x \xi_x(P) + M_y \xi_y(P) + M_z \xi_z(P) \,, \tag{10.9}$$

where ξ_x, ξ_y, and ξ_z are integral terms involving the geometry of the seamount. Assuming the total-field anomaly is measured at N discrete locations, equation 10.9 can be written in matrix notation as

$$\begin{bmatrix} \Delta T_1 \\ \Delta T_2 \\ \vdots \\ \Delta T_N \end{bmatrix} = \begin{bmatrix} \xi_{1x} & \xi_{1y} & \xi_{1z} \\ \xi_{2x} & \xi_{2y} & \xi_{2z} \\ \vdots & \vdots & \vdots \\ \xi_{Nx} & \xi_{Ny} & \xi_{Nz} \end{bmatrix} \begin{bmatrix} M_x \\ M_y \\ M_z \end{bmatrix}. \qquad (10.10)$$

The three columns of matrix ξ_{ij} in equation 10.10 represent the total-field anomaly at the various field locations, assuming unit magnetizations in the x, y, and z directions, respectively. Each element of the matrix can be calculated using forward-calculation methods (Chapter 9), but to do so requires approximation of the shape of the body with a simplified model, for example, an aggregate of rectangular blocks (Vacquier [285]), a stack of laminas (Talwani [275]), or a stack of layers (Plouff [229]). Then equation 10.10 represents N equations with only three unknowns, namely, the three components of magnetization, and these can be estimated by straightforward least-squares techniques (Claerbout [60]); that is, find M_x, M_y, and M_z that minimize the quantity

$$E^2 = \sum_{i=1}^{N} (\Delta T_i - \Delta T_i')^2 \,,$$

where $\Delta T_i'$, $i = 1, 2, \ldots, N$, are the measured anomaly values. In addition, we may wish to subtract a regional field $F(P)$ so that equation 10.9 becomes

$$\Delta T(P) = M_x \xi_x(P) + M_y \xi_y(P) + M_z \xi_z(P) - F(P) \,.$$

If $F(P)$ is a linear surface, for example, equation 10.10 simply would be expanded to six unknowns.

Various quantities can be calculated to evaluate the ability of this simple model to fit the measured data. The residual field, $e_i = \Delta T_i - \Delta T_i'$, for example, provides a spatial view of the quality of the model, and the "goodness-of-fit" parameter

$$r = \frac{\sum_{i=1}^{N} |\Delta T_i|}{\sum_{i=1}^{N} |e_i|}$$

provides an overall evaluation.

Nonuniform Magnetization

Seamount magnetization has proven to be a powerful tool in understanding plate tectonics, but the method has limitations. Perhaps the most serious is the very assumption of uniform magnetization. Least-squares analysis requires that the residual errors e_i be randomly distributed, but residuals resulting from the assumption of uniform magnetization often show broad regions of positive and negative misfit, reflecting in part variable magnetizations. There are many reasons to suspect nonuniform magnetization. Seamounts may form over times during which the geomagnetic field has varied significantly. Chemistry and mineralogy may not be completely uniform throughout the seamount because of changing magma sources during formation and because of subsequent low-temperature oxidation and hydrothermal circulation.

Several studies have sought to allow for variable magnetization in seamounts (Harrison [115], Francheteau et al. [91], Sager et al. [249], Emilia and Massey [85], Ueda [284]) and arc-related volcanoes (Kodama and Uyeda [149], Blakely and Christiansen [25]) by dividing the topographic edifice into discrete compartments. McNutt [181], for example, modeled several seamounts located on the Cocos plate by dividing their topographic edifices into two or three compartments, assuming each compartment was uniformly magnetized, and using least-squares methods to simultaneously solve for the direction of magnetization within each compartment.

In each of these studies, the boundaries between uniformly magnetized parts of the topographic features were determined subjectively, usually by inspection of the residual magnetic anomaly. An entirely different approach was proposed by Parker et al. [210]. Rather than pursuing a single magnetization direction, they sought a magnetization with both uniform and nonuniform components such that the nonuniform component was as small as possible. Their main points are summarized here.

First a few definitions are in order. The total-field anomaly of the seamount is given by equation 10.3,

$$\Delta T(P) = -C_{\mathrm{m}} \hat{\mathbf{F}} \cdot \nabla_P \int_R \mathbf{M}(Q) \cdot \nabla_Q \frac{1}{r} \, dv,$$

which can be written more simply in a form like equation 10.4,

$$\Delta T_i = \int_R \mathbf{M}(Q) \cdot \mathbf{G}_i(Q) \, dv, \qquad i = 1, 2, \ldots, L, \tag{10.11}$$

where the subscript i denotes a discrete observation point. The function $\mathbf{M}(Q)$ in this equation is a vector-valued function of position and can be treated as an infinite-dimensional vector. It represents one *element* of an infinite variety of possible magnetizations that might occur in R. The set of all such magnetizations forms an infinite-dimensional space called a *Hilbert space*. The Green's function $\mathbf{G}_i(Q)$ in equation 10.11 is analogous to coordinate vectors in a three-dimensional vector space (e.g., $\hat{\mathbf{i}}$, $\hat{\mathbf{j}}$, and $\hat{\mathbf{k}}$ in the cartesian coordinate system) and is called a *coordinate function*. The *inner product* of two elements $\mathbf{A}(Q)$ and $\mathbf{B}(Q)$ of the Hilbert space is given by

$$(\mathbf{A}, \mathbf{B}) = \int_R \mathbf{A}(Q) \cdot \mathbf{B}(Q)\, dv,$$

analogous to the dot product of two vectors. Hence, equation 10.11 is the inner product

$$\Delta T_i = (\mathbf{M}, \mathbf{G}_i),\qquad\qquad (10.12)$$

and each observation ΔT_i is said to be a linear functional of $\mathbf{M}(Q)$. The "size" of an element is measured by its *norm*

$$||\mathbf{A}|| = (\mathbf{A}, \mathbf{A})^{\frac{1}{2}}$$

$$= \left[\int_R |\mathbf{A}|^2\, dv \right]^{\frac{1}{2}},$$

and the difference between two elements is given by $||\mathbf{A} - \mathbf{B}||$.

In the method described by Parker et al. [210], the magnetization $\mathbf{M}(Q)$ in equation 10.11 is represented by both uniform and nonuniform components,

$$\mathbf{M}(Q) = \mathbf{M}_0 + \mathbf{M}_N(Q),$$

where $\mathbf{M}_0$ is a vector constant. Equation 10.12 becomes

$$\Delta T_i = (\mathbf{M}_0, \mathbf{G}_i) + (\mathbf{M}_N, \mathbf{G}_i),\qquad i = 1, 2, \ldots, L.\qquad (10.13)$$

The element $\mathbf{M}(Q)$ having minimum $\mathbf{M}_N(Q)$ is the magnetization that is most nearly uniform for any given $\mathbf{M}_0$.

First consider that $\mathbf{M}_0$ is known. Then an element $\mathbf{M}_N(Q)$ of smallest norm is required obeying the L inner-product constraints given by equation 10.13, where all quantities are known except $\mathbf{M}_N(Q)$. Parker et al. [210] showed that this magnetization can be represented by an expansion in terms of Green's functions,

$$\mathbf{M}_{\mathrm{N}}(Q) = \sum_{j=1}^{L} \alpha_j \mathbf{G}_j(Q), \qquad (10.14)$$

where the expansion coefficients α_j, $j = 1, 2, \ldots, L$, are to be determined. Substituting equation 10.14 into 10.13 yields

$$\Delta T_i = (\mathbf{M}_0, \mathbf{G}_i) + \sum_{j=1}^{L} \alpha_j (\mathbf{G}_j, \mathbf{G}_i)$$

$$= (\mathbf{M}_0, \mathbf{G}_i) + \sum_{j=1}^{L} \alpha_j \Gamma_{ij}, \qquad (10.15)$$

where the inner product Γ_{ij} is called a *Gram matrix*. All quantities are known in equation 10.15 except the L expansion coefficients, and they can be determined uniquely because Γ_{ij} is nonsingular.

The vector constant $\mathbf{M}_0$ is not known in application to seamount magnetization, of course, and in fact is the main point of the calculation. The problem is the same, however: Find $\mathbf{M}_0$ and $\mathbf{M}_{\mathrm{N}}(Q)$ that cause $\mathbf{M}_{\mathrm{N}}(Q)$ to have the smallest norm. Being a constant, $\mathbf{M}_0$ belongs to a three-dimensional subspace of the Hilbert space and can be written, analogously to equation 10.14, as

$$\mathbf{M}_0 = \sum_{k=1}^{3} \beta_k \mathbf{X}_k, \qquad (10.16)$$

where $\mathbf{X}_k$, $k = 1, 2, 3$, are unit magnetizations in each of three orthogonal directions. Substituting this summation into equation 10.15 yields

$$\Delta T_i = \sum_{k=1}^{3} \beta_k (\mathbf{X}_k, \mathbf{G}_i) + \sum_{j=1}^{L} \alpha_j \Gamma_{ij}. \qquad (10.17)$$

Hence, we wish to find parameters $\beta_1, \beta_2, \beta_3, \alpha_1, \alpha_2, \ldots, \alpha_L$ that minimize

$$\|\mathbf{M} - \mathbf{M}_0\| = \left(\sum_{j=1}^{L} \alpha_j \mathbf{G}_j, \sum_{k=1}^{L} \alpha_k \mathbf{G}_k \right)^{\frac{1}{2}} \qquad (10.18)$$

subject to the constraints described by equation 10.17. All quantities are known in equations 10.17 and 10.18 except the $L + 3$ parameters $\beta_1, \beta_2, \beta_3, \alpha_1, \alpha_2, \ldots, \alpha_L$. Parker et al. [210] described how these could be determined by the introduction of Lagrange multipliers. (See Claer-bout [60] for a general discussion of least-squares inversion subject to constraints.) With these $L + 3$ parameters, the most uniform direction of magnetization is given by equation 10.16. Notice that if magnetization is indeed uniform, equation 10.18 vanishes, and equation 10.17 reduces to equation 10.9.

10.3 Nonlinear Inverse Problem

The potential field on the left side of equation 10.5 is a linear functional of the distribution of mass or magnetic material. Doubling the intensity of magnetization, for example, doubles the amplitude of the total-field anomaly, whereas tripling the magnetization would triple the amplitude of the anomaly. In general terms, a system is said to be linear if it satisfies the following test: If $f_1(P)$ is the field caused by source distribution $s_1(Q)$, and $f_2(P)$ is the field caused by another distribution $s_2(Q)$, then the field caused by $as_1(Q) + bs_2(Q)$ is simply $af_1(P) + bf_2(P)$, where a and b are constants.

Exercise 10.3 Prove the previous sentence using equation 10.5.

The same cannot be said for other parameters that define the source. The potential field is not a linear functional of, for example, depth, thickness, or shape of the source. All of these parameters are contained within $\psi(P, Q)$ and in the limits of integration implied by volume R in equation 10.5. Inverse methods that attempt to estimate these nonlinear parameters are called *nonlinear* methods, but in fact, most nonlinear methods entail simplifying assumptions that in effect render the problem linear.

10.3.1 Shape of Source

This difficult class of inverse problem is addressed here by describing several examples from the literature. By nature, these methods are highly model dependent. They must make simplifying assumptions about the source distribution, and the validity of the calculated shapes naturally depends on whether or not the true source behaves in accordance with the assumptions. For example, one of the methods subsequently discussed finds the most compact mass satisfying a given set of gravity measurements. It is easy to imagine geologic situations in which mass would not be distributed in this way.

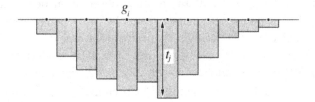

Fig. 10.5. Cross-sectional model of a sedimentary basin, as used in the method described by Bott [35]. Basin is assumed to be infinitely extended perpendicular to profile. Basin is divided into rectangular blocks, one block per field point.

Iterative Methods

As discussed in Chapter 9, forward models are developed by a three-step process. An anomaly is calculated from a model, the calculated anomaly is compared with the observed anomaly, and the model is adjusted in order to improve the comparison. The three-step process is repeated until the modeler is satisfied with the results. A number of computer-based algorithms use the same logical process, but we will consider them inverse methods here because the model is derived automatically with minimal control by the modeler.

An early example was described by Bott [35] to estimate the cross-sectional shape of sedimentary basins. In this method, the basin is assumed to be infinitely extended in one direction and to have uniform density contrast $\Delta\rho$ with respect to surrounding rocks. The basin is divided into N rectangular blocks infinitely extended parallel to the basin and extending to depths t_j, $j = 1, 2, \ldots, N$, as shown in Figure 10.5. Only N field points, g_i, $i = 1, 2, \ldots, N$, along a profile perpendicular to the basin are considered, and each field point is centered above a block. An initial guess is made for the thickness of each block by assuming each block is a slab infinite in all horizontal dimensions. Equation 3.27 provides the thickness of an infinite slab based on a single gravity measurement,

$$t_j^{(1)} = \frac{g_j}{2\pi\gamma\Delta\rho}, \qquad j = 1, 2, \ldots, N.$$

The superscript indicates the level of iteration, the first iteration in this case. Then a three-step procedure is conducted to iteratively modify block thickness. The steps are as follows, where k denotes the number of the iteration:

1. The field $g_j^{(k)}$ is calculated at each observation point due to all blocks, assuming thicknesses from the previous iteration. In the original work of Bott [35], this calculation was done in an elaborate way in order to save computer time. With modern computers, algorithms that implement equation 9.2.2 would be appropriate.

1. The residual $g_j - g_j^{(k)}$ is found at each observation point.

2. The infinite-slab approximation is used again to estimate a new set of thicknesses. The correction to each block is calculated under the assumption that the block is an infinite slab of thickness required to accommodate the residual; that is, the new thickness is

$$t_j^{(k+1)} = \frac{(g_j - g_j^{(k)})}{2\pi\gamma\Delta\rho} + t_j^{(k)}.$$

These three steps are repeated until the modeler is satisfied that convergence is met.

Cordell and Henderson [71] improved on this method in a number of ways. They employed data measured on or interpolated to a rectangular grid so that sources could be investigated in three dimensions. Sources are modeled as a bundle of rectangular blocks, one block per gravity value, as shown in Figure 10.6. Block thickness t_j, $j = 1, 2, \ldots, N$, is defined relative to a reference surface, which could represent, for example,

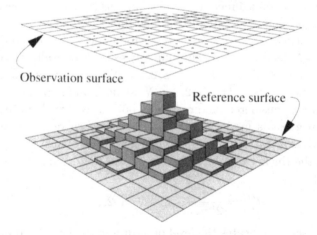

Observation surface

Reference surface

Fig. 10.6. Three-dimensional model for the iterative method of Cordell and Henderson [71]. Block thicknesses are relative to a common reference surface. Observed gravity is measured on a rectangular grid.

the top or bottom of all blocks. Similar to the method of Bott [35], initial block thickness is estimated by assuming each block to be an infinite slab. However, the ratio

$$\frac{t_j^{(k+1)}}{t_j^{(k)}} = \frac{g_j}{g_j^{(k)}}$$

is used to revise block thickness rather than an infinite-slab approximation. As before, the three-step procedure of calculation, comparison, and adjustment is carried out automatically at each iteration. This algorithm has been implemented in a Fortran program described by Cordell [65], and a similar version is available in a form compatible with microcomputers (Cordell, Phillips, and Godson [73]).

A somewhat different approach was described by Jachens and Moring [137]. Like the two previous methods, their method estimates the shapes of basins filled with low-density deposits, but their method takes into account the possibility that underlying basement rocks may have variable density. Their method proceeds by separating gravity measurements into two components: the component caused by the basins themselves and the component due to variations in density of underlying basement. Let g represent observed gravity after regional fields are removed (isostatic residual gravity (Chapter 7) would be an appropriate starting point) and let $g = g_b + g_d$, where g_b is the anomaly caused by underlying basement and g_d is the anomaly caused by low-density deposits. Then the following steps are conducted:

1. The first iteration assumes that g_b is defined by just those stations located on basement outcrops and calculates a smooth surface through just these data, as shown by the dashed line in Figure 10.7. This constitutes the first approximation $g_b^{(1)}$ to the basement field g_b; it is only a crude approximation because stations will still include the effects of nearby basins. These effects are to be removed in subsequent iterations.

2. The first approximation to g_d is found by subtracting $g_b^{(1)}$ from observed gravity g. This new residual $g_d^{(1)}$ is used to find a first approximation to basement depth using the infinite slab approximation, similar to the method of Bott [35].

3. The gravitational effect of the basins can then be calculated by a variety of methods. Jachens and Moring [137] used the method of Parker [204], to be discussed in Chapter 11. This result is subtracted

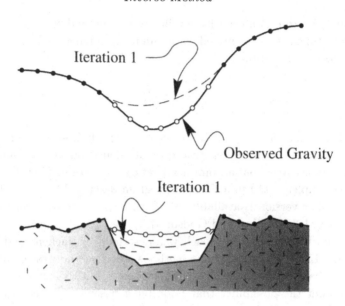

Fig. 10.7. Separation of residual gravity into two components, the component caused by density variations within basement and the component caused by basin fill. Closed dots signify measurements made on basement outcrops, open circles on sedimentary or volcanic cover.

from basement gravity stations to produce the next approximation for basement gravity $g_b^{(2)}$.

These three steps are repeated until the solution converges to the satisfaction of the modeler. Two products result: the shape of low-density basins and the gravitational attraction of basement without the effects of the basins. The method was applied to the entire state of Nevada by Jachens and Moring [137] in order to analyze the shape and distribution of basins in this part of the Basin and Range (Blakely and Jachens [31]), and a similar method was used by Saltus [250] to estimate the thickness of concealed sedimentary deposits beneath the Columbia River Basalt Group in Washington State.

Linearizing the Nonlinear

Although potential fields depend nonlinearly on certain source parameters, this dependence is nearly linear with respect to sufficiently small changes in those parameters. For example, the potential field of a polygonal prism is related to the coordinates of the corners of the polygon

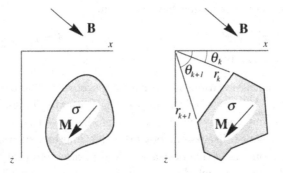

Fig. 10.8. Approximation of a two-dimensional source by an infinitely extended prism with polygonal cross section. Vectors **M** and **B** are projections of magnetization and ambient magnetic field, respectively, onto the x, z plane.

by way of arctangents and logarithms (see equations 9.2.2, 9.27, and 9.28), that is, the field is a nonlinear function of the coordinates of the polygon. Such nonlinear relationships can be rendered linear, however, by considering only very small changes in the parameters. For example, the gravity or magnetic field due to a set of polygonal prisms can be expanded in a Taylor's series based on changes in the positions of the coordinates of the polygons. If changes in the coordinates are small, the Taylor's series can be truncated, and the functional dependence on these changes thus becomes linear. An algorithm then could be devised to determine the best set of prisms for a given anomaly, where the cross-sectional shapes of the prisms are iteratively changed by small amounts through linear least-squares techniques.

This approach has been described in various forms for both gravity anomalies (Corbato [64], Al-Chalabi [2], Coles [62]) and magnetic anomalies (Johnson [144], McGrath and Hood [180], Rao and Babu [236]). Webring [290] has developed a computer program that implements both gravity and magnetic cases simultaneously and allows the user to intervene as necessary. This program also is available for microcomputers (Cordell et al. [73]). The following discussion summarizes the method described by Johnson [144].

Equations 9.2.2 and 9.30 provide the vertical attraction of gravity and the total-field anomaly, respectively, of a prism infinitely extended in one direction, with uniform density or magnetization, and with cross-sectional shape defined by an N-sided polygon (Figure 10.8). Let A_i represent one of L discrete measurements of the gravity or magnetic anomaly. Both equations then have the form

$$A_i = A(x'_1, z'_1, x'_2, z'_2, \ldots, x'_N, z'_N, x_i, z_i), \qquad i = 1, 2, \ldots, L$$
$$= A(x_i, z_i, \mathbf{w}),$$

where (x_i, z_i) is the location of the ith measurement. The primed coordinates are the N corners of the polygon, represented in shorthand by the $2N$-dimensional array $\mathbf{w}$. Array $\mathbf{w}$ includes only the body coordinates for the sake of discussion, but other parameters, such as density, the three components of magnetization, and parameters that describe a regional field, could be included as well. Furthermore, the following discussion treats all corners of the polygon as free to move in any direction. A practical algorithm would have the flexibility to analyze only specified corners.

Let A_i and $\bar{A}_i$ represent the observed and calculated anomalies, respectively, at one observation point. We wish to find the vector $\mathbf{w}$ such that the squared error

$$E^2 = \sum_{i=1}^{L} \left[A_i - \bar{A}_i(\mathbf{w}) \right]^2 \tag{10.19}$$

is as small as possible. Because $\bar{A}_i$ is a nonlinear function of $\mathbf{w}$, we cannot use the usual techniques of least-squares analysis. Instead, we will change the elements of $\mathbf{w}$ only small amounts over a number of iterations. If the changes in $\mathbf{w}$ are small, $\bar{A}_i$ will be nearly a linear function of those changes.

A Taylor's series is a convenient way to do this. For example, the value of a function $f(x, y)$ can be extrapolated to location $(x + \Delta x, y + \Delta y)$ by the series

$$f(x + \Delta x, y + \Delta y) = f(x, y) + \frac{\partial f(x, y)}{\partial x} \Delta x + \frac{\partial f(x, y)}{\partial y} \Delta y + \cdots,$$

where the higher-order terms have been dropped because Δx and Δy are taken sufficiently small.

Similarly, we let $\mathbf{w}^{(k)}$ represent the values of $(x'_1, z'_1, x'_2, z'_2, \ldots, x'_N, z'_N)$ after the kth iteration. Then, the Taylor's series expansion of the anomaly at point i is

$$\bar{A}_i(\mathbf{w}^{(k+1)}) \approx \bar{A}_i(\mathbf{w}^{(k)}) + \sum_{m=1}^{2N} \frac{\partial}{\partial w_m} \bar{A}_i(\mathbf{w}^{(k)}) \Delta w_m^{(k)}, \qquad i = 1, 2, \ldots, L,$$
$$\tag{10.20}$$

where $\Delta w_m^{(k)} = w_m^{(k+1)} - w_m^{(k)}$. Notice that $\bar{A}_i(\mathbf{w}^{(k+1)})$ is a linear function of $\Delta w_m^{(k)}$, where $m = 1, 2, \ldots, 2N$, that is, we have "linearized" a
nonlinear problem.

Now substitute equation 10.20 into equation 10.19 to get

$$E^2 = \sum_{i=1}^{L} \left[A_i - \bar{A}_i(\mathbf{w}^{(k)}) - \sum_{m=1}^{2N} \frac{\partial}{\partial w_m} \bar{A}_i(\mathbf{w}^{(k)}) \Delta w_m^{(k)} \right]^2 .$$

Expressions for $\bar{A}_i(\mathbf{w})$ have already been derived, namely, equations 9.2.2
and 9.30, and we could just as easily derive expressions for their partial
derivatives. Thus, given an initial array of parameters $\mathbf{w}^{(k)}$, the only
unknowns in the above equation are $\Delta w_m^{(k)}$, $m = 1, 2, \ldots, 2N$. To find
the set of these parameters that provides the smallest E^2, we calculate
the partial derivative of E^2 with respect to w_j, $j = 1, 2, \ldots, 2N$, and set
each equation equal to zero:

$$\sum_{i=1}^{L} \left[A_i - \bar{A}_i(\mathbf{w}^{(k)}) - \sum_{m=1}^{2N} \frac{\partial}{\partial w_m} \bar{A}_i(\mathbf{w}^{(k)}) \Delta w_m^{(k)} \right] \cdot \left[\frac{\partial}{\partial w_j} \bar{A}_i(\mathbf{w}^{(k)}) \right] = 0,$$

$$j = 1, 2, \ldots, 2N,$$

where again we have dropped higher-order terms. This complicated equation is just a matrix expression of the form

$$\alpha_j = \sum_{m=1}^{2N} G_{mj} \Delta w_m^{(k)}, \tag{10.21}$$

where

$$\alpha_j = \sum_{i=1}^{L} [A_i - \bar{A}_i(\mathbf{w}^{(k)})] \frac{\partial}{\partial w_j} \bar{A}_i(\mathbf{w}^{(k)}) \tag{10.22}$$

$$G_{mj} = \sum_{i=1}^{L} \left[\frac{\partial}{\partial w_m} \bar{A}_i(\mathbf{w}^{(k)}) \frac{\partial}{\partial w_j} \bar{A}_i(\mathbf{w}^{(k)}) \right]. \tag{10.23}$$

The steps in the solution then are as follows:

1. Pick an initial set of corner coordinates $\mathbf{w} = (x_1', z_1', x_2', z_2', \ldots)$. The
   foregoing discussion has assumed that all corners of the polygon are
   free to move. In practice, most corners would be fixed and only a few
   corners allowed to move during a given calculation.
2. Calculate $\bar{A}_i(\mathbf{w}^{(k)})$ and $\frac{\partial}{\partial w_m} \bar{A}_i(\mathbf{w}^{(k)})$ for $i = 1, 2, \ldots, L$ and $m = 1, 2, \ldots, 2N$.

3. Calculate α_j and G_{mj} in accordance with equations 10.22 and 10.23.
4. Invert the matrix in equation 10.21 in order to find $\Delta \mathbf{w}^{(k)}$.
5. Adjust $\mathbf{w}$ accordingly.

Steps 2 through 5 are repeated until the solution converges, that is, until E^2 is reduced to a specified level or elements of Δw_m, $m = 1, 2, \ldots, 2N$, become sufficiently small.

Johnson [144], following the algorithm of Marquardt [174], suggested weighting the diagonal of the matrix in equation 10.21 in order to control the convergence of the iterations. As discussed by Marquardt [174], equation 10.21 becomes

$$\alpha_j = \sum_{m=1}^{M} G_{mj} \Delta w_m^{(k)} (1 + \delta_{mj}\lambda) , \qquad (10.24)$$

where

$$\delta_{mj} = \begin{cases} 0, & \text{if } m \neq j; \\ 1, & \text{if } m = j, \end{cases}$$

and where λ is positive or zero. If $\lambda = 0$, equation 10.24 reduces to equation 10.21. For $\lambda > 0$, the new values of $\mathbf{w}^{(k+1)}$ are restricted to a neighborhood about $\mathbf{w}^{(k)}$; as $\lambda \to \infty$, equation 10.24 becomes the method of steepest descent.

Compact Bodies

Any solution for the distribution of mass or magnetic material from its corresponding gravity or magnetic field should strive to satisfy equation 10.5. Equation 10.5 is the principal constraint equation. We may also be willing to add other constraints, for example, that density should everywhere be positive (or everywhere negative) with respect to surrounding rocks. In addition to the constraints, we might strive to minimize or maximize some scalar property of the source. Last and Kubik [161], for example, described how to find the body with minimum volume that satisfies the constraint equations. They divided the source into N rectangular blocks, either two- or three-dimensional, as shown in Figure 10.9 for the two-dimensional case, and assumed that the density of each block ρ_j is uniform. Given measurements of gravity at L discrete locations, the constraint equations from equation 10.5 can be written as

$$g_i = \sum_{j=1}^{N} \psi_{ij} \rho_j + e_i, \qquad i = 1, 2, \ldots, L , \qquad (10.25)$$

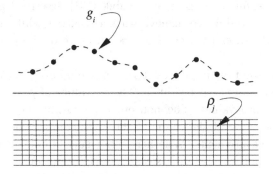

Fig. 10.9. Source region divided into rectangular blocks, where the number of blocks exceeds the number of measurements. As shown here, the source is considered to be two-dimensional.

where e_i, $i = 1, 2, \ldots, L$, are errors associated with each measurement. If $L > N$, we could use least-squares techniques and find a density for each block, as was discussed earlier in this chapter for the magnetic case. Instead, we select $N > L$, and require that the nonzero part of the source region be as small as possible, thereby providing an estimate of the shape of the most compact body that satisfies the L gravity measurements.

A single parameter representing the volume (or cross-sectional area in the two-dimensional case) of the nonzero source region is required. If we note that

$$\lim_{\epsilon \to 0} \frac{\rho_j^2}{\rho_j^2 + \epsilon} = \begin{cases} 0, & \text{if } \rho_j = 0; \\ 1, & \text{if } \rho_j \neq 0, \end{cases}$$

the volume of the body having nonzero density is given by

$$V = \Delta V \lim_{\epsilon \to 0} \sum_{j=1}^{N} \frac{\rho_j^2}{\rho_j^2 + \epsilon},$$

where ΔV is the volume of an individual block (or the area of an individual rectangle in the two-dimensional case). The algorithm then is required to minimize the scalar

$$q = \sum_{j=1}^{N} \frac{\rho_j^2}{\rho_j^2 + \epsilon} + \sum_{i=1}^{L} w_i e_i^2 \qquad (10.26)$$

subject to the L constraints expressed by equation 10.25, where ϵ is chosen to be sufficiently small and where w_i, $i = 1, 2, \ldots, L$, is a noise-weighting function. Last and Kubik [161] described an iterative, least-squares procedure to achieve such a solution while allowing the weights and densities to depend on the outcome of the previous iteration.

Hence, the solution minimizes both the volume of the body and a weighted sum of squared residuals. The N values of ρ_j will have both zero and nonzero values. The nonzero cells within the source region provide an estimate of the shape of the most compact body satisfying the gravity measurements. Whether or not such a body is geologically reasonable is another matter; it is easy to imagine geological situations in which masses are not expected to be packed into the smallest of volumes.

10.3.2 Depth to Source

Methods of estimation of the depth to magnetic or gravity sources can be classed into two categories: those that analyze a single, isolated anomaly and those that analyze a profile over many sources. A few examples of each type are described here. In the next chapter, we will visit the subject again in the Fourier domain.

Peters's Method

One early and useful depth rule was described by Peters [219]. If a magnetic anomaly is caused by a two-dimensional body with vertical sides, uniform and nearly vertical magnetization, and great depth extent, then the depth to the top of the body can be found approximately by the following graphical procedure (Figure 10.10). Draw two parallel lines with slope equal to one half of the maximum gradient of the anomaly, one line tangent to the peak of the anomaly and the other tangent to the minimum part of the anomaly. The horizontal separation of the two lines is proportional to the depth to the top of the body. The proportionality constant is 1.2 for very thin bodies and 2.0 for very thick bodies; a value of 1.6 is often used.

Peters's method requires many simplifying assumptions about the body (Skillbrei [262]). Yet it provides, with just graph paper and a pencil, a quick and rough estimate of depth to source in many geological situations at high latitudes. It amounts to finding the horizontal distance over which the anomaly is roughly a linear function of distance. Therefore, it is amenable to contour maps of gravity or magnetic anomalies; the distance over which the contours are roughly equally spaced is proportional to the depth to the source.

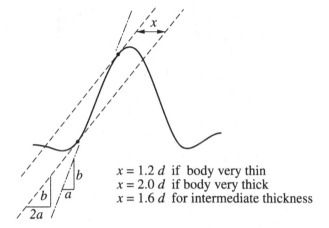

$x = 1.2\ d$ if body very thin
$x = 2.0\ d$ if body very thick
$x = 1.6\ d$ for intermediate thickness

Fig. 10.10. Illustration of Peters's method.

Maximum Depth

Interpretation of gravity or magnetic anomalies is complicated by the fact that many geologically reasonable density or magnetization solutions may perfectly satisfy the observed anomaly. Fortunately, some unequivocal parameters can be gotten directly from the anomalies without appeal to numerous assumptions about the source distribution. We will delve into this subject in more detail in the upcoming section on ideal bodies, but here we list some handy rules derived by Smith [263] and Bott and Smith [41] that quickly provide the maximum depth of causative sources. These rules are based on the first, second, and third derivatives of gravity and magnetic anomalies as measured along profiles (Figure 10.11); they are especially useful because they require no assumptions about the shape of the causative body.

The limiting depths from Smith [263] and Bott and Smith [41] are summarized below for both gravity and magnetic anomalies. In these inequalities, $A(x)$ signifies a profile across either a gravity or magnetic anomaly. In the gravity case, $A(x)$ represents the vertical attraction of gravity. In the magnetic case, $A(x)$ represents the component of magnetic field parallel to $\hat{\mathbf{r}} = (\hat{r}_x, \hat{r}_y, \hat{r}_z)$ (e.g., if $A(x)$ is the total-field anomaly, $\hat{\mathbf{r}}$ is in the direction of the ambient field). Parameter d is depth to a plane below which the entire source distribution lies. The following abbreviations are used: $A = A(x)$, $A' = \frac{dA(x)}{dx}$, $A'' = \frac{d^2 A(x)}{dx^2}$, and $A''' = \frac{d^3 A(x)}{dx^3}$. Subscript "max" signifies the maximum value obtained along the x axis.

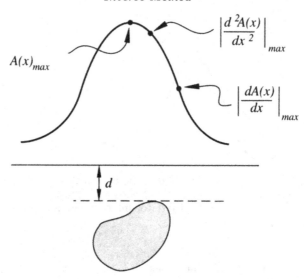

Fig. 10.11. Maximum depth to causative sources based on first, second, and third derivatives of their anomalies. Profile $A(x)$ represents either a magnetic or gravity anomaly.

The parameter Δw is the half-width of a symmetric anomaly, that is, the distance between positions where the anomaly reaches its maximum and half-maximum values. The maximum value of density or magnetization is denoted by $\rho_{\max}$ or $M_{\max}$, respectively. Note that these relationships were derived in the emu system and should be used accordingly.

Three-Dimensional Gravity Anomalies

Density of both signs:

$$d \leq 5.40 \ \frac{\gamma \rho_{\max}}{|A''|_{\max}},$$

$$d^2 \leq 6.26 \ \frac{\gamma \rho_{\max}}{|A'''|_{\max}}.$$

Density entirely positive (or entirely negative if signs are changed appropriately):

$$d \leq 1.5 \ \frac{A}{|A'|} \qquad \text{for all } x,$$

$$d^2 \leq -3 \ \frac{A}{A''} \qquad \text{for all } x \text{ for which } A'' \text{ is negative,}$$

$$d \leq 0.86 \quad \frac{A_{\max}}{|A'|_{\max}},$$

$$d \leq 2.70 \quad \frac{\gamma \rho_{\max}}{|A''|_{\max}},$$

$$d^2 \leq 3.13 \quad \frac{\gamma \rho_{\max}}{|A'''|_{\max}}.$$

Two-dimensional Gravity Anomalies

Density entirely positive (or entirely negative if signs are changed appropriately):

$$d \leq \frac{A}{|A'|} \qquad \text{for all } x,$$

$$d^2 \leq -2 \quad \frac{A}{A''} \qquad \text{for all } x \text{ for which } A'' \text{ is negative,}$$

$$d \leq 0.65 \quad \frac{A_{\max}}{|A'|_{\max}},$$

$$d \leq \Delta w \qquad \text{(symmetric anomaly).}$$

Three-Dimensional Magnetic Anomalies

No restrictions on magnetization:

$$d \leq 6.28 (4\hat{r}_x^2 + 3\hat{r}_y^2 + 3\hat{r}_z^2)^{\frac{1}{2}} \quad \frac{M_{\max}}{|A'|_{\max}},$$

$$d^2 \leq 9.73 (3\hat{r}_x^2 + 2\hat{r}_y^2 + 2\hat{r}_z^2)^{\frac{1}{2}} \quad \frac{M_{\max}}{|A''|_{\max}}.$$

Magnetization everywhere parallel and same sense:

$$d \leq 3.14 (4\hat{r}_x^2 + 3\hat{r}_y^2 + 3\hat{r}_z^2)^{\frac{1}{2}} \quad \frac{M_{\max}}{|A'|_{\max}},$$

$$d^2 \leq 4.87 (3\hat{r}_x^2 + 2\hat{r}_y^2 + 2\hat{r}_z^2)^{\frac{1}{2}} \quad \frac{M_{\max}}{|A''|_{\max}}.$$

*Vertical $\hat{r}$ and vertical **M**:*

$$d \leq 5.18 \quad \frac{M_{\max}}{|A'|_{\max}},$$

$$d^2 \leq 6.28 \quad \frac{M_{\max}}{|A''|_{\max}}.$$

Vertical $\hat{\mathbf{r}}$ and vertical $\mathbf{M}$; $\mathbf{M}$ everywhere of same sense:

$$d \leq 2.59 \quad \frac{M_{\max}}{|A'|_{\max}},$$

$$d^2 \leq 3.14 \quad \frac{M_{\max}}{|A''|_{\max}}.$$

Two-Dimensional Magnetic Anomalies

No restrictions on magnetization:

$$d \leq 8(\hat{r}_x^2 + \hat{r}_z^2)^{\frac{1}{2}} \quad \frac{M_{\max}}{|A'|_{\max}},$$

$$d^2 \leq 9.42(\hat{r}_x^2 + \hat{r}_z^2)^{\frac{1}{2}} \quad \frac{M_{\max}}{|A''|_{\max}}.$$

$\mathbf{M}$ *everywhere parallel and same sense:*

$$d \leq 4(\hat{r}_x^2 + \hat{r}_z^2)^{\frac{1}{2}} \quad \frac{M_{\max}}{|A'|_{\max}},$$

$$d^2 \leq 4.71(\hat{r}_x^2 + \hat{r}_z^2)^{\frac{1}{2}} \quad \frac{M_{\max}}{|A''|_{\max}}.$$

Each depth rule has specific applicability. Some require knowledge of a significant part of the anomaly in order to find maximum amplitudes and gradients. Others require only parts of the anomaly. In practice, as many depth rules as seem applicable should be tried to determine which provides the smallest value for maximum depth.

Euler's Equation

The previous methods of depth estimation are best suited for anomalies caused by single, isolated sources. Another class of techniques uses a somewhat different strategy, namely, to consider magnetic or gravity anomalies to be caused by many relatively simple sources. Such a strategy is amenable for long profiles or large surveys with many anomalies. For example, a technique that can estimate the location of a simple body (monopole, dipole, thin sheet, etc.) from only a few measurements of the magnetic or gravity field could be applied to a long profile of measurements by dividing the profile into windows of consecutive measurements, each window providing a single estimate of source location.

When all such determinations are plotted in cross section, they may tend to cluster around magnetization or density contrasts of geologic interest.

Euler's equation, discussed in Section 6.4 (equation 6.19), has led to one such method. Euler's equation in its general form is given by

$$\mathbf{r} \cdot \nabla f = -nf \, .$$

Functions f that satisfy Euler's equation are said to be homogeneous; if they also satisfy Laplace's equation, they can be represented in spherical coordinates as a sum of spherical surface harmonics. Any spatial derivative of a homogeneous function is also homogeneous. For example, taking the partial derivative with respect to x of both sides of Euler's equation yields

$$\frac{\partial}{\partial x}[\mathbf{r} \cdot \nabla f] = \frac{\partial}{\partial x} f + \mathbf{r} \cdot \nabla \frac{\partial}{\partial x} f$$

$$= -n \frac{\partial}{\partial x} f,$$

$$\mathbf{r} \cdot \nabla \left[\frac{\partial}{\partial x} f \right] = -(n+1) \frac{\partial}{\partial x} f,$$

and consequently $\frac{\partial}{\partial x} f$ is homogeneous with degree $n+1$.

It is easily shown that $f = \frac{1}{r}$ satisfies Euler's equation with $n = 1$, so clearly the potential of a point mass (or uniform sphere) must also. Because potential fields arising from other point sources (dipoles, wires, and so forth) involve spatial derivatives of $\frac{1}{r}$, they too should satisfy Euler's equation, but with their own characteristic integer values for n. The total-field anomaly of a magnetic dipole (or uniformly magnetized sphere), for example, is given by

$$\Delta T = C_{\mathrm{m}} \hat{\mathbf{b}} \cdot \nabla \left(\mathbf{m} \cdot \frac{1}{r} \right),$$

where $\hat{\mathbf{b}}$ is the unit vector parallel to the ambient field, and $\mathbf{m}$ is the dipole moment. It is easily shown that ΔT thus defined satisfies Euler's equation with $n = 3$. The parameter n in Euler's equation is referred to as the *structural index* or *attenuation rate*. Table 10.1 shows the structural index for various sources.

Exercise 10.4 Use Euler's equation to verify the values of n in Table 10.1.

Euler's equation has been used by a number of authors for analyzing both magnetic anomalies (Thompson [280], Barongo [12], Reid et al. [242]) and gravity anomalies (Marson and Klingele [176]). Consider, for example, the total-field anomaly over a simple body, such as a sphere

Table 10.1. *"Structural index" for various gravity and magnetic sources.*

n	Type of Body
1	line of mass
2	line of dipoles
2	point mass (sphere of uniform density)
3	point dipole (sphere of uniform magnetization)

or cylinder. Let ΔT_i be the ith point of a magnetic survey over a simple body, such as a sphere or cylinder, with the point of measurement at (x, y, z), and the center of the body at (x_0, y_0, z_0). Substituting into equation 6.19 provides

$$\left[\frac{\partial}{\partial x} \Delta T_i \quad \frac{\partial}{\partial y} \Delta T_i \quad \frac{\partial}{\partial z} \Delta T_i \right] \begin{bmatrix} x - x_0 \\ y - y_0 \\ z - z_0 \end{bmatrix} = n \Delta T_i \,.$$

Assuming we have some way of measuring or calculating horizontal and vertical gradients of the total-field anomaly, this equation has only four unknowns: x_0, y_0, z_0, and n, the first three of which provide the location of the body. We can obtain as many equations as necessary by writing this equation for various measurement locations,

$$\begin{bmatrix} \dfrac{\partial}{\partial x} \Delta T_1 & \dfrac{\partial}{\partial y} \Delta T_1 & \dfrac{\partial}{\partial z} \Delta T_1 \\[2mm] \dfrac{\partial}{\partial x} \Delta T_2 & \dfrac{\partial}{\partial y} \Delta T_2 & \dfrac{\partial}{\partial z} \Delta T_2 \\[2mm] \vdots & \vdots & \vdots \end{bmatrix} \begin{bmatrix} x - x_0 \\ y - y_0 \\ z - z_0 \end{bmatrix} = n \begin{bmatrix} \Delta T_1 \\ \Delta T_2 \\ \vdots \end{bmatrix} \,,$$

and use least-squares methods to solve for the unknowns. If the position of the body is known, we can solve for n and learn something about the kind of body involved (Barongo [12]). If on the other hand we suspect the nature of the body (e.g., we might believe that the body is spherical in shape), we can select an appropriate n and solve for the body's position.

Although Euler's equation provides a useful way to locate bodies of ideal shape, such as spheres and cylinders, the method has definite limitations when applied to more typical, distributed sources. In these more realistic cases, n may not be a constant with respect to depth and position of the source (Steenland [272]; LaFehr, as quoted by Steenland

[272]; Ravat [238]), because f is no longer simply a derivative of $\frac{1}{r}$ but rather an integration over the entire source distribution. Reid et al. [242] showed that anomalies over certain extended bodies, such as thin dipping sheets, do satisfy Euler's equation, but such is not the case in general. Ravat [238] concluded that the method is strictly valid only when the anomaly has constant attenuation rate with respect to distance from the source.

In spite of these theoretical limitations, Euler's method seems to provide useful results in practical applications. Thompson [280] applied the Euler method to profile data. In this application, the source distribution is assumed to be two dimensional, and the derivative with respect to y is not needed. The profile is divided into groups of points, and a source location (x_0, y_0) is calculated for each group using the previous equation and least-squares methods. Source locations, when plotted in cross section, tend to cluster around presumed magnetization contrasts. Rather than solving for n simultaneously, Thompson [280] recommended specifying various values of n. Reid et al. [242] and Marson and Klingele [176] expanded this general technique to two-dimensional magnetic and gravity surveys, respectively, over three-dimensional source distributions. In this case, the survey is divided into square, perhaps overlapping windows, and for a specified value of n, each window is analyzed for the location of a causative source.

It is clear that Euler's method requires not only the anomaly but also its gradient in three directions. We might be fortunate enough to have actual gradient measurements at our disposal (Hood [129]), but more likely the gradients would have to be calculated from anomaly measurements, a subject that is left until Section 12.2.

Werner Deconvolution

The magnetic field of a thin sheetlike body has a very simple form that directly depends in part on its location and depth. It is possible in principle to estimate the location of the top of the body and its magnetization from only four measurements of the total-field anomaly. A profile of total-field measurements over the body could be divided into groups of four or more measurements, each group providing an estimate of the source location. When locations are plotted in cross section, individual depth estimates tend to cluster around the true location of the sheetlike body. If the true geologic section can be modeled appropriately by a collection of such bodies, then analyzing numerous groups of consecutive points along the profile may indicate their locations.

This technique was discussed by Werner [291] and later implemented in a practical way by Hartman, Teskey, and Friedberg [120]. Groups of consecutive points are treated as a "window" sliding along the profile. The method is similar in some respects to *deconvolution* in seismic interpretation, where seismic waves are transformed into impulses representing individual reflectors. Hence the method is called *Werner deconvolution*.

Vertical dikes may seem like a highly specialized application. However, as shown by Hartman et al. [120] and Hansen and Simmonds [113], it is relatively straightforward to generalize the method to other hypothetical bodies. For example, the anomaly over a very thin dike is equivalent to the horizontal gradient of a semi-infinite half-space. (Proof of this relationship is left for the problem set at the end of this chapter.) Hence the horizontal gradient of the anomaly can be analyzed in precisely the same way, thereby providing a way to detect abrupt offsets in magnetic layers, as well as dikelike bodies. Klitgord and Behrendt [148] described an application of Werner deconvolution to a detailed aeromagnetic survey along the entire Atlantic seaboard of the United States. Ku and Sharp [150] combined the Werner deconvolution method with the Marquardt inversion scheme (discussed in a previous section) in order to analyze the shapes of two-dimensional magnetic sources. An algorithm to perform Werner deconvolution on total-field data was written by L. Cordell and R. Godson and is available in a form suitable for microcomputers (Cordell et al. [73]).

Ku and Sharp [150] provided an excellent discussion of Werner deconvolution, which is summarized here. A magnetic sheet of vanishing thickness is equivalent to a sheet of dipoles, as shown in Figure 10.12. To find the anomaly of a semi-infinite sheet, we begin with the equations for a single line of dipoles and integrate from depth d to infinity. The horizontal and vertical components of a horizontal line of dipoles are given by equations 5.18. Dipole moment per unit length $\mathbf{m}$ is equivalent to magnetization $\mathbf{M}$ times the cross-sectional area. We let the width of the sheet be Δx and let $\mathbf{m} = \mathbf{M}\,\Delta x\,dz$. The x and z components of the magnetic field due to a line of dipoles and observed at the origin then are given by

$$B_x = \frac{2C_{\mathrm{m}}M\Delta x}{(x'^2 + z'^2)^2}\left[\hat{M}_x(x'^2 - z'^2) + 2\hat{M}_z x' z'\right]\,dz',$$

$$B_z = \frac{2C_{\mathrm{m}}M\Delta x}{(x'^2 + z'^2)^2}\left[\hat{M}_z(z'^2 - x'^2) + 2\hat{M}_x x' z'\right]\,dz',$$

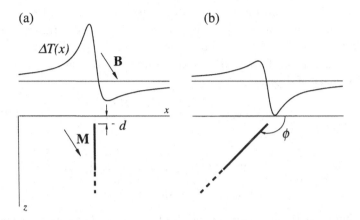

Fig. 10.12. Total-field anomalies over sheets of dipoles. (a) A single vertical sheet; (b) a single sheet with dip ϕ.

and the total-field anomaly of the line source is given approximately by

$$\Delta T_l = \hat{\mathbf{F}} \cdot \mathbf{B}$$

$$= \frac{2C_m M \Delta x}{(x'^2 + z'^2)^2} \left[\alpha(x'^2 - z'^2) + 2\beta x' z'\right] dz', \qquad (10.27)$$

where $\hat{\mathbf{F}}$ is a unit vector parallel to the ambient field and where

$$\alpha = \hat{F}_x \hat{M}_x - \hat{F}_z \hat{M}_z,$$

$$\beta = \hat{F}_x \hat{M}_z + \hat{F}_z \hat{M}_x.$$

To find the anomaly of a vertical sheet, we integrate equation 10.27 from depth d to infinity to get

$$\Delta T_s = 2C_m M \Delta x \int\limits_{d}^{\infty} \frac{\alpha x'^2 - \alpha z'^2 + 2\beta x' z'}{(x'^2 + z'^2)^2} \, dz'$$

$$= -2C_m M \Delta x \frac{\alpha d - \beta x'}{x'^2 + d^2}.$$

This equation remains valid under a rotation of the coordinate system. Rotating the x, z coordinate system by an angle θ produces a new coordinate system u, w, where the rotation is accomplished by letting

$$x = u \cos \theta + w \sin \theta,$$

$$z = -u \sin \theta + w \cos \theta.$$

But rotating the coordinate system by an angle θ is equivalent to rotating the thin sheet by an amount $-\theta$. Applying the coordinate transformation and letting the observation point be located at $(x, 0)$ rather than $(0, 0)$ provides the general equation for the total-field anomaly over a semi-infinite sheet of dipoles

$$\Delta T_s(x) = \frac{A(x - x') + Bd}{(x - x')^2 + d^2},\qquad(10.28)$$

where

$$A = -2C_m M \Delta x (\alpha \cos \phi + \beta \sin \phi),$$

$$B = 2C_m M \Delta x (-\alpha \sin \phi + \beta \cos \phi),$$

and where ϕ is the dip angle of the sheet below the horizontal, as shown in Figure 10.12.

For any given dip ϕ, equation 10.28 has four unknowns, A, B, x', and d, and these can be determined from four or more measurements of the total-field anomaly. Equation 10.28 can be rewritten as

$$x^2 \Delta T_i = a_0 + a_1 x + b_0 \Delta T_i + b_1 x \Delta T_i,$$

or in matrix notation as

$$x^2 \Delta T_i = \begin{bmatrix} 1 & x & \Delta T_i & x\Delta T_i \end{bmatrix} \begin{bmatrix} a_0 \\ a_1 \\ b_0 \\ b_1 \end{bmatrix},\qquad(10.29)$$

where ΔT_i is the ith point of the profile and

$$a_0 = -Ax' + Bd,$$

$$a_1 = A,$$

$$b_0 = -x'^2 - d^2,$$

$$b_1 = 2x'.$$

The four unknowns are contained within a_0, a_1, b_0, and b_1; that is,

$$x' = b_1/2,$$

$$d = \sqrt{-b_0 - \frac{b_1^2}{4}},$$

$$A = a_1,$$

$$B = \frac{2a_0 + a_1 b_1}{2d},$$

so solving for a_0, a_1, b_0, and b_1 provides an estimate of x', d, A, and B.

Four consecutive measurements of the total-field anomaly provide four equations like equation 10.29, or in matrix notation,

$$
\begin{bmatrix} x_i^2 \Delta T_i \\ x_{i+1}^2 \Delta T_{i+1} \\ x_{i+2}^2 \Delta T_{i+2} \\ x_{i+3}^2 \Delta T_{i+3} \end{bmatrix} = \begin{bmatrix} 1 & x_i & \Delta T_i & x_i \Delta T_i \\ 1 & x_{i+1} & \Delta T_{i+1} & x_{i+1} \Delta T_{i+1} \\ 1 & x_{i+2} & \Delta T_{i+2} & x_{i+2} \Delta T_{i+2} \\ 1 & x_{i+3} & \Delta T_{i+3} & x_{i+3} \Delta T_{i+3} \end{bmatrix} \begin{bmatrix} a_0 \\ a_1 \\ b_0 \\ b_1 \end{bmatrix}. \quad (10.30)
$$

Application of equation 10.30 to four consecutive measurements of the total-field anomaly over a single sheet of dipoles in principle should yield the four unknowns concerning the sheet, namely, x', d, A, and B. If the geologic section can be modeled appropriately by many such sheets, a four-point operator such as equation 10.30 can be moved through the anomaly profile, each operation potentially yielding the location of a hypothetical sheet. The resulting locations when plotted in cross section may assist the interpreter in identifying geologic structure.

A regional field could be determined simultaneously with unknowns a_0, a_1, b_0, and b_1. If the observed total-field anomaly is composed of both the effects of the thin sheet and a regional field given by $c_0 + c_1 x + c_2 x^2$, for example, then equation 10.30 could be modified to include three additional unknowns, namely, c_0, c_1, and c_2.

As mentioned earlier, the Werner deconvolution method can be extended by noting that the total-field anomaly over a thin sheet is equivalent to the horizontal gradient of the total-field anomaly over a semi-infinite half-space (see problem set). The method described here is typically applied to both the total-field anomaly and the horizontal gradient of the total-field anomaly (e.g., Ku and Sharp [150]). Hansen and Simmonds [113] discussed an extension of Werner deconvolution to polygonal sources. In their method, deconvolution is applied to the analytic signal of the anomaly, to be discussed in Chapter 12, rather than to the anomaly itself. Hansen and Simmonds [114] also discussed the similarities between Werner deconvolution and CompuDepth™,† another widely used algorithm developed by O'Brien [196].

In the previous discussion, it was assumed that consecutive points of the profile are used for each estimate, but this may not be appropriate. If consecutive points are too close together relative to the depth to the body, they may fall on relatively linear parts of the anomaly and lead to depth estimates that are too deep. On the other hand, if the consecutive

† CompuDepth™ is a registered trademark of TerraSense, Inc.

points are too far apart, they may sample more than one anomaly leading to erroneous results. In practice, profiles can be analyzed numerous times with successive decimation of the measured total-field profile.

10.3.3 Ideal Bodies

The previous sections of this chapter discussed a variety of techniques to find various details about source distributions from their gravity or magnetic fields. These methods sometimes take a cavalier approach in confronting the problems of nonuniqueness and model construction. The iterative methods, for example, iterate until the calculated anomaly satisfies some measure of goodness of fit; the model is deemed successful if it fits the geologic intuition of the modeler. Measures of goodness of fit can be misleading. The Taylor's series method, for example, tries to find the minimum squared error E^2 in a multi-dimensional parameter space $\mathbf{w}$ by changing the elements of $\mathbf{w}$ by small amounts. The method may converge rapidly to acceptably low levels of E^2, but there is no guarantee that $\mathbf{w}$ will seek the true rather than a local minimum of E^2. Moreover, there is no guarantee of uniqueness even if the fit is perfect. (The depth rules of the previous section are an exception; they provide limiting depths to the top of a body with only minor assumptions about the body.)

Rather than attempting to find detailed information about the causative source, we could look instead for fundamental properties that are common to the entire infinite set of reasonable sources. It should be clear that such fundamental properties are available; the total excess mass of a body, for example, can be found from its gravity anomaly using Gauss's law (Chapter 3) with no assumptions about the shape or density of the mass.† "Reasonable" is, of course, a subjective term. A reasonable mass distribution, for example, would have a density distribution that nowhere exceeds the greatest density contrast expected on the basis of rock property studies.

Parker [205, 206] discussed how to find the greatest lower bound on maximum density ρ_l consistent with a set of gravity measurements. If such a bound could be found, then every body that produces the observed anomaly must somewhere have a density contrast equal to or greater than ρ_l. This parameter is of some interest in itself. For

---

† But even this calculation has basic assumptions: The gravity anomaly must be isolated from the effects of all other sources, a difficult proposition in practical applications. See Grant and West [99] for further discussion on this calculation.

example, it might help the interpreter to determine the predominant rock type of the anomalous mass (Goodacre [98]). Unconsolidated sedimentary deposits have much different densities than granitic plutons, but both lithologies are usually less dense than pre-Cenozoic sedimentary and metamorphic rocks. A sedimentary basin within pre-Cenozoic crust typically produces a negative gravity anomaly; calculation of ρ_1 might help decide whether the gravity anomaly is caused entirely by the sedimentary basin or in part by underlying granitic rocks.

The primary application of ρ_1, however, is in determining other bounding parameters (Parker [205, 206]). For example, ρ_1 is a function of depth to source; the deeper that all bodies are constrained to lie, the greater must be the limiting density contrast. It stands to reason that ρ_1 must increase monotonically as the limiting depth increases. For a given gravity anomaly, a series of ρ_1 could be calculated for various source depths and plotted as shown in Figure 10.13. For any depth, the curve in Figure 10.13 provides the greatest lower bound on maximum density; that is, any source located at that depth, including the one true source, must somewhere have a density contrast at least as large as the corresponding

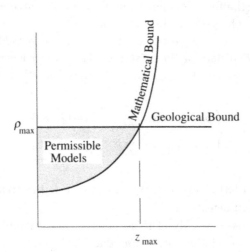

Fig. 10.13. Trade-off curve for ideal body. The curve represents the greatest lower bound on maximum density. The horizontal line represents the maximum density based on geologic constraints. The intersection of the curve and line provide the maximum depth to the body.

ρ_1. The concave region above this "trade-off curve" represents the range of acceptable solutions based on mathematical grounds.

Given some knowledge about the geologic setting, we might be able to specify the largest density contrast of any reasonable source causing the anomaly. This provides a *maximum* density based on geologic grounds, shown by the horizontal line in Figure 10.13. The two bounding conditions, one mathematical and the other geological, constrain the permissible depths and densities of all possible sources. In particular, they determine the maximum depth of any geologically reasonable source that can cause the anomaly.

To find ρ_1, we begin with gravity measurements at N discrete locations. Equation 10.5 provides

$$g_i = \int_V \rho(Q)\psi_i(Q)\,dv, \qquad i = 1, 2, \ldots, N, \qquad (10.31)$$

where g_i represents the anomaly at point i, V is the region within which the body is confined, $\rho(Q)$ is the density contrast at Q, and $\psi_i(Q)$ is the Green's function (the field at point i due to a point source at Q). The problem is to find the greatest lower bound on $\rho(Q)$ that satisfies the N constraint equations implied by equation 10.31. Parker [205] showed that a sufficient condition for the existence of a greatest lower bound is the existence of $N+1$ constants $\rho_0, \alpha_1, \alpha_2, \ldots, \alpha_N$ such that $\rho_0 > 0$ and

$$\rho(Q) = \begin{cases} \rho_0 & \text{where} \quad \sum_{i=1}^{N} \alpha_i\psi_i(Q) > 0, \\ 0 & \text{where} \quad \sum_{i=1}^{N} \alpha_i\psi_i(Q) \leq 0 \end{cases} \qquad (10.32)$$

satisfies equation 10.31. Moreover, the constant ρ_0 is the greatest lower bound on maximum density ρ_1.

The body composed entirely of this minimum density and satisfying equation 10.31 is called the *ideal body*. The ideal body is unique; for any higher density, there are infinitely many bodies that satisfy equation 10.31; for any lower density, there are none. The only assumption about the mass distribution is that density contrast must be of one sign. The shape of the ideal body most likely will not reflect the true shape of the source. This is not important; the ideal-body density is the only important property in this application.

If no restrictions are placed on the sign of $\rho(Q)$, then a sufficient condition is the existence of parameters ρ_0, α_1, $\alpha_2, \ldots, \alpha_N$ such that

$$\rho(Q) = \begin{cases} \rho_0 & \text{where} \quad \sum_{i=1}^{N} \alpha_i \psi_i(Q) > 0, \\ 0 & \text{where} \quad \sum_{i=1}^{N} \alpha_i \psi_i(Q) = 0, \\ \rho_0 & \text{where} \quad \sum_{i=1}^{N} \alpha_i \psi_i(Q) < 0 \end{cases}$$

satisfies equation 10.32.

As it turns out, the only analytical solutions for ideal bodies are for $N = 1$ and $N = 2$. To incorporate more measured data, we must settle for an approximation. Parker [206] described how to use the techniques of linear programming (e.g., Gass [93]). The source region is divided into many compartments similar to Figure 10.9, and a uniform density ρ_j is assigned to each compartment. The objective then is to minimize all ρ_j subject to the constraints

$$g_i = \sum_{j=1}^{M} \rho_j \psi_{ij} + e_i, \qquad i = 1, 2, \ldots, N,$$

$$-|e_{\max}| \le e_i \le |e_{\max}|, \qquad i = 1, 2, \ldots, N,$$

$$\rho_j \ge 0, \qquad j = 1, 2, \ldots, M,$$

where e_i, $i = 1, 2, \ldots, N$, are the errors associated with each measurement, and $e_{\max}$ is the maximum expected error. Minimizing the maximum value of ρ_j is equivalent to finding the smallest value of ρ_1 such that $\rho_j \le \rho_1$ for all j, a problem suitable for the methods of linear programming. Huestis and Ander [131] describe a Fortran algorithm to calculate ideal two-dimensional bodies, and Ander and Huestis [4] provide a Fortran algorithm for the three-dimensional case.

Figures 10.14 and 10.15 show an example of this method. A linear gravity gradient in the Cascade Range of the northwestern United States (Figure 10.14) was suspected of being caused by a magma chamber at mid-crustal depths (Blackwell et al. [19]). This hypothesis can be tested (Blakely [24]) by calculating trade-off curves for various profiles across the gravity gradient (Figure 10.15). If the maximum density contrast of partially melted crustal materials is, say, -300 kg·m^{-3}, then the trade-off curves indicate that the source can be no deeper than about 2 km (Figure 10.15). But this limiting depth is too shallow to be directly related to pervasive partial melting, and we can safely reject the hypothesis that the gradient is caused directly by partial melting.

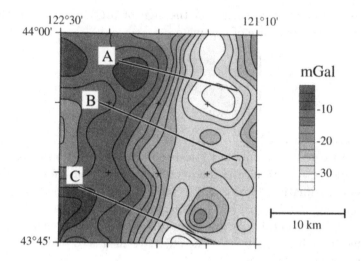

Fig. 10.14. Gravity gradient in the Cascade Range of the northwestern United States suspected to be caused by partial melting at mid-crustal depths. Profiles A, B, and C were analyzed by ideal-body theory, and the results are shown in Figure 10.15. Modified from Blakely [24].

In the previous discussion, ρ_l was found as a function of increasing depth to the top of the source region. We could also investigate the behavior of ρ_l as a function of the thickness or lateral extent of the source region, thereby constraining the size of all bodies that satisfy the gravity measurements. Ander and Huestis [4] discuss how this might be done for the gravity case, and Huestis and Parker [132] apply the method to the magnetic case. The steps are similar to the foregoing discussion: That is, choose a layer thickness, calculate ρ_l (or an intensity of magnetization) for that source region, and repeat for a range of thicknesses. At some minimum thickness, ρ_l will exceed the geologically acceptable value; this thickness is then the lower bound on all models that fit the anomaly. The source may be distributed in a thicker layer, but not entirely within a thinner layer.

In the gravity case, we minimize some scalar parameter of the source, namely, its density. The magnetic case presents a particular problem because we are dealing with a vector quantity. However, as discussed earlier, norms are scalar parameters that represent the "size" of vector quantities. We have various choices, as discussed by Huestis and

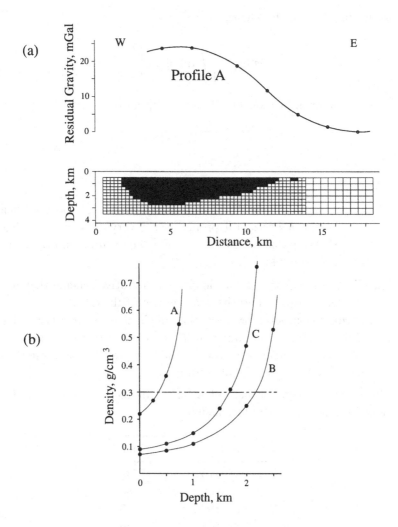

Fig. 10.15. Ideal-body theory applied to the gradient of Figure 10.14. (a) The ideal body for profile A assuming that the mass distribution is entirely below a depth of 0.5 km; (b) trade-off curves for the three profiles. If maximum density contrast is 0.3 g·cm^{-3}, the maximum depth is approximately 2 km. Modified from Blakely [24].

Parker [132]. Possible norms include the 1-norm

$$||m||_1 = \int\limits_V |\mathbf{m}|\, dv$$

used in this section, the 2-norm

$$\|m\|_2 = \left[\int_V \mathbf{m} \cdot \mathbf{m} \, dv \right]^{\frac{1}{2}}$$

discussed earlier in Section 10.2.2, or the sup norm

$$\|m\|_\infty = \max_{x,z} |\mathbf{m}| \, .$$

10.4 Problem Set

1. Suppose that you have N discrete measurements of the total-field anomaly along a horizontal profile over a very thin, vertical dike. Assume that you know the direction of magnetization and the depth to the top of the dike.

   (a) Write the equations and describe an iterative scheme that uses Taylor's series to find the depth extent of the dike.

   (b) Write a computer program that implements your scheme. Use Subroutine B.15 in Appendix B, or one like it, to calculate the total-field anomaly over various thin, vertical dikes and use these anomalies to test your algorithm.

2. The vertical magnetic field of a buried dipole is measured at five locations on the $z = 0$ plane as follows:

Coordinates, m	B_z, nT
(0,0,0)	1730
(−10,0,0)	385
(0,−10,0)	285
(0,10,0)	20
(10,0,0)	−75

   Assume that the dipole is located at coordinates (0,0,10) and calculate its inclination, declination, and magnitude in the sense of least squares.

3. Show that the following bodies satisfy the inequalities discussed in Section 10.3.2 of this chapter.

   (a) A sphere of radius a and uniform density ρ buried so that its center is at a depth d $(d > a)$.

(a)   (b)

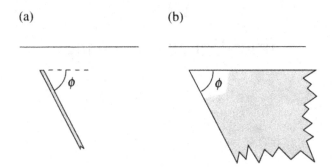

Fig. 10.16. (a) A very thin dike with dip ϕ; (b) a semi-infinite half space with dip ϕ.

  (b) A sphere of radius a and uniform magnetization **M**. Magnetization is induced in an ambient field with inclination I and declination D.

  (c) A horizontal cylinder of density ρ, radius a, and center at depth d $(d > a)$.

4. Consider total-field anomalies over the two bodies shown in Figure 10.16. Anomaly (A) is over a very thin dike, and anomaly (B) is over a semi-infinite half-space. Show that anomaly (A) is the horizontal gradient of anomaly (B).

11
Fourier-Domain Modeling

If your experiment needs statistics, you ought to have done a better experiment.

(Ernest Rutherford)

Fourier is a mathematical poem.

(W. Thompson and P. G. Tait)

The relationship between the dominant wavelengths of a potential-field anomaly and the size, depth, and shape of its causative source was a continuing theme of previous chapters. Figure 4.9, for example, shows that the width of an anomaly produced by a dipole is fundamentally related to the depth of the dipole. It stands to reason that Fourier analysis, a methodology that maps functions of space (or time) into functions of wavenumber (or frequency), might provide insights into the relationship between potential fields and causative sources. We will see, for example, that Fourier transformation of the gravity or magnetic anomaly caused by a layered source immediately separates the anomaly into two multiplicative factors: a function that describes the depth and thickness of the layer and a function that describes the distribution of density or magnetization within the layer. The source distribution can be determined in this case, at least in principle, by simply dividing the Fourier transform of the anomaly by a simple function that depends on the depth and thickness of the layer.

Tsuboi and Fuchida [282, 283] were apparently the earliest to apply Fourier analysis to the interpretation of potential-field anomalies; they used Fourier series to show the relationship between gravity anomalies and mass distributions, both two and three dimensional, confined to horizontal planes. In the 1960s, a number of authors used Fourier analysis in the interpretation of marine magnetic anomalies, notably

Gudmundsson [106, 107], Heirtzler and Le Pichon [122], and Neidell [195]. Harrison [116] has provided a good review of this general topic. At about the same time, Bhattacharyya published several key papers [14, 15, 16] on Fourier analysis of gravity and magnetic anomalies. Perhaps his most significant contribution in this regard [16] was his recognition that many operations, such as upward continuation and reduction to the pole, are relatively simple linear relationships in the Fourier domain.

We begin here with a brief review of Fourier transforms and linear systems analysis, and then derive the Fourier transforms of anomalies caused by simple sources, such as dipoles, monopoles, and line sources. These simple Fourier transforms provide a foundation for the treatment of more complicated magnetic and gravity sources.

11.1 Notation and Review

A brief review of Fourier analysis and linear systems analysis is presented here. For more complete discussions, the interested reader is referred to textbooks by Lee [162], Bracewell [42], and Papoulis [202].

11.1.1 Fourier Transform

Most textbooks devoted to Fourier analysis focus on functions of time. Consequently they employ time-related terms such as "frequency," "period," and "time domain." In this chapter, however, we will be concerned primarily with the analysis of functions of distance, such as the behavior of gravity or magnetic fields as measured on specified surfaces. To minimize confusion, we will use terms relevant to functions of distance (e.g., wavenumber, wavelength, space domain) and reserve time-related terms for when temporal analogies are required.

The Fourier series was discussed briefly in Chapter 6. In short, a periodic function can be synthesized by an infinite sum of weighted sinusoids, where the weights of the sinusoids are determined through analysis of the periodic function. For example, if $f(x)$ is a function that repeats itself over an interval X, it can be represented as

$$f(x) = \sum_{n=-\infty}^{\infty} F_n e^{ik_n x}, \qquad (11.1)$$

where $k_n = \frac{2\pi n}{X}$ and $i = \sqrt{-1}$. The weights F_n in this summation are complex numbers and can be determined with the integral

$$F_n = \frac{1}{X} \int\limits_{x_0}^{x_0+X} f(x) e^{-ik_n x} \, dx \,. \tag{11.2}$$

Exercise 11.1 Show that equation 11.1 and equation 6.1 are equivalent representations of $f(x)$, where $F_n = \frac{1}{2}(a_n - ib_n)$.

Now suppose that $f(x)$ does not repeat itself over a finite segment of the x axis. Instead we require that $f(x)$ be reasonably well behaved and have its variation confined to a finite-length segment of the x axis. In other words, we will require that

$$\int\limits_{-\infty}^{\infty} |f(x)| \, dx < \infty \,. \tag{11.3}$$

It stands to reason that gravity and magnetic anomalies satisfy this property if the survey extends well beyond the lateral extent of all causative bodies. Then letting $X \to \infty$ in equation 11.2 provides the Fourier transform of an aperiodic function $f(x)$,

$$F(k) = \int\limits_{-\infty}^{\infty} f(x) e^{-ikx} \, dx \,. \tag{11.4}$$

Exercise 11.2 Use Gauss's law to show that gravity and magnetic anomalies measured along a horizontal profile satisfy inequality 11.3, given that all sources are restricted to a localized region below the profile.

The variable k in equation 11.4 is called *wavenumber* and has units of inverse distance; it is analogous to angular frequency in time-domain Fourier transforms, which has units of inverse time. Wavenumber is inversely related to wavelength λ, that is,

$$k = \frac{2\pi}{\lambda} \,.$$

Notice from equation 11.4 that the Fourier transform of a function $f(x)$ evaluated at $k = 0$ is simply the average of $f(x)$ over the entire x axis, that is,

$$F(0) = \int\limits_{-\infty}^{\infty} f(x) \, dx \,.$$

The Fourier transform $F(k)$ is, in general, a complex function with real and imaginary parts, that is, $F(k) = \mathrm{Re}F(k) + i\,\mathrm{Im}F(k)$. It also can be written as

$$F(k) = |F(k)|e^{i\Theta(k)} ,$$

where

$$|F(k)| = \left[(\mathrm{Re}F(k))^2 + (\mathrm{Im}F(k))^2\right]^{\frac{1}{2}} ,$$

$$\Theta(k) = \arctan \frac{\mathrm{Im}F(k)}{\mathrm{Re}F(k)} .$$

The functions $|F(k)|$ and $\Theta(k)$ are called the amplitude and phase spectrum, respectively. The total energy of $f(x)$ is given by

$$E = \int_{-\infty}^{\infty} |F(k)|^2 dx ,$$

and $|F(k)|^2$ is called the *energy-density spectrum*.

Of particular importance is the fact that the Fourier transform has an inverse operation. Analogous to equation 11.1, the inverse Fourier transform is given by

$$f(x) = \frac{1}{2\pi} \int_{-\infty}^{\infty} F(k)e^{ikx} \, dk . \tag{11.5}$$

If $f(x)$ satisfies inequality 11.3, then the Fourier transform $F(k)$ exists and satisfies both equations 11.4 and 11.5.

The discussion thus far has dealt with a function of one variable, but the Fourier transform can be extended easily to functions of two variables. The Fourier transform of $f(x,y)$ and its inverse Fourier transform, for example, are given by

$$F(k_x, k_y) = \int_{-\infty}^{\infty} \int_{-\infty}^{\infty} f(x,y)e^{-i(k_x x + k_y y)} \, dx \, dy, \tag{11.6}$$

$$f(x, y) = \frac{1}{4\pi^2} \int_{-\infty}^{\infty} \int_{-\infty}^{\infty} F(k_x, k_y)e^{i(k_x x + k_y y)} \, dk_x \, dk_y , \tag{11.7}$$

where k_x and k_y are inversely related to wavelengths in the x and y directions, respectively: $k_x = 2\pi/\lambda_x$ and $k_y = 2\pi/\lambda_y$.

It is important to note that $f(x)$ and $F(k)$ are simply different ways of

looking at the same phenomenon. The Fourier transform *maps* a function
from one domain (space or time) into another domain (wavenumber or
frequency). Consequently, the following discussion will refer to the *space
domain* and the *Fourier domain* as two different frameworks to view the
same phenomenon.

In the remainder of this chapter and in the subsequent chapter, the
Fourier transform of a function $f(x)$ will sometimes be denoted by the
shorthand notation $\mathcal{F}[f]$, that is,

$$\mathcal{F}[f] = \int\limits_{-\infty}^{\infty} f(x)e^{-ikx}\,dx\,.$$

11.1.2 Properties of Fourier Transforms

The Fourier transform has a number of important properties that will
be particularly useful in later discussions and derivations. Some of these
properties are described subsequently. The proofs are straightforward
and left to the exercises. As a shorthand notation in the following,
$f(x) \leftrightarrow F(k)$ should be read "$f(x)$ has a Fourier transform given by
$F(k)$."

Symmetries

If $f(x) \leftrightarrow F(k)$, and if $f(x)$ is a real function, then $F(k)$ has a real
part that is symmetric and an imaginary part that is antisymmetric
about $k = 0$; that is, if $f(x)$ is real, then $F(k) = F^*(-k)$, where the
asterisk denotes complex conjugation, and the Fourier transform of a
real function is said to be *Hermitian*. Moreover, if $F(k) = F^*(-k)$,
then $f(x)$ must be a real function; that is, the Hermitian property is a
necessary and sufficient condition for $f(x)$ to be real.

Linearity

The Fourier transform is a linear operation. For example, if $f_1(x) \leftrightarrow F_1(k)$ and $f_2(x) \leftrightarrow F_2(k)$, then

$$[a_1 f_1(x) + a_2 f_2(x)] \leftrightarrow [a_1 F_1(k) + a_2 F_2(k)]\,,$$

where a_1 and a_2 are arbitrary constants.

Scaling

If $f(x) \leftrightarrow F(k)$, then

$$f(ax) \leftrightarrow \frac{1}{|a|} F\left(\frac{k}{a}\right),$$

where a is an arbitrary constant. This implies that the x interval containing most of the energy of $f(x)$ is inversely related to the bandwidth containing most of the energy of $F(k)$. In terms of gravity or magnetic anomalies, the scaling property shows that a broad anomaly will have a narrower amplitude spectrum than will a narrow anomaly. Because the width of an anomaly is directly related to the depth of its source, we can expect that the narrowness of a Fourier-transformed anomaly also will be related to depth of source.

Shifting

Shifting a function along the x axis in the space domain is equivalent to adding a linear phase factor to the function's Fourier transform; that is, if $f(x) \leftrightarrow F(k)$, then

$$f(x - x_0) \leftrightarrow F(k)e^{-ix_0 k}.$$

Note that the amplitude spectrum and the energy-density spectrum of $f(x)$ are unaffected by a shift of $f(x)$ along the x axis.

Differentiation

Differentiation of a function in the space domain is equivalent to multiplication by a power of wavenumber in the Fourier domain. For example, if $f(x) \leftrightarrow F(k)$, then

$$\frac{d^n}{dx^n} f(x) \leftrightarrow (ik)^n F(k). \tag{11.8}$$

If the function depends on two variables and if $f(x, y) \leftrightarrow F(k_x, k_y)$, then

$$\frac{\partial^n}{\partial x^n} \frac{\partial^m}{\partial y^m} f(x, y) \leftrightarrow (ik_x)^n (ik_y)^m F(k_x, k_y). \tag{11.9}$$

We will have much use for this theorem in subsequent discussions.

Exercise 11.3 Prove each of the preceding properties (symmetry, linearity, scaling, shifting, and differentiation). In each case, start with the defining equations 11.4 and 11.6.

11.1.3 Random Functions

Suppose $r(x)$ is a random function of x extending to infinity in both the $+x$ and $-x$ directions. By a random function, we mean that the value of $r(x)$ at any particular x is defined by a random process and cannot be precisely predicted in advance. For example, the magnetization of an infinitely extended, horizontal slab would constitute a random function if the magnetization at any x could be described by, say, a Gaussian probability distribution. In such cases, $r(x)$ does not satisfy inequality 11.3 and does not have a Fourier transform. Still we should be able to describe how the power of such a random function is distributed in the Fourier domain.

Wiener Theorem

To see how this might be done, we need the definition of the *autocorrelation function*. First consider a function $f(x)$ that *does* satisfy inequality 11.3 and does have a Fourier transform $F(k)$. The autocorrelation in this case is

$$\phi(x) = \int\limits_{-\infty}^{\infty} f(x + x')\, f(x')\, dx'$$

(Lee [162]), and the Fourier transform of the autocorrelation is given by $F(k)F^*(k) = |F(k)|^2$. Hence, if $f(x)$ has a Fourier transform, its energy-density spectrum and its autocorrelation are Fourier transform pairs.

This relationship between autocorrelation and the energy-density spectrum can be extended to random functions. Although the Fourier transform of a random function may not exist, the autocorrelation of the random function in many cases can be derived and will have a Fourier transform. Hence, we can find a representation for the energy of $r(x)$, analogous to the energy-density spectrum, by first calculating an autocorrelation appropriate for random functions,

$$\phi(x) = \lim_{X \to \infty} \frac{1}{2X} \int\limits_{-X}^{X} r(x')\, r(x + x')\, dx' \tag{11.10}$$

(Lee [162]), and then Fourier transforming the autocorrelation

$$\Phi(k) = \int\limits_{-\infty}^{\infty} \phi(x) e^{-ikx}\, dx\,. \tag{11.11}$$

This relationship is called the *Wiener theorem* for autocorrelation, and the function $\Phi(k)$ is called the *power-density spectrum* of the random function.

Ensemble Average

Of course, we never will have a complete description of $r(x)$ with which to solve completely equation 11.10, but there is another way to estimate the autocorrelation if we know its probabilistic behavior. To this end, assume that $r(x)$ describes a stationary process and let $P_r(\rho)$ describe the probability that $r(x)$ has the value ρ at any given x. Suppose first that we wanted to find the average value of $r(x)$. The average of $r(x)$ can be estimated in two ways: We could simply average $r(x)$ over a long section of x, or we could statistically average over all possible realizations of the random process. The equivalence of these two averages is expressed by

$$\lim_{X \to \infty} \frac{1}{2X} \int\limits_{-X}^{X} r(x)\, dx = \int\limits_{-\infty}^{\infty} \rho\, P_r(\rho)\, d\rho. \qquad (11.12)$$

The right-hand side of equation 11.12 is an *ensemble average* and expresses the expected value of $r(x)$.

Now note from equation 11.10 that the autocorrelation is simply a spatial average of $r(x)\, r(x' + x)$. We should be able to equate this to an ensemble average analogous to the right-hand side of equation 11.12. Lee [162] showed that if $P_{r_1,r_2}(\rho_1, \rho_2; \chi)$ is the joint probability that $r(x) = \rho_1$ at any x and $r(x) = \rho_2$ a distance χ away from x, then the autocorrelation is given by

$$\phi(\chi) = \int\limits_{-\infty}^{\infty} \int\limits_{-\infty}^{\infty} \rho_1\, \rho_2\, P_{r_1,r_2}(\rho_1, \rho_2; \chi)\, d\rho_1\, d\rho_2. \qquad (11.13)$$

Equations 11.10 and 11.13, therefore, show two ways to derive the autocorrelation of a random function. The first averages $r(x)\, r(x + x')$ over all space, the second is an average over all members of the ensemble. Most important, the power-density spectrum of a random function can be derived from a model of its probabilistic behavior through the use of equations 11.13 and 11.11.

Exercise 11.4 Assume that $r(x)$ is an uncorrelated function with an amplitude at any x predicted by a Gaussian normal distribution with zero mean. What is the autocorrelation and power-density spectrum of $r(x)$?

11.1.4 Generalized Functions

Earlier chapters have made much use of the impulse $\delta(x)$ function. In Section 2.3, for example, an impulse was used to represent a point mass in Poisson's equation. We have treated it much like an ordinary function, but it really is not a function in the usual sense. Because $\delta(x)$ has unit area and because $\delta(x) = 0$ when $x \neq 0$, the impulse is not only discontinuous but also has infinite magnitude at $x = 0$. Indeed, the impulse is defined in terms of other more ordinary functions, as in the definition

$$f(x_0) = \int\limits_{-\infty}^{\infty} f(x)\,\delta(x - x_0)\,dx, \qquad (11.14)$$

where $f(x)$ is an integrable function. In other words, when the impulse is employed as a weighting function for $f(x)$, it provides a value of $f(x)$ at a single value of x. It follows from equation 11.14 that

$$f(0) = \int\limits_{-\infty}^{\infty} f(x)\,\delta(x)\,dx, \qquad (11.15)$$

and

$$\int\limits_{-\infty}^{\infty} \delta(x)\,dx = 1.$$

It can be seen from these relationships that the Fourier transform of an impulse is given by

$$\mathcal{F}[\delta(x)] = \int\limits_{-\infty}^{\infty} \delta(x)e^{-ikx}\,dx$$

$$= 1.$$

11.1.5 Convolution

The convolution of two functions $f(x)$ and $g(x)$ is given by the integral

$$h(x) = \int\limits_{-\infty}^{\infty} f(x')\,g(x - x')\,dx', \qquad (11.16)$$

or, for two-dimensional functions,

$$h(x,y) = \int\limits_{-\infty}^{\infty} \int\limits_{-\infty}^{\infty} f(x',y')\, g(x - x', y - y')\, dx'\, dy'\,. \qquad (11.17)$$

The convolution integral has a simple physical meaning which can be seen by considering an electrical circuit. Suppose that we have the ability to submit a current to one part of this circuit and measure the resulting voltage at another part, as shown schematically in Figure 11.1. Suppose further that a current pulse of very short duration (i.e., an impulse) submitted to the circuit at time $t = 0$ results in a voltage $g(t)$. The response of the circuit to an impulsive input is called its *impulse response*. Electrical circuits are approximately linear systems, so it stands to reason that another impulse submitted t_0 seconds later would produce an identical response delayed by exactly t_0 seconds; that is, the voltage would be $g(t - t_0)$. A series of impulses at times $t_1, t_2, \ldots$ and weighted by $a_1, a_2, \ldots$ would result in the output $a_1 g(t - t_1) + a_2 g(t - t_2) + \cdots$. In other words, the response to a series of impulses is simply a linear combination of impulse responses. As the impulses are made arbitrarily close together, the input to the circuit becomes a continuously varying

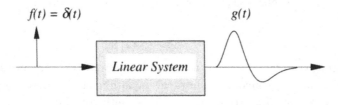

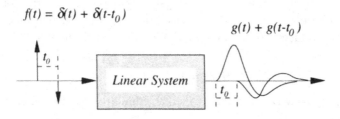

Fig. 11.1. An electrical circuit is approximately a linear system. If an impulsive input results in a response $g(t)$, then a series of impulses will produce a linear combination of $g(t)$.

current. It should be clear that, at this limit, the output voltage will be a continuous combination of impulse responses.

The convolution integral, equation 11.16, is the mathematical expression of this linear relationship between input and output, where $f(x)$ is the input and $h(x)$ is the output. Accordingly, if $f(x)$ is an impulse in equation 11.16, we see from the definition of an impulse (equation 11.14) that

$$\int_{-\infty}^{\infty} \delta(x')\, g(x - x')\, dx' = g(x)\,,$$

and $g(x)$ is the impulse response. It is also easy to show that if the impulse is displaced by a distance x_0 and amplified by a_0, the impulse response is similarly delayed and amplified,

$$\int_{-\infty}^{\infty} a_0 \delta(x' - x_0) g(x - x')\, dx' = a_0 g(x - x_0)\,,$$

which is exactly the property of the electrical circuit. In general, a series of weighted impulses produces an output consisting of a series of weighted impulse responses,

$$\int_{-\infty}^{\infty} (a_1 \delta(x' - x_1) + a_2 \delta(x' - x_2) + \cdots) g(x - x')\, dx'$$

$$= a_1 g(x - x_1) + a_2 g(x - x_2) + \cdots\,.$$

In the limit, the series of weighted impulses becomes a continuous function $f(x)$, and the convolution integral (equation 11.16) represents the continuous output $h(x)$ resulting from this continuous input.

Fourier-Convolution Theorem

A very important property of linear systems can be seen by transforming the convolution integral to the Fourier domain. If the convolution of $f(x)$ and $g(x)$ produces a function $h(x)$, the Fourier transform of $h(x)$ is given by

$$H(k) = F(k)\, G(k)\,, \qquad (11.18)$$

where $f(x) \leftrightarrow F(k)$ and $g(x) \leftrightarrow G(k)$. Similarly, for two-dimensional functions,

$$H(k_x, k_y) = F(k_x, k_y)\, G(k_x, k_y)\,. \qquad (11.19)$$

Exercise 11.5 Prove equation 11.18.

Thus convolution in the space (or time) domain transforms to multiplication in the Fourier domain.

This relationship between convolution in the space (or time) domain and multiplication in the Fourier domain is important for several reasons. First, suppose that $g(x)$ is the impulse response of a linear system; when $f(x)$ is submitted to the system, we can expect the output to be $h(x)$. Further suppose that $g(x)$ is known on theoretical grounds so that $G(k)$ can be derived. We might be very interested in finding the input to the system that resulted in some measured output, and equation 11.18 in principle provides a straightforward way to make this analysis: (1) Fourier transform $h(x)$ to get $H(k)$, (2) divide $H(k)$ by $G(k)$ to get $F(k)$, and (3) inverse Fourier transform $F(k)$ to get $f(x)$. Second, equation 11.18 (or equation 11.19) provides a straightforward description of how the wavenumber (or frequency) content of any input will be transformed by the linear system. The energy-density spectrum of any input will be multiplied by $|G(k)|^2$.

Just as the convolution of two functions of space has a Fourier transform given by the product of the Fourier transforms of the two functions, it is easy to show that a similar relation holds for convolution in the Fourier domain. If $f(x)$, $g(x)$, and $h(x)$ have Fourier transforms given by $F(k)$, $G(k)$, and $H(k)$, respectively, and if $h(x) = f(x) \, g(x)$, then

$$H(k) = \frac{1}{2\pi} \int\limits_{-\infty}^{\infty} F(k') \, G(k - k') \, dk' \, . \tag{11.20}$$

A similar relationship can be written for two-dimensional functions.

Exercise 11.6 Prove equation 11.20.

Parseval's Formula

The total energy of a real function $f(x)$ can be found by either integrating $f^2(x)$ over all space or by integrating the energy-density spectrum of $f(x)$ over all k. This relationship is expressed by *Parseval's formula*: If $f(x)$ is real and if $f(x) \leftrightarrow F(k)$, then

$$2\pi \int\limits_{-\infty}^{\infty} |f(x)|^2 \, dx = \int\limits_{-\infty}^{\infty} |F(k)|^2 \, dk \, .$$

Parseval's formula is easily derived from equation 11.20 by letting $g(x) = f(x)$.

11.1.6 Discrete Fourier Transform

The preceding discussion has treated convolution and Fourier transforms of continuous functions. In practice, we must deal with sampled data, and this limitation has profound effects on the kind of information available through Fourier analysis. Bracewell [42] provides a general review of this aspect of Fourier transforms, and Cordell and Grauch [69] and Ricard and Blakely [244] have discussed its limitations in the context of potential-field data. Appendix C briefly reviews sample theory; here we summarize the most important results.

The Fourier transform of sampled data is known as the *discrete Fourier transform*. It has limitations at both the longest and shortest wavelengths. It should be clear, for example, that wavelengths less than twice the sample interval cannot be represented adequately by the discrete Fourier transform. This limitation is expressed in the Fourier domain in an interesting way: The discrete Fourier transform is periodic with a period inversely proportional to the sample interval.

Consider N sequential samples of $f(x)$ evenly spaced at Δx intervals. If we assume that $f(x)$ is zero beyond these N samples, then we can consider N to be effectively infinite. In this case, the discrete Fourier transform $F_D(k)$ is related to the true Fourier transform $F(k)$ by the summation

$$F_D(k) = \frac{1}{\Delta x} \sum_{j=-\infty}^{\infty} F\left(k - \frac{2\pi j}{\Delta x}\right)$$

(see Appendix C). At any given k_0, we obviously would like for $F_D(k_0)$ to equal $F(k_0)$. Unfortunately, according to the previous equation $F_D(k_0)$ actually equals $F(k_0)$ plus $F(k)$ evaluated at an infinite number of other wavenumbers. This "self-contamination" is known as *aliasing*. The period of the discrete Fourier transform is $k_s = 2\pi/\Delta x$, and k_s is called the *sampling wavenumber*; half of the sampling wavenumber ($\pi/\Delta x$) is called the *Nyquist* wavenumber. Because the discrete Fourier transform repeats itself each $2\pi/\Delta x$, all unique information lies between $\pm\pi/\Delta x$. Hence, the Nyquist wavenumber is the largest wavenumber at our disposal. Note that it has a wavelength of twice the sample interval.

As we will see shortly, potential-field anomalies, like many physical phenomena, can be considered to be *band-limited*, that is, they have Fourier transforms that decay with increasing wavenumber. Hence, the contaminating high-wavenumber terms in the foregoing summation may be relatively small, especially if the sample interval is made sufficiently small relative to the significant wavelengths of $f(x)$. These principles are important considerations in developing a digitizing strategy.

Numerous algorithms are available to perform the discrete Fourier transform. Many of these employ a manipulation called *doubling* (Claerbout [60]) that makes them computationally efficient; such algorithms are called *fast Fourier transforms*. Appendix B provides two such subroutines. Subroutine B.16 is from Claerbout [60] and performs the one-dimensional discrete Fourier transform. Subroutine B.17 is from Press et al. [233] and performs multi-dimensional transforms.

11.2 Some Simple Anomalies

Using the principles of the previous section, we now can derive the Fourier transforms of potential fields caused by a variety of simple sources, such as dipoles, monopoles, lines, and ribbons. This effort will be well spent; these transforms will form the foundation for more complex gravity and magnetic sources, eventually leading to a wide variety of applications, including forward and inverse calculations, upward continuation, and depth-to-source estimations.

Let r describe the distance between point P located at (x, y, z) and point Q at point (x', y', z'). The Fourier transform of $1/r$ is the cornerstone of this discussion because potential fields depend on various derivatives of $1/r$. With the Fourier transform of $1/r$ in hand and with the aid of the differentiation theorem (Section 11.1.2), subsequent derivations will be relatively straightforward. Spector and Bhattacharyya [267] used the same sort of strategy in deriving energy-density spectra and autocorrelation functions for dipole and line-source anomalies.

We will confine P to a horizontal plane at height z_0 and, for the moment, consider Q to be fixed and located on the z axis at $(0, 0, z')$, where $z' > z_0$ (Figure 11.2). The two-dimensional Fourier transform of $1/r$ is given by

$$\mathcal{F}\left[\frac{1}{r}\right] = \int\limits_{-\infty}^{\infty} \int\limits_{-\infty}^{\infty} \frac{1}{\sqrt{x^2 + y^2 + (z_0 - z')^2}} e^{-i(k_x x + k_y y)} \, dx \, dy \, .$$

We can simplify this equation considerably by noting that the function $1/r$ is cylindrically symmetrical about the z axis and by converting the integral to polar coordinates. If we let

$$x = a \cos \theta,$$

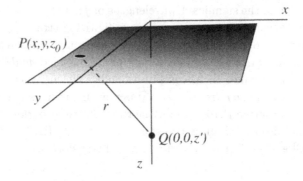

Fig. 11.2. Coordinate system for the derivation of Fourier transformed anomalies caused by point sources. Field is measured on a horizontal surface at z_0, and source is located on the z axis at z'.

$$y = a \sin \theta,$$

$$k_x = k \cos \phi,$$

$$k_y = k \sin \phi,$$

$$a = \sqrt{x^2 + y^2},$$

$$k = \sqrt{k_x^2 + k_y^2},$$

$$w = z_0 - z',$$

the two-dimensional Fourier transform of $1/r$ becomes

$$\mathcal{F}\left[\frac{1}{r}\right] = \int_0^{2\pi} \int_0^{\infty} \frac{1}{\sqrt{a^2 + w^2}} e^{-iak\cos(\theta - \phi)} \, a \, da \, d\theta$$

$$= \int_0^{\infty} \frac{1}{\sqrt{a^2 + w^2}} \left[\int_0^{2\pi} e^{-iak\cos\theta} \, d\theta \right] a \, da.$$

The integral over θ has the form of a zeroth-order Bessel function,

$$J_0(z) = \frac{1}{2\pi} \int_0^{2\pi} e^{-iz\cos\theta} \, d\theta,$$

Table 11.1. *Fourier transforms of anomalies caused by simple sources.*[a]

Type of Source	Fourier Transform	Eqn.
Vertical attraction of gravity		
Monopole (uniform sphere)	$2\pi\gamma\mu e^{\lvert k\rvert(z_0-z')}$	11.22
Vertical line	$\frac{2\pi\gamma\lambda}{\lvert k\rvert}e^{\lvert k\rvert z_0}\left(e^{-\lvert k\rvert z_1}-e^{-\lvert k\rvert z_2}\right)$	11.23
Horizontal line (uniform cylinder)	$2\pi\gamma\lambda e^{\lvert k\rvert(z_0-z')}$	11.27
Vertical ribbon	$\frac{2\pi\gamma\sigma}{\lvert k\rvert}e^{\lvert k\rvert z_0}\left(e^{-\lvert k\rvert z_1}-e^{-\lvert k\rvert z_2}\right)$	11.28
Total-field anomaly		
Dipole (uniform sphere)	$2\pi C_m m\,\Theta_m\Theta_f\lvert k\rvert e^{\lvert k\rvert(z_0-z')}$	11.25
Vertical line	$2\pi C_m m'\,\Theta_m\Theta_f e^{\lvert k\rvert z_0}\left(e^{-\lvert k\rvert z_1}-e^{-\lvert k\rvert z_2}\right)$	11.26
Horizontal line (uniform cylinder)	$2\pi C_m m'\,\Theta_m'\Theta_f'\lvert k\rvert e^{\lvert k\rvert(z_0-z')}$	11.29
Vertical ribbon	$2\pi C_m m''\,\Theta_m'\Theta_f' e^{\lvert k\rvert z_0}\left(e^{-\lvert k\rvert z_1}-e^{-\lvert k\rvert z_2}\right)$	11.30

[a]Note: Observation surface is $z = z_0$. Point sources are located at $(0,0,z')$; vertical lines and ribbons extend from $(0,0,z_1)$ to $(0,0,z_2)$. μ is mass, λ is mass per unit length, and σ is mass per unit area; m is dipole moment, m' is dipole moment per unit length, and m'' is dipole moment per unit area. See text for definitions of Θ_m, Θ_f, Θ_m', and Θ_f'. In each case, $z_0 < z'$, $z_0 < z_1$, and $z_1 < z_2$.

and making this substitution into the Fourier transform produces a Hankel transform of zeroth order

$$\mathcal{F}\left[\frac{1}{r}\right] = 2\pi \int\limits_0^\infty \frac{1}{\sqrt{a^2 + w^2}}\, J_0(ak)\, a\, da\,.$$

This integral illustrates a general result: The two-dimensional Fourier transform of a cylindrically symmetrical function reduces to a Hankel transform. The solution of this particular Hankel transform is given by Bracewell [42]:

$$\mathcal{F}\left[\frac{1}{r}\right] = 2\pi\frac{e^{\lvert k\rvert(z_0-z')}}{\lvert k\rvert}, \qquad z' > z_0,\ \lvert k\rvert \neq 0\,. \tag{11.21}$$

Bhattacharyya [15] found the same expression using a different derivation. From equation 11.21, we can easily derive the Fourier transform of the potential fields caused by a number of simple sources. The derivations follow, and the important results are repeated in Table 11.1.

We should note in passing that, although we have found a suitable expression for $\mathcal{F}\left[\frac{1}{r}\right]$, $1/r$ does not satisfy inequality 11.3 because the integral

$$\int\limits_{-\infty}^{\infty} \int\limits_{-\infty}^{\infty} \frac{dx\,dy}{\sqrt{x^2 + y^2 + (z_0 - z')^2}} = \int\limits_{0}^{2\pi} \int\limits_{0}^{\infty} \frac{a\,da\,d\theta}{\sqrt{a^2 + (z_0 - z')^2}}$$

$$= 2\pi \int\limits_{0}^{\infty} \frac{a\,da}{\sqrt{a^2 + (z_0 - z')^2}}$$

is not finite. Hence, the Fourier transform of $1/r$ does not exist in the rigorous sense of inequality 11.3, a fact expressed in equation 11.21 by the undefined nature of $\mathcal{F}\left[\frac{1}{r}\right]$ at infinite wavelength ($|k| = 0$). We can take some solace in the fact that any spatial derivative of $1/r$ (and therefore gravity and magnetic anomalies in general) does satisfy inequality 11.3.

Exercise 11.7 Show that any horizontal derivative of $1/r$ has a Fourier transform.

11.2.1 Three-Dimensional Sources

In this section and the next, the Fourier transforms of several simple two- and three-dimensional sources will be discussed. The terminology herein can be a source of confusion. The anomaly of a three-dimensional body (e.g., a sphere) is appropriately measured on a two-dimensional surface, and to study the anomaly in the Fourier domain requires a two-dimensional Fourier transform. Likewise, a two-dimensional source (e.g., a cylinder) is measured along a profile that requires the one-dimensional Fourier transform. We will continue to refer to sources as either two- or three-dimensional in accordance with their geometries, with the understanding that their anomalies are analyzed in the Fourier domain in either one or two dimensions, respectively.

Monopole

Consider the Fourier transform of the gravitational potential observed on a horizontal plane at $z = z_0$ and caused by a point mass (equivalent of course to a spherical mass with uniform density) located below the plane. The Fourier transform of this potential can be written immediately from equation 11.21. The gravitational potential of a point mass μ is given by

$U = \gamma\mu/r$, where γ is the gravitational constant; the Fourier transform of this potential observed on a horizontal plane is simply

$$\mathcal{F}[U] = \gamma\mu\mathcal{F}\left[\frac{1}{r}\right]$$

$$= 2\pi\gamma\mu\frac{e^{|k|(z_0-z')}}{|k|}, \qquad z' > z_0.$$

Gravitational acceleration **g** is related to the potential by the equation **g** $= \nabla_P U$, so any component of **g** is simply a directional derivative of U. In particular, the vertical attraction of gravity due to a point mass is the vertical derivative of $\gamma\mu/r$, that is,

$$g_z = \gamma\mu\frac{\partial}{\partial z}\frac{1}{r}.$$

Observed on a horizontal plane, this field has a Fourier transform given by

$$\mathcal{F}[g_z] = \gamma\mu\mathcal{F}\left[\frac{\partial}{\partial z}\frac{1}{r}\right]$$

$$= \gamma\mu\frac{\partial}{\partial z}\mathcal{F}\left[\frac{1}{r}\right]$$

$$= 2\pi\gamma\mu e^{|k|(z_0-z')}, \qquad z' > z_0. \tag{11.22}$$

A number of important characteristics of gravity anomalies can be seen from equation 11.22 and are illustrated by Figure 11.3(a). Maximum energy of the gravitational field occurs at $|k| = 0$, and the value of the energy-density spectrum at $|k| = 0$ is proportional to the total mass; the proof of this assertion is left to the problem set at the end of this chapter. Energy decreases exponentially with increasing wavenumber; that is, the energy at each wavelength dominates the energy at all shorter wavelengths. Moreover, the rate of decrease in energy with respect to wavenumber depends on the depth to the mass; the deeper the mass, the less significant are short wavelengths as compared with longer wavelengths of the anomaly. In other words, equation 11.22 and Figure 11.3 show that the gravity anomaly is approximately band limited; although all wavenumbers contribute to the anomaly, the largest wavenumbers are relatively insignificant.

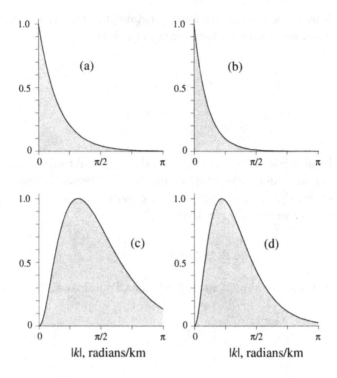

Fig. 11.3. Energy-density spectra of anomalies caused by some simple bodies. (a) Monopole at depth 1 km; (b) vertical line mass with top at 1 km and bottom at 2 km; (c) dipole at depth 1 km; (d) vertical line of dipoles with top at 1 km and bottom at 2 km. Vertical axes are normalized to one.

Vertical Line Mass

Consider the gravitational attraction observed on a horizontal plane and caused by a vertical wire extending along the z axis from $(0, 0, z_1)$ to $(0, 0, z_2)$, where $z_2 > z_1$, as shown by Figure 11.4. Let the mass of one element of the wire be $\mu = \lambda \, dz$, where λ is mass per unit length. The Fourier transform of this field is found by integrating equation 11.22 along the z axis from z_1 to z_2,

$$\mathcal{F}[g_z] = 2\pi\gamma\lambda \int_{z_1}^{z_2} e^{|k|(z_0 - z')} \, dz'$$

$$= \frac{2\pi\gamma\lambda}{|k|} e^{|k|z_0} \left(e^{-|k|z_1} - e^{-|k|z_2} \right), \quad z_2 > z_1, \; z_1 > z_0 . \quad (11.23)$$

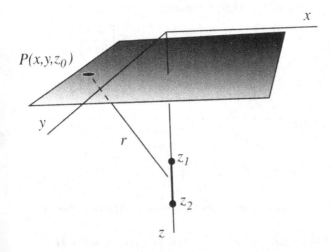

Fig. 11.4. Vertical line source located along the z axis between z_1 and z_2 and observed on a horizontal surface $z_0 < z_1$.

Figure 11.3(b) shows the energy-density spectrum of this Fourier transform; it has a shape very similar to that of an isolated point mass.

Exercise 11.8 Notice in Figure 11.3 that the energy of the field caused by a vertical wire decays more rapidly with increasing k than for a point mass. Why is this so?

Magnetic Dipole

The magnetic potential of a dipole with dipole moment $\mathbf{m} = m\,\hat{\mathbf{m}}$ is given by equation 4.13,

$$V = -C_{\mathrm{m}}\mathbf{m} \cdot \nabla_P \frac{1}{r}$$

$$= -C_{\mathrm{m}}m\left(\hat{m}_x\frac{\partial}{\partial x}\frac{1}{r} + \hat{m}_y\frac{\partial}{\partial y}\frac{1}{r} + \hat{m}_z\frac{\partial}{\partial z}\frac{1}{r}\right),$$

where C_{m} is a constant as discussed in Chapter 4. This is, of course, identical to the potential of a uniformly magnetized sphere with magnetization equal to $\mathbf{m}$ divided by the volume of the sphere. The Fourier transform of this potential, as observed on a horizontal plane (Figure 11.2), is given by application of equation 11.9 to equation 11.21:

$$\mathcal{F}[V] = -C_{\mathrm{m}}m\left(\hat{m}_x\mathcal{F}\left[\frac{\partial}{\partial x}\frac{1}{r}\right] + \hat{m}_y\mathcal{F}\left[\frac{\partial}{\partial y}\frac{1}{r}\right] + \hat{m}_z\mathcal{F}\left[\frac{\partial}{\partial z}\frac{1}{r}\right]\right)$$

$$= -C_{\mathrm{m}}m\left(\hat{m}_x ik_x\mathcal{F}\left[\frac{1}{r}\right] + \hat{m}_y ik_y\mathcal{F}\left[\frac{1}{r}\right] + \hat{m}_z\frac{\partial}{\partial z}\mathcal{F}\left[\frac{1}{r}\right]\right)$$

$$= -2\pi C_{\mathrm{m}}m\,\Theta_{\mathrm{m}}\,e^{|k|(z_0-z')}, \qquad z' > z_0, \tag{11.24}$$

where

$$\Theta_{\mathrm{m}} = \hat{m}_z + i\frac{\hat{m}_x k_x + \hat{m}_y k_y}{|k|}$$

is a complex function of k_x and k_y that depends only on the orientation of the dipole.

The magnetic field is related to the potential by the equation $\mathbf{B} = -\nabla_P V$, so any component of $\mathbf{B}$ can be found by deriving a directional derivative of V. For example, the total-field anomaly is given approximately by

$$\Delta T = -\hat{\mathbf{f}} \cdot \nabla_P V$$

$$= -\hat{f}_x\frac{\partial}{\partial x}V - \hat{f}_y\frac{\partial}{\partial y}V - \hat{f}_z\frac{\partial}{\partial z}V,$$

where $\hat{\mathbf{f}}$ is the unit vector parallel to the ambient field. Therefore, observed on a horizontal plane, the Fourier transform of the total-field anomaly is

$$\mathcal{F}[\Delta T] = -\hat{f}_x\mathcal{F}\left[\frac{\partial}{\partial x}V\right] - \hat{f}_y\mathcal{F}\left[\frac{\partial}{\partial y}V\right] - \hat{f}_z\mathcal{F}\left[\frac{\partial}{\partial z}V\right]$$

$$= -i\hat{f}_x k_x\mathcal{F}[V] - i\hat{f}_y k_y\mathcal{F}[V] - \hat{f}_z\frac{\partial}{\partial z}\mathcal{F}[V],$$

and combining this equation with equation 11.24 yields

$$\mathcal{F}[\Delta T] = 2\pi C_{\mathrm{m}}m\,\Theta_{\mathrm{m}}\Theta_{\mathrm{f}}\,|k|\,e^{|k|(z_0-z')}, \qquad z' > z_0, \tag{11.25}$$

where

$$\Theta_{\mathrm{f}} = \hat{f}_z + i\frac{\hat{f}_x k_x + \hat{f}_y k_y}{|k|}.$$

Exercise 11.9 As discussed in Section 11.1.2, the Fourier transform of any real function must be Hermitian. Show that this is indeed the case for the Fourier transform of the total-field anomaly of a dipole.

In equation 11.25, note that the orientations of vectors **m** and **f** are contained entirely within Θ_m and Θ_f, respectively, whereas the depth of the dipole is contained exclusively within the exponential term. This separation of source distribution and source location into multiplicative factors is an important attribute of Fourier-transformed potential fields that will be exploited later in this chapter.

The functions Θ_m and Θ_f behave in interesting ways. Although they both are variables of k_x and k_y, they assume constant values along any ray projected from the origin. This can be seen by converting k_x and k_y to polar coordinates. In general, the value along each ray differs from neighboring rays, and rays on opposite sides of the origin are complex conjugates of one another. Consequently, the imaginary parts of Θ_m and Θ_f are discontinuous through the origin. The average of Θ_m and Θ_f along any circle concentric about the origin is given by

$$\frac{1}{2\pi} \int_0^{2\pi} \Theta_m \, d\phi = \hat{m}_z$$

and

$$\frac{1}{2\pi} \int_0^{2\pi} \Theta_f \, d\phi = \hat{f}_z \,,$$

respectively. Hence, although Θ_m and Θ_f are not radially symmetric in general, they have average values along any concentric circle that are independent of the radius of the circle. It follows that the shape of the amplitude spectrum (or energy-density spectrum) of the anomaly is identical, to within a multiplicative constant, when viewed along any ray projected from the origin, and the shape of the spectrum along any ray is proportional to the radially averaged spectrum. Consequently, the shape of the amplitude spectrum as a function of $|k|$ depends only on the exponential term of equation 11.25 which in turn depends only on the depth of the dipole; the shape is independent of the orientation of the dipole and the orientation of the ambient field.

The energy-density spectrum of the total-field anomaly over a vertical dipole is shown in Figure 11.3(c). Unlike point masses, the maximum energy of the total-field anomaly does not occur at $|k| = 0$, but rather at a value of $|k|$ that depends on the depth of the dipole.

Exercise 11.10 At what wavenumber, in terms of z_0 and z', does the maximum of equation 11.25 occur? What does this imply about the shape of the anomaly as a function of depth to the dipole?

At wavenumbers greater than that maximum, the energy-density spectrum decays monotonically and approaches exponential decay at high wavenumbers. Hence, as for the gravity case, any given wavenumber greater than the maximum wavenumber will dominate all higher wavenumbers. Note that the energy-density spectrum of the dipole anomaly falls off much slower with increasing $|k|$ than does the monopole anomaly. This is to be expected: The field caused by a dipole has shorter wavelengths than the field caused by a monopole.

The fact that the energy-density spectrum approaches zero at $|k| = 0$ is a reflection of Gauss's law for magnetic sources. As discussed in Section 11.1.1, the value of the Fourier transform at $|k| = 0$ equals the average of the space-domain function over all x and y, and Gauss's law tells us that this average must be zero for a magnetic anomaly caused by any localized source.

Vertical Line of Dipoles

To find the Fourier transform of the anomaly over a vertical line of dipoles, we simply integrate equation 11.25 along the z axis. Let the top of the line source be at $(0, 0, z_1)$ and the bottom at $(0, 0, z_2)$, as in Figure 11.4, and let each element of the line have a dipole moment given by $\mathbf{m} = \mathbf{m}' dz$, where $\mathbf{m}'$ is dipole moment per unit length. Then the Fourier transform of the total-field anomaly is given by

$$\mathcal{F}[\Delta T] = 2\pi C_m m' \,\Theta_m \Theta_f |k| \int_{z_1}^{z_2} e^{|k|(z_0 - z')} \, dz'$$

$$= 2\pi C_m m' \,\Theta_m \Theta_f \, e^{|k| z_0} \left(e^{-|k| z_1} - e^{-|k| z_2} \right), \quad z_2 > z_1, \; z_1 > z_0 \,.$$

$$(11.26)$$

The energy-density spectrum of this Fourier transform is shown in Figure 11.3(d).

Exercise 11.11 Consider the vertical magnetic field observed on a horizontal plane and caused by a vertically magnetized, vertical wire that extends from z_1 to infinite depth. Show that the Fourier transform of this field is proportional to the Fourier transform of a gravity anomaly observed on the same plane and caused by a point mass located at $z = z_1$. Explain why this is so.

11.2.2 Two-Dimensional Sources

For infinitely extended line sources, equivalent to wires or uniform cylinders, we can consider just single profiles above and perpendicular to the line source, as shown by Figure 11.5. The one-dimensional Fourier transform of the profile is appropriate in these cases, but the derivations and results are very similar to the previous section.

Horizontal Line Mass

The vertical attraction at (x, z_0) of a wire mass (equivalent to a uniform horizontal cylinder) infinitely extended parallel to the y axis and passing through the z axis at $z = z'$ (Figure 11.5) is given by equation 3.17,

$$g_z(x) = -2\gamma\lambda\frac{(z_0 - z')}{r^2},$$

where λ is mass per unit length and $r = \sqrt{x^2 + (z_0 - z')^2}$ is the perpendicular distance from the wire to the observation point. This function has a Fourier transform given by

$$\mathcal{F}[g_z] = -2\gamma\lambda(z_0 - z')\mathcal{F}\left[\frac{1}{r^2}\right]$$

$$= 2\pi\gamma\lambda\, e^{|k|(z_0 - z')}, \qquad z_0 < z'. \tag{11.27}$$

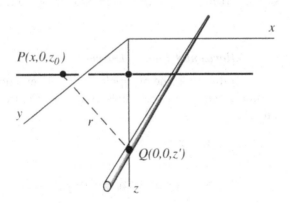

Fig. 11.5. Coordinate system and geometrical arrangement of line sources as used for derivations of the Fourier transforms of their anomalies. Line source is parallel to the y axis and intersects the z axis at $z = z'$. Field is measured along a horizontal line above and perpendicular to the line source.

Note the similarity between equations 11.27 and 11.22. In fact, the radial energy-density spectrum of the field caused by a monopole, shown in Figure 11.3(a), is identical in shape to that of an infinitely extended, horizontal wire mass. All of the comments made earlier, concerning the Fourier-domain characteristics of the anomaly caused by an isolated monopole, also hold for the anomaly caused by a wire.

Vertical Ribbon Mass

A vertical ribbon infinitely extended in the horizontal direction is equivalent to a vertical stack of horizontal wires all of infinite length. Let the ribbon extend from z_1 to z_2 in the vertical direction and to infinity in the $+y$ and $-y$ directions. The field is observed along a single line above and perpendicular to the ribbon. The Fourier transform of the vertical gravity field can be found by integrating equation 11.27 along the z axis:

$$\mathcal{F}[g_z] = 2\pi\gamma\sigma \int_{z_1}^{z_2} e^{|k|(z_0 - z')} dz'$$

$$= \frac{2\pi\gamma\sigma}{|k|} e^{|k|z_0}(e^{-|k|z_1} - e^{-|k|z_2}), \qquad z_1 < z_2,\ z_2 < z_0, \quad (11.28)$$

where σ is mass per unit area of the ribbon. The energy-density spectrum of the field caused by this ribbon source is identical in shape to the energy-density spectrum of a vertical wire extending from z_1 to z_2 (Figure 11.3(b)).

Horizontal Line of Dipoles

The potential at (x, z_0) due to a line of dipoles infinitely extended parallel to the y axis and intersecting the z axis at $z = z'$ (Figure 11.5) is given by equation 5.14

$$V = 2C_{\mathrm{m}} \frac{\mathbf{m}' \cdot \mathbf{r}}{r^2}$$

$$= 2C_{\mathrm{m}} m'(\hat{m}_x x + \hat{m}_z(z_0 - z')) \frac{1}{r^2},$$

where $\mathbf{m}' = m'\hat{\mathbf{m}}$ is dipole moment per unit length. Note that vector $\mathbf{r}$ lies within the x, z plane. This potential is, of course, identical to the potential of a uniformly magnetized cylinder with axis at $z = z'$. The Fourier transform of the potential, as observed along a horizontal line above and perpendicular to the source, is given by

$$\mathcal{F}[V] = 2C_{\mathrm{m}}m' \left(\hat{m}_x \mathcal{F} \left[\frac{x}{r^2} \right] + \hat{m}_z(z_0 - z')\mathcal{F} \left[\frac{1}{r^2} \right] \right)$$

$$= -2\pi C_{\mathrm{m}}m'\Theta_{\mathrm{m}}m'e^{|k|(z_0-z')} ,$$

where $\Theta'_{\mathrm{m}} = \hat{m}_z + i\hat{m}_x \operatorname{sgn} k$. The total-field anomaly is given by

$$\Delta T = -\hat{\mathbf{f}} \cdot \nabla_P V$$

$$= -\hat{f}_x \frac{\partial V}{\partial x} - \hat{f}_z \frac{\partial V}{\partial z} ,$$

and the Fourier transform of the anomaly is found by application of equation 11.8 to the previous equation,

$$\mathcal{F}[\Delta T] = 2\pi C_{\mathrm{m}}m' \, \Theta'_{\mathrm{m}}\Theta'_{\mathrm{f}} \, |k| \, e^{|k|(z_0-z')} , \qquad (11.29)$$

where $\Theta'_{\mathrm{f}} = \hat{f}_z + i\hat{f}_x \operatorname{sgn} k$.

Again note the similarity between this result and the Fourier transform of the anomaly caused by a single dipole (equation 11.25). Hence, Figure 11.3(c) is applicable to the magnetized wire, and many of the comments made earlier for isolated dipoles also pertain to the magnetized wire.

Vertical Ribbon of Dipoles

Finally, we can use equation 11.29 to find the Fourier transform of the total-field anomaly caused by a vertical magnetic ribbon extending infinitely in the $+y$ and $-y$ directions and from z_1 to z_2 in the vertical direction. We simply integrate along the z axis from z_1 to z_2,

$$\mathcal{F}[\Delta T] = 2\pi C_{\mathrm{m}}m'' \, \Theta'_{\mathrm{m}}\Theta'_{\mathrm{f}}e^{|k|z_0}(e^{-|k|z_1} - e^{-|k|z_2}) , \qquad (11.30)$$

where m'' is dipole moment per unit area of the ribbon. As before, the energy-density spectrum of the total-field anomaly caused by a magnetic ribbon is identical in form to that of a vertical line of dipoles (Figure 11.3(d)).

11.3 Earth Filters

The previous section was motivated by more than just an interest in anomalies caused by simple sources. It also will provide the groundwork to apply Fourier transforms to more general gravity and magnetic sources, eventually leading to both forward and inverse calculations and other applications. The keystone in this discussion is the Fourier-convolution theorem of Section 11.1.5.

Equation 10.5 shows the relationship between a potential field f and the distribution of source material s:

$$f(P) = \int_R s(Q)\,\psi(P,Q)\,dv\,, \qquad (11.31)$$

where R is the region occupied by source material, P is the point of observation, Q is one point of the distribution, and $\psi(P,Q)$ is the Green's function. Recall that the Green's function depends on the geometrical placement of P and Q and is simply the potential field at point P due to a single element of the source located at Q. If $f(P)$ represents vertical attraction of gravity, for example, then $s(Q)$ is density and $\psi(P,Q)$ is the vertical attraction at P due to a monopole at Q. If $f(P)$ represents the total-field anomaly, then $s(Q)$ is magnetization and $\psi(P,Q)$ is the total-field anomaly of a single dipole.

Now suppose that the source distribution is confined to a horizontal layer with top at z_1 and bottom at z_2. As usual, we will orient the z axis down so that $z_1 < z_2$. We also will require the source distribution to vary in only the x and y directions, which will cause $s(Q)$ to be constant along any vertical line through the layer. With these restrictions, equation 11.31 becomes

$$f(x,y,z) = \int_{-\infty}^{\infty} \int_{-\infty}^{\infty} \int_{-\infty}^{\infty} s(x',y')\,\psi(x-x',y-y',z-z')\,dx'\,dy'\,dz'$$

$$= \int_{-\infty}^{\infty} \int_{-\infty}^{\infty} s(x',y') \int_{z_1}^{z_2} \psi(x-x',y-y',z-z')\,dz'\,dx'\,dy'$$

$$= \int_{-\infty}^{\infty} \int_{-\infty}^{\infty} s(x',y')\,\xi(x-x',y-y')\,dx'\,dy'\,, \qquad (11.32)$$

which is the two-dimensional convolution integral. The Green's function ξ in this case represents the field at (x,y) due to a single element of the layer, namely, a vertical line element extending from (x',y',z_1) to (x',y',z_2).

We now assume that the field is measured on a horizontal plane at height $z = z_0$, where $z_0 < z_1$, and Fourier transform both sides of equation 11.32. The Fourier-convolution theorem leads to

$$\mathcal{F}[f] = \mathcal{F}[s]\,\mathcal{F}[\xi]\,; \qquad (11.33)$$

that is, the Fourier transform of the potential field is equal to the Fourier transform of the source distribution multiplied by $\mathcal{F}[\xi]$, the Fourier transform of the Green's function. In the previous section, several Fourier transforms were derived that will serve for $\mathcal{F}[\xi]$. Equation 11.23, for example, describes the Fourier transform of the vertical attraction of a vertical line, and substituting into equation 11.33 yields

$$\mathcal{F}[g_z] = \mathcal{F}[\rho]\left\{\frac{2\pi\gamma}{|k|}\,e^{|k|z_0}\left(e^{-|k|z_1} - e^{-|k|z_2}\right)\right\}, \qquad z_0 < z_1, \quad z_1 < z_2,$$

$$(11.34)$$

where ρ is the density of the slab, a function of x and y only. Hence, in the Fourier domain, the vertical attraction of a horizontal layer is equal to two multiplicative factors: the Fourier transform of the density and a function that depends on the depth and thickness of the layer.

Exercise 11.12 Show that equation 11.34 reduces to the "infinite slab formula," $g_z = 2\pi\gamma\rho(z_2 - z_1)$, if z_1, z_2, and ρ are constants. Hint: $f(x)\,\delta(x) = f(0)\,\delta(x)$.

For the total-field anomaly, $\mathcal{F}[\xi]$ comes from equation 11.26, and equation 11.33 becomes

$$\mathcal{F}[\Delta T] = \mathcal{F}[M]\left\{2\pi C_m \Theta_m \Theta_f\, e^{|k|z_0}\left(e^{-|k|z_1} - e^{-|k|z_2}\right)\right\},$$

$$z_0 < z_1, \quad z_1 < z_2, \qquad (11.35)$$

where M is magnetization, a function of x and y only. As before, the Fourier transform has separated the total-field anomaly into two factors: the magnetization and a function that depends on other attributes of the layer, namely, its depth, thickness, and direction of magnetization.

Exercise 11.13 Use equation 11.35 to show that a uniformly magnetized, infinite slab produces no magnetic field.

If the magnetization or density varies in only one horizontal direction, then equation 11.31 reduces to a single convolution integral,

$$f(x) = \int\limits_{-\infty}^{\infty} s(x')\,\xi(x - x')\,dx'\,,$$

where $s(x)$ is the source distribution of a horizontal layer. The Green's function in this case is the field caused by a vertical ribbon extending

from z_1 to z_2 in the vertical direction and infinitely extended parallel
to y. In the Fourier domain, therefore, $\mathcal{F}[\xi]$ is given by equation 11.28
for the two-dimensional gravity case and by equation 11.30 for the two-
dimensional magnetic case.

Exercise 11.14 What form would equations 11.34 and 11.35 take if density
and magnetization were confined to a vanishingly thin plane?

The function $\mathcal{F}[\xi]$ in equation 11.33 relates the spectrum of the potential field to the spectrum of the causative source in a very simple way.
It not only provides insights into the spectral relationship between field
and source, but it also facilitates a simple calculation of one from the
other. Two landmark papers by Schouten [252] and Schouten and Mc-
Camy [253] discussed these principles in detail for the two-dimensional
magnetic case. They called $\mathcal{F}[\xi]$ the *earth filter*; here we will apply the
term generally to both gravity and magnetic fields and to both two- and
three-dimensional situations.

Forward Calculation

Equations 11.34 and 11.35 permit a calculation of the vertical gravity
anomaly or total-field magnetic anomaly from a given distribution of
density or magnetization, respectively, when the distribution is confined
to a horizontal layer and varies in only horizontal directions. The steps
are to (1) Fourier transform $s(x, y)$, (2) multiply by the earth filter, and
(3) inverse Fourier transform the product. Algorithms B.18 and B.19 in
Appendix B show the application of these three steps to gravity and magnetic sources, respectively. These subroutines treat the three-dimensional
case and employ the two-dimensional discrete Fourier transform. Similar algorithms could be developed for the two-dimensional case using
the one-dimensional discrete transform (Subroutine B.16).

The spectrum of the field caused by a layer source, therefore, is equivalent to the spectrum of the source distribution after "shaping" by the
earth filter. The earth filter in this case is merely the Fourier transform
of the anomaly caused by a vertical line element, so all of the earlier
comments regarding these line-element transforms are applicable here.
Moreover, the energy-density spectra shown on Figure 11.3 also reflect
the spectra of their respective earth filters. For example, Figure 11.3(a)
shows that the contributions of $\rho(x, y)$ to a gravity anomaly $g_z(x, y)$
will be attenuated exponentially as a function of increasing wavenumber, and the rate of that attenuation will increase with increasing depth
to the layer. It follows that the $k = 0$ component of $\rho(x, y)$ will be least

attenuated. Put another way, the gravity anomaly will always be smoother than the density distribution, in the sense that all wavenumbers (except $|k| = 0$) of the gravity anomaly are attenuated relative to the density distribution; the higher the wavenumber (shorter the wavelength), the greater the attenuation.

Exercise 11.15 Show that when $\rho(x, y)$ is a constant, equation 11.34 reduces to the infinite-slab formula (equation 3.27). Hint: L'Hospital's rule is helpful.

Likewise, many of the comments of Section 11.2.1 concerning vertical lines of dipoles are relevant to the earth filter for a magnetic layer. In particular, the earth filter for a magnetic layer has its maximum value at a wavenumber k_{max} that depends on the depth and thickness of the layer (Figures 11.3(c) and 11.6). Hence, the $|k| = k_{max}$ component of magnetization will be less attenuated than all other wavenumbers in the total-field anomaly. Notice in Figure 11.6 that k_{max} shifts to higher wavenumbers (shorter wavelengths) as depth of the layer decreases. Wavenumbers higher than k_{max} are each attenuated less than all higher wavenumbers, and this attenuation approaches exponential decay at high wavenumbers. The rate of attenuation with increasing wavenumber increases with increasing depth to the magnetic layer. The $k = 0$ component of magnetization is eliminated entirely by the earth filter.

Exercise 11.16 Use equation 11.35 to show that a uniformly magnetized slab produces no magnetic anomaly.

Parameters that determine the location and shape of the source (i.e., the depth and thickness of the layer) are restricted to the exponential terms of equation 11.35; they influence the behavior of the earth filter as a function of $|k|$ only. On the other hand, the direction of magnetization and the direction of the regional field are restricted entirely to the functions Θ_m and Θ_f, respectively. The direction of magnetization and the direction of the regional field, therefore, have no effect on the behavior of the earth filter as a function of $|k|$, but they completely control its phase. For example, if the magnetization and regional field are directed vertically ($\hat{m}_x = \hat{m}_y = \hat{f}_x = \hat{f}_y = 0$), then Θ_m and Θ_f will be real constants. This implies that the phase of the source distribution and the phase of the anomaly will be identical when magnetization and regional field are vertical; for example, if $M(x, y)$ is symmetric about the origin, $\Delta T(x, y)$ also will be symmetric. On the other hand, if magnetization is horizontal ($\hat{m}_z = 0$) and the regional field is vertical ($\hat{f}_x = \hat{f}_y = 0$),

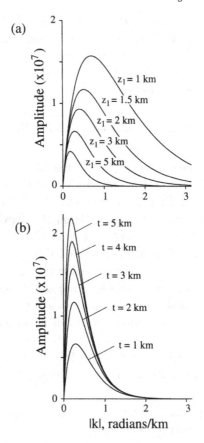

Fig. 11.6. Examples of the earth filter for a magnetic layer. Parameters z_1 and t represent depth and thickness of the layer, respectively, in km. Magnetization and regional field are vertical. (a) Various depths to the layer with $t = 1$ km; (b) various layer thicknesses with $z_1 = 3$ km.

Θ_m will be an imaginary function of k_x and k_y, and Θ_f will be a real constant. The earth filter in this case will be purely imaginary. In the space domain, this implies that if $M(x, y)$ is symmetric about the origin, $\Delta T(x, y)$ will be antisymmetrical.

Inverse Calculation

For a density or magnetization distribution confined to a layer, the forward calculation was a simple matter of multiplying the Fourier transform of the source distribution by the appropriate earth filter. It would

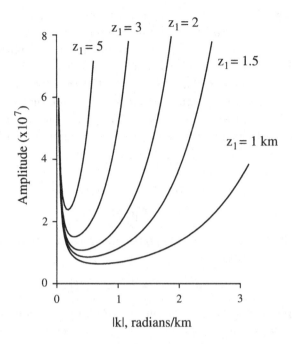

Fig. 11.7. The inverse earth filter for a magnetic layer at various depths. Thickness of layer is 1 km. Magnetization and regional field are vertical. Parameter z_1 indicates depth to top of layer.

seem that the inverse calculation would be just as simple; that is, dividing both sides of equation 11.33 by the earth filter yields

$$\mathcal{F}[s] = \mathcal{F}[f]\,\mathcal{F}^{-1}[\xi],\qquad(11.36)$$

where $\mathcal{F}^{-1}[\xi]$ is the *inverse earth filter*. To calculate density or magnetization from a measured anomaly, therefore, we might try the following steps: Fourier transform the potential field, multiply by $\mathcal{F}^{-1}[\xi]$, and inverse Fourier transform the product. Unfortunately, as discussed in detail by Schouten [252], this inverse method is not as straightforward as it seems.

Figure 11.7 shows the amplitude of the inverse earth filter $\mathcal{F}^{-1}[\xi]$ for a magnetic layer located at various depths. Notice that the amplitude approaches infinity at both high and low wavenumbers. In the case of gravity fields, the low wavenumbers pose no problem, but high wavenumbers similarly approach infinite amplitudes. Consequently, application of the inverse earth filter to measured gravity or magnetic fields will in general cause the highest wavenumbers to be greatly amplified. More important, any noise contained within the potential-field measurements will be similarly amplified, thereby producing high-amplitude, short-wavelength oscillations in the calculated source distribution.

Suppose, for example, that $\Delta T(x, y)$ is the true total-field anomaly caused by a layer with magnetization $M(x, y)$, but that our measurements of ΔT include a random noise $e(x, y)$. The magnetization $M'(x, y)$ that would be derived by inverting these contaminated data is given by

$$
\begin{aligned}
\mathcal{F}\left[M'\right] &= \mathcal{F}\left[\Delta T + e\right] \mathcal{F}^{-1}\left[\xi\right] \\
&= \mathcal{F}\left[\Delta T\right] \mathcal{F}^{-1}\left[\xi\right] + \mathcal{F}\left[e\right] \mathcal{F}^{-1}\left[\xi\right] \\
&= \mathcal{F}\left[M\right] + \mathcal{F}\left[e\right] \mathcal{F}^{-1}\left[\xi\right],
\end{aligned}
$$

and the difference between the determined magnetization and the true magnetization will be

$$
\mathcal{F}\left[M' - M\right] = \mathcal{F}\left[e\right] \mathcal{F}^{-1}\left[\xi\right].
$$

If the amplitudes of the measurement errors are randomly distributed about zero mean with variance σ^2, then the power spectrum of the error will be a constant proportional to the variance.† Hence, the error in the determined magnetization will be a function of wavenumber. Indeed, it will be proportional to the inverse earth filter itself. As shown by Figure 11.7, the amplified error will be most dramatic at shortest wavelengths and can reach extraordinary amplifications at high wavenumbers relative to the mid-range of the spectrum. Consequently, a small random error added to the measurements can cause large-amplitude, short-wavelength, and unrealistic fluctuations in the calculated source distribution. In short, the solution is unstable.

In practice, we are concerned only with the Fourier transform of measured field values out to the Nyquist wavenumbers ($k_x = \pm\pi/\Delta x$ and $k_y = \pm\pi/\Delta y$, where Δx and Δy are the intervals at which the field was sampled). Consequently, discrete values of $\mathcal{F}^{-1}\left[\xi\right]$ similarly are needed only out to $k_x = \pm\pi/\Delta x$ and $k_y = \pm\pi/\Delta y$. Nevertheless, if the depth of the layer is large relative to the sample intervals, the Nyquist wavenumbers will be sufficiently large to cause extreme amplification of both signal and noise. Figure 11.8 shows this relationship for a magnetic layer with $\Delta x = \Delta y$ and with a thickness equal to $2\Delta x$.

---

† Strictly speaking, the power spectrum of a finite segment of a random function such as $e(x, y)$ will not be a constant but will be σ^2 plus a random fluctuation. See Claerbout [60, pp. 76–80].

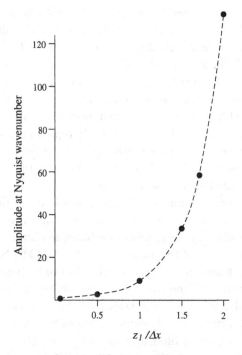

Fig. 11.8. High-wavenumber amplification of the inverse earth filter (magnetic case) as a function of attempted resolution. Each point on the curve represents the ratio of the amplitude spectrum at the Nyquist wavenumber divided by the minimum amplitude. Abscissa represents resolution, that is, depth to the top of the layer divided by sample interval.

Figure 11.8 is reminiscent of an earlier discussion in Section 10.2.1. An inverse method was discussed in that section in which the magnetic source was modeled in the space domain with a horizontal layer. The layer was divided into N horizontal prisms, each with vertical sides and width W, as shown by Figure 10.1. The total-field anomaly in this case is given by

$$\Delta T_i = \sum_{j=1}^{N} M_j \psi_{ij}, \qquad i = 1, 2, \ldots, L, \qquad (11.37)$$

where ΔT_i is the ith observation point, M_j is the magnetization of the jth cell, and ψ_{ij} is the field at point i due to cell j with unit magnetization. Solving for the N values of M_j from L measured values of ΔT_i constitutes the linear inverse problem, and if $L > N$, a least-squares technique would be an appropriate strategy to find the M_j. However,

Bott and Hutton [38] showed that this calculation is unstable if too much resolution is attempted, that is, if the width of the blocks is made too small relative to the depth of the layer (Figure 10.3).

Now if the cells in Figure 10.1 all have the same top and bottom depths, equation 11.37 is just the discrete form of the one-dimensional convolution, where cell width W is equivalent to the sample interval Δx. Hence, equation 11.37 is the space-domain equivalent of equation 11.33. It follows that Figures 10.3 and 11.8 both represent the same phenomenon. If the inverse calculation attempts too much resolution, small errors in the measured field will generate large and unrealistic oscillations in the calculated source distribution.

In practice, we can assume that the calculated source distribution is not meaningful at wavenumbers outside of a certain range. Multiplying the inverse earth filter by an appropriate bandpass filter will eliminate all wavenumbers outside of this range (Schouten and McCamy [253]), in effect forcing the determined source distribution to be band-limited. By adjusting the bandwidth, the combined filter can be "tuned" depending on the depth and thickness of the layer and the sample interval of the measurements. Caution is still required, however, because the combination of the inverse earth filter and the band-pass filter has a narrow, peaked shape that can generate unrealistic oscillations in determinations of magnetization (Blakely and Schouten [32]).

11.3.1 Topographic Sources

The earth filters discussed previously are limited by an overly simple model for the source distribution. In order to apply the Fourier-convolution theorem, equation 11.31 had to be rendered into a convolution (equation 11.32), and this was possible only because the source distribution was confined between horizontal surfaces z_1 and z_2 and was allowed to vary in only the horizontal directions. Parker [204] showed how the first of these two assumptions could be relaxed; that is, he developed a model consisting of a source layer with uneven top and uneven bottom surfaces. The following derivation is similar to that of Parker's three-dimensional magnetic case; the derivation for gravity anomalies follows the same lines and is left to the exercises.

The total-field anomaly is measured on a horizontal surface at altitude z_0, and all magnetic material will be confined between two surfaces $z_1(x,y)$ and $z_2(x,y)$, as in Figure 11.9. We will require $z_1(x,y) > z_0$ and $z_2(x,y) > z_1(x,y)$ for all x and y. Note that z_1 and z_2 are no longer

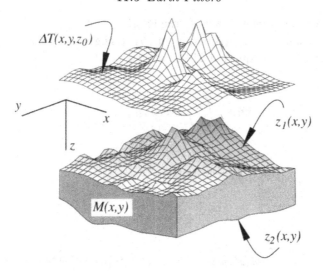

Fig. 11.9. Calculation of the magnetic field caused by a magnetic layer with uneven top and bottom surfaces. The source distribution is confined between two surfaces z_1 and z_2, both functions of x and y. The total-field anomaly ΔT is measured on a horizontal surface at $z = z_0$.

constants but rather are functions of x and y. From equation 11.31, the total-field anomaly is given by

$$\Delta T(x, y, z_0) = \int\limits_{-\infty}^{\infty} \int\limits_{-\infty}^{\infty} M(x', y') \int\limits_{z_1}^{z_2} \psi(x - x', y - y', z_0 - z') \, dz' \, dx' \, dy' ,$$

(11.38)

where ψ is the field at (x, y, z_0) due to a point source at (x', y', z'). Fourier transforming both sides of equation 11.38 yields

$$\mathcal{F}[\Delta T] = \mathcal{F} \left[\int\limits_{-\infty}^{\infty} \int\limits_{-\infty}^{\infty} M(x', y') \int\limits_{z_1}^{z_2} \psi(x - x', y - y', z_0 - z') \, dz' \, dx' \, dy' \right]$$

$$= \int\limits_{-\infty}^{\infty} \int\limits_{-\infty}^{\infty} M(x', y') \int\limits_{z_1}^{z_2} \mathcal{F}[\psi(x - x', y - y', z_0 - z')] \, dz' \, dx' \, dy' .$$

(11.39)

The two-dimensional Fourier transform of $\psi(x - x', y - y', z_0 - z')$ in equation 11.39 poses no problem because we already know the Fourier transform of $\psi(x, y, z)$; it is simply the Fourier transform of the anomaly caused by a single dipole (equation 11.25). Using the shifting property (Section 11.1.2) and equation 11.25, we get

$$\mathcal{F}[\Delta T] = 2\pi C_m \, \Theta_m \Theta_f |k| \int\limits_{-\infty}^{\infty} \int\limits_{-\infty}^{\infty} M(x', y')$$

$$\int\limits_{z_1}^{z_2} e^{|k|(z_0 - z')} \, e^{-i(k_x x' + k_y y')} \, dz' \, dx' \, dy'$$

$$= 2\pi C_m \, \Theta_m \Theta_f e^{|k|z_0}$$

$$\int\limits_{-\infty}^{\infty} \int\limits_{-\infty}^{\infty} M(x', y') \{ e^{-|k|z_1} - e^{-|k|z_2} \} e^{-i(k_x x' + k_y y')} \, dx' \, dy' \, .$$

The top and bottom surfaces of the magnetic layer are carried in this equation by the bracketed exponentials. Parker [204] suggested replacing these exponential terms with their equivalent power series. The double integral then is converted to a sum of two-dimensional Fourier transforms, each taken with respect to x', y' coordinates. Making these modifications and collecting terms leads to the equation

$$\mathcal{F}[\Delta T] = 2\pi C_m \, \Theta_m \Theta_f \, e^{|k|z_0} \, \mathcal{F} \left[M \sum_{n=0}^{\infty} \frac{(-|k|)^n}{n!} (z_1^n - z_2^n) \right]$$

$$= 2\pi C_m \, \Theta_m \Theta_f \, e^{|k|z_0} \sum_{n=1}^{\infty} \frac{(-|k|)^n}{n!} \mathcal{F}[M(z_1^n - z_2^n)] \, . \qquad (11.40)$$

Hence, the Fourier transform of the total-field anomaly caused by a layer with uneven top and bottom surfaces is represented as a summation; each element of the summation includes a Fourier transform of the magnetization weighted by a power of either the top or bottom surface. After the summation is completed, the inverse Fourier transform will provide the total-field anomaly. Notice that if z_1 and z_2 are both constant, equation 11.40 reduces to equation 11.35. Parker [207] showed that this summation converges most rapidly if the origin is selected so that $z = 0$ midway between the minimum values of z_1 and z_2.

The gravitational equivalent of equation 11.40 is

$$\mathcal{F}[g_z] = 2\pi\gamma\, e^{|k|z_0} \sum_{n=1}^{\infty} \frac{(-|k|)^{n-1}}{n!} \mathcal{F}[\rho(z_1^n - z_2^n)] , \qquad (11.41)$$

where the mass is confined between surfaces $z_1(x, y)$ and $z_2(x, y)$ and is described by density $\rho(x, y)$.

Exercise 11.17 Derive equation 11.41 along the same lines as done for equation 11.40.

The forward calculation of equation 11.40 has been particularly useful in investigating magnetic anomalies over magnetic terrain or bathymetry (e.g., Macdonald [170]; Grauch [101]; Blakely [22]; Blakely and Grauch [28]; Hildenbrand, Rosenbaum, and Kauahikaua [128]). The gravitational equivalent of equation 11.40 (equation 11.41) has been used in a variety of applications, such as calculating isostatic residual gravity anomalies (Simpson, Jachen, and Blakely [260]) and estimating the gravitational effects of sedimentary basins (e.g., Jachens and Moring [137], Saltus [250]). Subroutine B.23 in Appendix B is a Fortran implementation of equation 11.40.

Equation 11.40 also provides a framework for the inverse problem; that is, to calculate the magnetization M from measurements of ΔT (Parker [204], Parker and Huestis [208]). This calculation requires that the thickness of the layer $t = z_2 - z_1$ remain uniform. If t is a constant, then equation 11.40 can be written

$$\mathcal{F}[\Delta T] = 2\pi C_m \Theta_m \Theta_f\, e^{|k|z_0} (1 - e^{-|k|t}) \sum_{n=0}^{\infty} \frac{(-|k|)^n}{n!} \mathcal{F}[M z_1^n] . \quad (11.42)$$

Isolating the $n = 0$ term of the summation on the left side of equation 11.42 yields

$$\mathcal{F}[M] = \frac{\mathcal{F}[\Delta T]}{2\pi C_m \Theta_m \Theta_f\, e^{|k|z_0} (1 - e^{-|k|t})} - \sum_{n=1}^{\infty} \frac{(-|k|)^n}{n!} \mathcal{F}[M z_1^n] . \quad (11.43)$$

Equation 11.43 can be solved iteratively by (1) making an initial guess at the magnetization on the right side of equation 11.43, (2) solving the equation for a revised magnetization, (3) moving this revised magnetization to the right-hand side, and (4) repeating steps 2 and 3 until the solution converges. As discussed earlier, step 2 involves a summation that converges rapidly if the origin is midway between the topographic extremes. The iterations, on the other hand, converge most rapidly if the origin is at a different level, namely, at the minimum value of topography

(Parker and Huestis [208]). The change in origin can be accommodated through the exponential term containing z_0.

The solution to equation 11.43 is susceptible to the same instabilities discussed earlier in this and the previous chapter, and the higher wavenumber components must be filtered at each iteration. Nevertheless, this inverse method has proved useful in many applications, particularly in the analysis of marine magnetic anomalies (e.g., Macdonald [170]; Macdonald et al. [171]) where the magnetic part of ocean crust can sometimes be considered uniform in thickness.

Equation 11.43 can be used to find the annihilator for a particular layer with constant thickness and a specified upper topographic surface (Parker and Huestis [208]). Recall that the annihilator for a volume R is the source distribution $a(Q)$ that satisfies the equation

$$\int_R a(Q)\psi(P, Q)\, dv = 0\,,$$

where $\psi(P, Q)$ is the appropriate Green's function. A magnetization $a(Q)$ that satisfies

$$\sum_{n=0}^{\infty} \frac{(-|k|)^n}{n!} \mathcal{F}\left[az_1^n\right] = \delta(k_x)\delta(k_y) \tag{11.44}$$

will produce no anomaly when confined to the layer because substituting equation 11.44 into 11.42 causes the total-field anomaly to vanish. This can be verified by comparing the preceding equations with equation 11.15. Hence, we rewrite equation 11.44 as

$$\mathcal{F}[a] = \delta(k_x)\delta(k_y) - \sum_{n=1}^{\infty} \frac{(-|k|)^n}{n!} \mathcal{F}\left[az_1^n\right]\,, \tag{11.45}$$

which can be solved iteratively for the annihilator $a(x, y)$ as was done earlier for magnetization $M(x, y)$. The annihilator represents the non-uniqueness of the solution for M. Any amount of the annihilator can be added to the magnetization without affecting the calculated total-field anomaly; that is, the anomaly caused by $M(x, y) + \alpha a(x, y)$ is independent of α.

Exercise 11.18 Use equation 11.44 to show that the annihilator for an infinite slab with flat top and flat bottom is a constant.

11.3.2 General Sources

At this point, we return to equation 11.31 and derive general relationships in the Fourier domain between a gravity or magnetic anomaly measured on a horizontal surface and its causative source distribution located entirely below the surface. As before, we assume that the field is measured on a horizontal plane at height $z = z_0$ and rewrite equation 11.31 in a cartesian coordinate system with z positive down,

$$f(x, y, z_0) = \int\limits_{z_0}^{\infty} \int\limits_{-\infty}^{\infty} \int\limits_{-\infty}^{\infty} s(x', y', z')\, \psi(x - x', y - y', z_0 - z')\, dx'\, dy'\, dz'\,.$$

It is assumed in this equation that $s(x', y', z')$ is zero outside of a region with finite dimensions and especially at all $z < z_0$. Applying the Fourier transform to both sides of the equation (and remembering that the Fourier transform involves integration with respect to x and y, not x' and y'), we have

$$\mathcal{F}[f] = \int\limits_{z_0}^{\infty} \int\limits_{-\infty}^{\infty} \int\limits_{-\infty}^{\infty} s(x', y', z')\, \mathcal{F}[\psi(x - x', y - y', z_0 - z')]\, dx'\, dy'\, dz'\,.$$

The shifting property of Fourier transforms (Section 11.1.2) allows a substitution for the Green's function in the previous equation, that is,

$$\mathcal{F}[\psi(x - x', y - y')] = \mathcal{F}[\psi(x, y)]\, e^{-i(k_x x' + k_y y')}\,.$$

In the gravitational case, the anomaly f is the vertical attraction of gravity g, the source distribution s is density ρ, and ψ is the field of a monopole with Fourier transform given by equation 11.22. Making these substitutions, we have

$$\mathcal{F}[g] = 2\pi\gamma\, e^{|k|z_0} \int\limits_{z_0}^{\infty} \int\limits_{-\infty}^{\infty} \int\limits_{-\infty}^{\infty} \rho(x', y', z')\, e^{-|k|z'}\, e^{-i(k_x x' + k_y y')}\, dx'\, dy'\, dz'\,.$$

Now the two inner integrals are simply another two-dimensional Fourier transform, in this case taken with respect to x' and y', so

$$\mathcal{F}[g] = 2\pi\gamma\, e^{|k|z_0} \int\limits_{z_0}^{\infty} \mathcal{F}[\rho(z')]\, e^{-|k|z'}\, dz'\,, \tag{11.46}$$

where the term $\mathcal{F}[\rho(z')]$ represents the two-dimensional Fourier transform of the density on one horizontal slice through the body at depth z'.

In the same way, we can derive a similar relationship for the magnetic case,

$$\mathcal{F}[\Delta T] = 2\pi C_m \Theta_m \Theta_f \, |k| \, e^{|k|z_0} \int_{z_0}^{\infty} \mathcal{F}[M(z')] \, e^{-|k|z'} \, dz' \, , \qquad (11.47)$$

where the term $\mathcal{F}[M(z')]$ represents the Fourier transform of the magnetization on one horizontal slice through the body at depth z'.

Equations 11.46 and 11.47 provide general relationships in the Fourier domain between arbitrary mass or magnetization distributions and the anomalies that they produce. The integral terms in each equation essentially consist of dividing the source into horizontal slices, Fourier transforming the density or magnetization of each slice, weighting the Fourier transform by an exponential term that depends on the depth of the slice, and summing the results over all slices. We will have use for these relationships in Chapter 12. Notice that equations 11.46 and 11.47 reduce to equations 11.34 and 11.35 if the source is confined to a horizontal layer with top at depth z_1, bottom at z_2, and density or magnetization that varies in only the horizontal directions.

11.4 Depth and Shape of Source

As we have seen, Fourier analysis under certain assumptions can greatly facilitate solutions to the linear inverse problem. If the source material is confined to a horizontal layer, for example, equation 11.36 provides a direct calculation of magnetization or density from the potential field of the layer, thus rendering the linear inverse problem for this particular case quite easy, at least conceptually.

Fourier transforms also can assist in determining the shape and location of potential-field sources. Oldenburg [200], for example, modified the method of Parker [204] in order to estimate the shape of a causative mass from its gravity anomaly. If the body has uniform density contrast and a flat bottom at $z = 0$, then the right side of equation 11.41 reduces to a summation of Fourier transforms of powers of the upper surface, that is,

$$\mathcal{F}[g_z] = 2\pi\gamma e^{|k|z_0} \rho \sum_{n=1}^{\infty} \frac{(-|k|)^{n-1}}{n!} \mathcal{F}[z_1^n] \, .$$

Isolating the first term of this summation, as was done for magnetization in Section 11.3.1, leads to an iterative scheme to estimate the shape of the upper surface. The free parameters ρ and z_0 in this equation, as well as the requisite bandpass filtering at each iteration, characterize the nonuniqueness of this inversion scheme; the solution for z_1 depends on the values selected for these parameters (Oldenburg [200]). Pilkington and Crossley [221, 222] described a similar technique for estimating the shape of magnetic basement or crustal interfaces from magnetic anomalies.

Hansen and Wang [114] simplified a Fourier-domain formulation, originally from Pedersen [216], that approximates potential-field sources by polyhedrons, similar to the space-domain models described by Bott [36] and Barnett [11]. The bodies are assumed to have uniform magnetization or density. The potential field $f(x, y)$ of a polyhedron can be expressed in the Fourier domain as a summation over the N vertices of the polyhedron,

$$\mathcal{F}[f] = \sum_{n=1}^{N} \alpha_n e^{-i(k_x x_n + k_y y_n) - |k| z_n} , \qquad (11.48)$$

where (x_n, y_n, z_n) are the coordinates of vertex n, and α_n depends on the orientation of each edge of each facet composing vertex n and, in the magnetic case, on the direction of magnetization and regional field. The coordinates of each vertex, therefore, are represented in the argument of a specific exponential term in the summation, so the forward problem is reasonably straightforward. Because the summation depends on the vertex coordinates, the polyhedron does not need to be decomposed into individual facets, as was done in Section 9.3.2 and in Subroutine B.10.

Wang and Hansen [288] showed how this formalism could be adapted to the inverse problem for magnetic anomalies, that is to estimate the locations in space of the vertices of a magnetic polyhedron from its measured anomaly. Here we use the total-field anomaly, but their discussion applies to any component of the anomalous field. The exponential in each term of summation 11.48 includes both an attenuation factor related to the depth of one vertex and a phase factor related to the horizontal position of the same vertex. The problem then is to analyze $\mathcal{F}[\Delta T]$ in order to resolve these various factors for each vertex. First let $F(k_x, k_y) = |k|^2 \mathcal{F}[\Delta T]$. We will show in the next chapter that $F(k_x, k_y)$ so defined is the Fourier transform of the second vertical derivative of ΔT. Wang and Hansen [288] defined N complex parameters

$$\delta_j = i \frac{k_x x_j + k_y y_j}{|k|} + z_j, \qquad j = 1, 2, \ldots, N,$$

and showed that the first N inward derivatives of $F(k_x, k_y)$ are related by a summation,

$$\frac{\partial^N F(k_x, k_y)}{\partial(-|k|)^N} = \sum_{n=0}^{N-1} \gamma_n \frac{\partial^n F(k_x, k_y)}{\partial(-|k|)^n}, \qquad (11.49)$$

where the various γ_n are factors in a complex polynomial equation,

$$\delta_j^N = -\sum_{n=0}^{N-1} \gamma_n \delta_j^n. \qquad (11.50)$$

The roots of equation 11.50 are the various δ_j, and these as defined previously are related to the vertex coordinates.

The algorithm of Wang and Hansen [288] essentially proceeds by computing the first N inward derivatives of $|k|^2 \mathcal{F}[\Delta T]$; using these in equation 11.49 to solve for γ_n, $n = 0, 1, 2, \ldots, N-1$, along rays of the k_x, k_y plane; and finally solving for δ_j, $j = 1, 2, \ldots, N$, in equation 11.50, also along rays of the k_x, k_y plane.

11.4.1 Statistical Models

The previous examples are applicable to potential fields caused by one or a few causative bodies. In the remainder of this section, we will focus on estimating the average depth of a large collection of gravity or magnetic sources from their statistical properties, an approach pioneered by Spector and Grant [268] and Treitel, Clement, and Kaul [281]. Magnetic fields are used here for illustration, but the discussion could easily be adapted to gravity fields as well.

To see how the statistical approach works in a general way, consider the total-field anomaly measured on a horizontal surface and caused by a horizontal layer with top at depth d and thickness t. From equation 11.35,

$$\mathcal{F}[\Delta T] = \mathcal{F}[M]\left\{2\pi C_m \Theta_m \Theta_f e^{-|k|d}(1 - e^{-|k|t})\right\}. \qquad (11.51)$$

The total-field anomaly is measured at discrete locations, and $\mathcal{F}[\Delta T]$ and Θ_f are easily calculated. If by some lucky happenstance we also know the distribution of magnetization, then only the exponential terms remain to be determined in equation 11.51. These could be transposed to one side of the equation, and curve fitting could be used to estimate d and t.

It is unlikely that we would know $M(x,y)$ in detail, but we may have some idea about how magnetization behaves statistically. For example, we might assume that the layer extends infinitely far in all horizontal directions and that $M(x,y)$ is a random function of x and y. In this case, $\mathcal{F}[M]$ and $\mathcal{F}[\Delta T]$ do not exist because inequality 11.3 is not satisfied. Equation 11.51 should be written instead as

$$\Phi_{\Delta T}(k_x, k_y) = \Phi_M(k_x, k_y) \cdot F(k_x, k_y), \qquad (11.52)$$

where $\Phi_{\Delta T}$ and Φ_M are power-density spectra of the total-field anomaly and the magnetization, respectively, and

$$F(k_x, k_y) = 4\pi^2 C_{\mathrm{m}}^2 |\Theta_{\mathrm{m}}|^2 |\Theta_{\mathrm{f}}|^2 e^{-2|k|d}(1 - e^{-|k|t})^2 \,.$$

If the probabilistic behavior of $M(x,y)$ is known or assumed, its power-density spectrum can be analytically derived as described in Section 11.1.3. Moreover, although we only have measurements of $\Delta T(x,y)$ over finite distances, a variety of ways are available to estimate power-density spectra from finite segments of a random function. Hence, with a suitable probabilistic model for $M(x,y)$, the remaining unknown function $F(k_x, k_y)$ can be analyzed in terms of d and t.

This analysis can be simplified greatly by noting that all terms, except $|\Theta_{\mathrm{m}}|^2$ and $|\Theta_{\mathrm{f}}|^2$, in equation 11.52 are radially symmetric.† Moreover, as discussed in Section 11.2.1, the radial averages of Θ_{m} and Θ_{f} are constants. Hence, the radial average of $\Phi_{\Delta T}$ is

$$\bar{\Phi}_{\Delta T}(|k|) = A\, \Phi_M(|k|)\, e^{-2|k|d}(1 - e^{-|k|t})^2 \,, \qquad (11.53)$$

where A is a constant that depends on the orientations of magnetization and regional field. We now need a suitable statistical model for $M(x,y)$. For example, if $M(x,y)$ is completely random and uncorrelated, $\Phi_M(k_x, k_y)$ is a constant, and equation 11.53 becomes

$$\bar{\Phi}_{\Delta T}(|k|) = B e^{-2|k|d}(1 - e^{-|k|t})^2 \,,$$

where B is a constant. Finally, taking the logarithm of both sides yields

$$\log \bar{\Phi}_{\Delta T}(|k|) = \log B - 2|k|d + 2\log(1 - e^{-|k|t}) \,. \qquad (11.54)$$

As shown by Figure 11.10, this equation at medium to high wavenumbers (wavelengths less than about twice the thickness of the layer) is approximately that of a straight line with slope equal to $-2d$. We could estimate

---

† We have assumed here that the probabilistic behavior of $M(x,y)$ is isotropic, that is, its statistical behavior is the same in all horizontal directions.

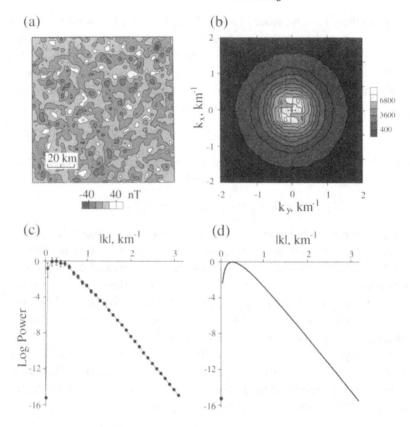

Fig. 11.10. Power-density spectrum of the anomaly due to a randomly magnetized layer. (a) Total-field anomaly caused by magnetic layer with top at 3 km depth, thickness 1 km, and random magnetization. Magnetization is described by uniform distribution ranging between +1 and −1 A/m. Directions of magnetization and regional field are vertical. (b) Amplitude spectrum of the anomaly; values smoothed for contouring. (c) Log of radial power-density spectrum. Dots indicate average power within rings concentric about the origin. Error bars indicate standard deviation of values within the ring, divided by the square root of the number of values encountered. (d) Log of the theoretical power-density spectrum.

d, therefore, by calculating the power-density spectrum of $\Delta T(x, y)$, radially averaging the spectrum within rings concentric about $|k| = 0$, and fitting a straight line through the high-wavenumber part of the radially averaged spectrum. One-half the slope of the line is an estimate of d (Figure 11.10).

Exercise 11.19 Show that the same result is achieved if $M(x, y) = M_0 + r(x, y)$, where M_0 is a constant and $r(x, y)$ is a random function of x and y with zero mean. In other words, show that a slab, principally uniform in magnetization but having a small random component, also satisfies $\Phi_M = $ constant.

Ensemble Sources

In a landmark paper, Spector and Grant [268] framed the random description of the source distribution in a more complete context. In their model, the anomaly is assumed to be produced by a large number of blocks. The parameters describing any one block (e.g., depth, thickness, width, length, magnetization) are assumed to obey probabilities common to the entire set. Such a distribution of sources constitutes an ensemble. Equation 11.12 showed that the expected value of a continuous random variable r can be expressed as an ensemble average,

$$\langle r \rangle = \int\limits_{-\infty}^{\infty} \rho P_r(\rho)\, d\rho \,,$$

where $P_r(\rho)$ is the probability that r takes the value ρ. Similarly, if g is a measurable function of various random variables $a_1, a_2, \ldots$, then g has an expected value given by

$$\langle g \rangle = \int\limits_{-\infty}^{\infty} \cdots \int\limits_{-\infty}^{\infty} g(\alpha_1, \alpha_2, \ldots) P_{a_1, a_2, \ldots}(\alpha_1, \alpha_2, \ldots)\, d\alpha_1\, d\alpha_2 \cdots \,,$$

where $P_{a_1, a_2, \ldots}(\alpha_1, \alpha_2, \ldots)$ is the joint probability distribution for the random variables of g.

Now consider the magnetic anomaly caused by an ensemble of sources, where the parameters (depth, thickness, and so forth) describing each element of the ensemble are each a random variable. We wish to frame the power-density spectrum of the anomaly in terms of an ensemble average, as shown by the previous equation. If $P_{d, t, \ldots}(\delta, \tau, \ldots)$ is the joint probability that depth d assumes the value δ, thickness t assumes the value τ, and so on, then the expected value of the power-density spectrum is given by

$$\langle \Phi_{\Delta T} \rangle = \int\limits_{-\infty}^{\infty} \cdots \int\limits_{-\infty}^{\infty} |\mathcal{F}[\Delta T]|^2 P_{d, t, \ldots}(\delta, \tau, \ldots)\, d\delta\, d\tau \cdots \,,$$

where $|\mathcal{F}[\Delta T]|^2$ is the spectrum of the anomaly caused by just one element of the ensemble having parameters δ, τ, and so on. Presumably the random parameters vary independently of one another, so the joint probability can be factored, that is,

$$\langle \Phi_{\Delta T} \rangle = \int\limits_{-\infty}^{\infty} \cdots \int\limits_{-\infty}^{\infty} |\mathcal{F}[\Delta T]|^2 \, P_d(\delta) \, P_t(\tau) \cdots d\delta \, d\tau \cdots . \qquad (11.55)$$

Next we need an expression for $|\mathcal{F}[\Delta T]|^2$ in equation 11.55. Spector and Grant [268] let each member of the ensemble be a rectangular parallelepiped. Equation 11.51 can provide the spectrum of the anomaly caused by a rectangular parallelepiped if we let $M(x,y) = 0$ except within a rectangular patch. Let the patch have dimensions $2a$ and $2b$ in the x and y directions, respectively, and let $M(x,y) = M_0$, a constant, within the patch. The parameters a, b, and M_0 become additional random variables in equation 11.55. From equation 11.51,

$$|\mathcal{F}[\Delta T]|^2 = 4\pi^2 C_{\mathrm{m}}^2 |\Theta_{\mathrm{m}}|^2 |\Theta_{\mathrm{f}}|^2 M_0^2 e^{-2|k|d}(1 - e^{-|k|t})^2 S^2(a,b) , \qquad (11.56)$$

where

$$S(a,b) = \frac{4 \sin k_x a \, \sin k_y b}{k_x k_y}$$

is the "shape factor."

Now we exploit a result that has recurred throughout this chapter. In the Fourier domain, the potential field separates into multiplicative factors, where each factor is a function of just one of the random variables. Substituting equation 11.56 into equation 11.55 yields

$$\begin{aligned} \langle \Phi_{\Delta T} \rangle &= 4\pi^2 C_{\mathrm{m}}^2 \int\limits_{-\infty}^{\infty} e^{-2|k|\delta} P_d(\delta) \, d\delta \int\limits_{-\infty}^{\infty} (1 - e^{-|k|\tau})^2 P_t(\tau) \, d\tau \cdots \\ &= 4\pi^2 C_{\mathrm{m}}^2 \, \langle e^{-2|k|d} \rangle \, \langle (1 - e^{-|k|t})^2 \rangle \\ &\quad \langle S^2(a,b) \rangle \, \langle M_0^2 \rangle \, \langle |\Theta_{\mathrm{m}}|^2 \rangle \, \langle |\Theta_{\mathrm{f}}|^2 \rangle . \end{aligned} \qquad (11.57)$$

Hence, the expected value of the power-density spectrum is factored into various ensemble averages, where each average involves just one of the random variables.

Consider the first of these factors. Suppose that the depth to the top of each element of the ensemble can lie with equal probability between depths $\bar{d} - \Delta d$ and $\bar{d} + \Delta d$. Then

$$\langle e^{-2|k|d} \rangle = \frac{1}{2\Delta d} \int_{\bar{d}-\Delta d}^{\bar{d}+\Delta d} e^{-2|k|\delta} \, d\delta$$

$$= \frac{1}{4|k|\Delta d} e^{-2|k|\bar{d}} \left(e^{2|k|\Delta d} - e^{-2|k|\Delta d} \right).$$

If Δd is taken small compared to $\bar{d}$, then

$$\langle e^{-2|k|d} \rangle = e^{-2|k|\bar{d}}.$$

Hence, we have the surprising result that the power-density spectrum of an entire ensemble depends on depth in the same way that an "average" member of the ensemble depends on depth. It is easy to show that $\langle |\Theta_m|^2 \rangle$ and $\langle |\Theta_f|^2 \rangle$ do not depend on $|k|$, but that $\langle S^2(a,b) \rangle$ will affect the shape of the radially averaged spectrum depending on the values of a and b.

To estimate $\bar{d}$, therefore, we could proceed as before: From the measured anomaly, calculate the logarithm of the radially averaged power-density spectrum; the slope of this function at medium to high wavenumbers is proportional to the depth of the ensemble. This implies, however, that all other factors of the ensemble (i.e., $\langle S^2(a,b) \rangle$) are accounted for. In order to investigate how depth varies throughout a large region, various authors (e.g., Connard, Couch, and Gemperle [63]; Okubo et al. [198]; Phillips [220]; Blakely and Hassanzadeh [29]; Blakely [21]) have divided magnetic surveys into overlapping cells (or segments of long profiles) and made calculations for each cell similar to those described here.

Fractal Source Distributions

The foregoing statistical models have assumed that magnetization (or density) has no spatial correlation. The method of Spector and Grant [268], for example, employs a model with infinitely many rectangular prisms each with random dimensions; the magnetization of such a model becomes spatially uncorrelated if the characteristic dimensions of the prisms are assumed to be small. Similarly, the model of Blakely and Hassanzadeh [29] assumes that the magnetization is an uncorrelated random function of just one horizontal dimension. Such models imply that any two samples taken from the source distribution will be totally independent of one another, regardless of their proximity.

There is mounting evidence from aeromagnetic data (Gregotski, Jensen, and Arkani-Hamed [104]; Pilkington and Todoeschuck [226]), from truck-mounted magnetometer surveys (Gettings, Bultman, and Fisher [94, 95]; Bultman and Gettings [46]), and from density and susceptibility logs (Pilkington and Todoeschuck [225, 226]) that magnetization and density are not completely uncorrelated but rather have a degree of self-similarity. These authors have argued that magnetization and density can be described as a form of fractal geometry, called scaling noise. Such random processes have power-density spectra proportional to some power of wavenumber. A one-dimensional process of this sort, for example, has a power-density spectrum given by $\Phi(k) = Ak^{\alpha}$, where A and α are constants, and k is wavenumber. If $\alpha = 0$, the process is uncorrelated, as in the foregoing statistical models. If $\alpha < 0$, on the other hand, the process is correlated, and the degree of correlation is indicated by the magnitude of α. The parameter α is called the *fractal dimension*; in a sense, it describes the "roughness" of the random process. Gettings et al. [94, 95] and Bultman and Gettings [46] have related α, determined from truck-mounted magnetometer profiles, to near-surface lithologies. Volcanic terrain might possess a different fractal dimension than plutonic terrain, for example, and determination of α along a magnetic profile might help to distinguish between these two lithologies when concealed beneath sediments.

Pilkington and Todoeschuck [226] compared magnetic anomalies over three-dimensional fractal models. They confined the magnetization to a layer and let the magnetization be fractal in both the horizontal and vertical directions. A magnetization with $\alpha = -4$ produces an anomaly that is similar in appearance to typical observed anomalies and in this sense is far superior to anomalies produced by an uncorrelated magnetization ($\alpha = 0$). They further argued that the fractal behavior of magnetization may not be isotropic; that is, the fractal dimension in the horizontal directions may be different from that in the vertical direction. These results have important implications for statistical models to estimate depth to source (Pilkington, Gregotski, and Todoeschuck [223]); the often-used assumption in these models, that the power-density spectrum of magnetization is constant, may be inaccurate and may lead to systematic errors in depth estimation.

11.4.2 Depth to Bottom

The depth extent of magnetic sources can be of considerable geologic interest. The limiting depth may be controlled, for example, by the Curie-temperature isotherm. If the Curie temperature of crustal rocks in the region of the magnetic survey is known from geologic and rock magnetic considerations, then an estimate of the depth to the Curie-temperature isotherm based on magnetic anomalies can help characterize the geothermal setting of the area. Unfortunately, this calculation ranks among the most difficult in potential-field inversion. As can be seen from equation 11.35, $e^{-|k|z_2}$ dominates $e^{-|k|z_1}$ throughout all wavelengths of the spectrum; that is, the contribution of the bottom of the source is dominated at all wavelengths by the contributions from shallower parts.

Several methods to estimate z_2 in the Fourier domain have been developed in recent years and have shown encouraging results in spite of the difficulties involved. Various workers (Connard et al. [63], Smith et al. [265], Shuey et al. [257], Miyazaki [185], and Blakely [23]) have used the shape of radially averaged spectra to estimate the depth extent of magnetic sources. In particular, the position k_{max} of the maximum along the $|k|$ axis is related to the depth to the bottom of the layer according to

$$k_{max} = \frac{\log z_2 - \log z_1}{z_2 - z_1}.$$

To estimate depths to the Curie-temperature isotherm in Oregon, for example, Connard et al. [63] divided a magnetic survey into overlapping cells and calculated for each cell a radially averaged power-density spectrum, as in Figure 11.10. From these spectra, they attempted to locate k_{max} and, using the preceding equation, to estimate z_2. Note that z_2 in this equation cannot be determined independently of z_1.

Bhattacharyya and Leu [18] determined depth to the bottom of individual sources by analyzing isolated anomalies. They calculated the location of the *centroid* of isolated magnetic sources from the moments of their respective anomalies. If the depth to the top of the source is also known, perhaps determined with the method of Spector and Grant [268], then the depth to the bottom of the source is easily calculated. This method requires isolating individual anomalies from all surrounding anomalies and regional fields, a challenging task in most geologic environments.

Okubo et al. [198, 199] expanded on the method of Bhattacharyya and Leu [18] by treating ensembles of sources. They reframed equation 11.57 into an expression dependent on the centroid of the ensemble average.

By making calculations for various patches of the magnetic survey, they were able to derive a map of the Curie-temperature isotherm for a part of the Japanese arc.

Each of these methods has fundamental limitations. First, the determination of the depth to the bottom of a source cannot be made without knowledge of the depth to its top, which in itself is a difficult inverse determination. Second, estimation of the bottom of potential-field sources, by their nature, must focus on the lowest wavenumber parts of the Fourier domain. This part of the spectrum is susceptible to noise from various sources, particularly from poorly known regional fields that may be unrelated to the bottom parts of the sources of concern. Moreover, discrete Fourier analysis provides relatively little information about the lowest wavenumbers of the power-density spectrum. With conventional techniques, the smallest wavenumber at our disposal is the fundamental wavenumber $k = 2\pi/L$, where L is the dimension of the magnetic survey. Connard et al. [63] reasoned that k_{max} should be at least twice the fundamental wavenumber in order to be able to resolve a peak in the spectrum. Thus,

$$L \geq \frac{4\pi(z_2 - z_1)}{\log z_2 - \log z_1}.$$

A magnetic survey conducted 1 km above the tops of magnetic sources, therefore, must have minimum dimensions on the order of 50 km in order to resolve depth extents of 10 km; the survey dimensions must be at least 160 km for sources extending to 50 km.

Finally, the shape of the power-density spectrum and, hence, depth determinations depend on the characteristic shapes of the magnetic bodies, expressed by the factor $S^2(a, b)$ in equation 11.57. At wavelengths large compared to the body dimensions, the ensemble average is equivalent to that of a random distribution of dipoles (Okubo et al. [198]). In such cases, the low-wavenumber part of the spectrum is independent of the exact model (blocks, cylinders, spheres, and the like) used for the elements of the ensemble. The true dimensions of the source, however, are not likely to be well understood in most geologic settings.

11.5 Problem Set

1. Consider a cartesian coordinate system with the z axis directed down, and let $\phi(x, y, z)$ be the potential of a source distribution located entirely below the plane $z = z_0$. The *Dirichlet integral* relates the

potential field observed on one plane $z = z_1$ to the potential observed on another plane $z = z_2$, where $z_2 < z_1$ and $z_1 < z_0$:

$$\phi(x, y, z_2) = \frac{z_1 - z_2}{2\pi} \int_{-\infty}^{\infty} \int_{-\infty}^{\infty} \frac{\phi(x', y', z_1)}{r^3} \, dx' \, dy', \qquad z_2 < z_1,$$

where $r = \sqrt{(x - x')^2 + (y - y')^2 + (z_1 - z_2)^2}$.

(a) Show for any z_3 that

$$\int_{-\infty}^{\infty} \int_{-\infty}^{\infty} \phi(x, y, z_3) \, dx \, dy = A,$$

where $z_3 < z_0$, and A is a constant independent of z_3.

(b) How does this result relate to Gauss's law for both the gravitational and magnetic potentials?

(c) Show that the total energy of the potential observed on a plane above the source decreases with the plane's distance from the source; that is, show that

$$\int_{-\infty}^{\infty} \int_{-\infty}^{\infty} \phi^2(x, y, z_2) \, dx \, dy < \int_{-\infty}^{\infty} \int_{-\infty}^{\infty} \phi^2(x, y, z_1) \, dx \, dy$$

for any $z_2 < z_1$.

2. Let g_z be the vertical gravitational attraction of a mass distribution located below a horizontal plane, and let $\mathcal{F}[g_z]$ be the Fourier transform of g_z observed on the plane.

(a) Show that the value of $\mathcal{F}[g_z]$ at $k = 0$ is proportional to the total mass of the distribution.

(b) More generally, show that the amplitude spectrum satisfies the inequality

$$|\mathcal{F}[g_z]| \leq 2\pi\gamma M \, e^{-|k|z_0},$$

where M is the total mass and z_0 is the depth to the top of the source distribution.

3. At the north magnetic pole, the total-field anomaly due to a given distribution of magnetization has a phase spectrum given by $\phi(k_x, k_y)$.

(a) What is the phase spectrum of the same magnetization distribution if it is relocated at the magnetic equator? Assume that magnetization is entirely induced.

(b) What is the phase spectrum at the south magnetic pole?

(c) What is the phase spectrum if the local field has an inclination of 63° and a declination of 18°?

4. A total-field anomaly $\Delta T(x, y)$ is measured on a horizontal surface at height z_0 above a uniformly magnetized, rectangular prism with magnetization M, thickness T, and horizontal dimensions D. The prism is at the north magnetic pole and has negligible Koenigsberger ratio.

(a) Describe the amplitude spectrum of $\Delta T(x, y)$ at all wavenumbers (including $k_x = k_y = 0$) in terms of z_0, T, D, and M. Hint: It might be helpful to consider the prism as part of an infinite slab.

(b) Discuss the feasibility of determining z_0, T, D, and M from the amplitude spectrum.

12

Transformations

The third, or depth, dimension of geology is a major frontier at present.

(Jack Oliver)

A theory has only the alternative of being right or wrong. A model has a third possibility: it may be right, but irrelevant.

(Paul R. Ehrlich)

Now we are ready to undertake the third category of potential field methods depicted in Figure 9.1, those methods that facilitate geologic interpretations by transforming measured data into some new form. These transformations, in general, do not directly define the distribution of sources, but they often provide insights that help to build a general understanding of the nature of the sources. Upward continuation, for example, is a method that transforms anomalies measured on one surface into those that would have been measured on some higher surface. The upward-continued anomalies do not provide direct information about the source, but they can be instructive nonetheless. In particular, the process of upward continuation tends to attenuate anomalies caused by local, near-surface sources relative to anomalies caused by deeper, more profound sources.

The aeromagnetic data shown in Figure 12.1 will serve to illustrate some of the methods discussed subsequently. These data are from a region of the Basin and Range geologic province in north-central Nevada. The mountain ranges in this area typically expose sedimentary rocks of Paleozoic age and volcanic tuffs of Tertiary age. The linear magnetic anomaly that trends north-northwest across the entire map (Figure 12.1(a)), however, is associated with exposures of basaltic flows and dikes of middle Miocene age, which are largely concealed. This association between the linear anomaly and basaltic dikes and flows has been interpreted as indicating a rift zone active during the middle Miocene

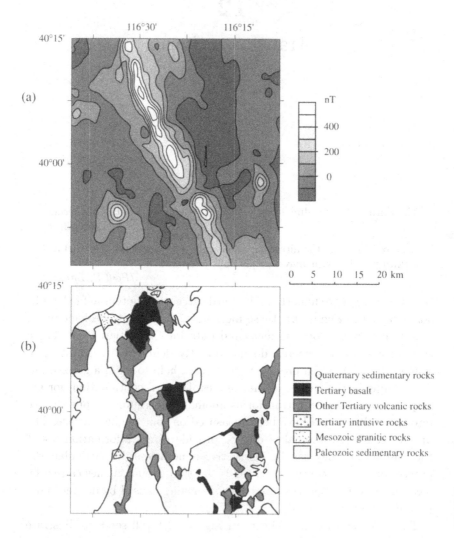

Fig. 12.1. Aeromagnetic data and geology of north-central Nevada. (a) Aero-magnetic compilation from Kucks and Hildenbrand [151]. Contour interval 100 nT. (b) Geologic map simplified from Stewart and Carlson [273]. Linear magnetic anomaly with north-northwest trend is associated with exposures of Tertiary basalt (solid black); isolated anomaly in southwest quadrant is caused by Tertiary intrusive rocks (plus pattern).

(Zoback and Thompson [295]). The isolated anomaly in the southwestern quadrant of the map, on the other hand, apparently is associated with an isolated granitic intrusion of Tertiary age (Grauch et al. [102]).

Some of the techniques discussed subsequently are accompanied by computer subroutines in Appendix B. In some cases, more comprehensive algorithms are available in the literature (e.g., Hildenbrand [127]; Blakely [20], Gibert and Galdeano [96]). Cordell et al. [73] have compiled in a form suitable for IBM-compatible computers a variety of interpretive programs, the most comprehensive, publicly available package of programs to date.

12.1 Upward Continuation

Upward continuation transforms the potential field measured on one surface to the field that would be measured on another surface farther from all sources. As we shall see, this transformation attenuates anomalies with respect to wavelength; the shorter the wavelength, the greater the attenuation. In this sense, the process of upward continuation degrades the measured data, and we might wonder why such a process would have any application at all. Several useful examples come to mind. First, it is sometimes necessary to compare or merge aerial surveys measured at disparate altitudes, and upward continuation provides a way to transform individual surveys onto a consistent surface. Second, upward continuation tends to accentuate anomalies caused by deep sources at the expense of anomalies caused by shallow sources. A magnetic survey over young volcanic terrain, for example, may be dominated by short-wavelength anomalies due to near-surface volcanic rocks; upward continuation can be used to attenuate the shallow-source anomalies in order to emphasize deeper, more profound sources, such as underlying plutonic rocks.

Green's third identity (equation 2.9) shows why upward continuation should be possible. If function U is harmonic, is continuous, and has continuous derivatives throughout a regular region R, then it follows from Green's third identity that the value of U at any point P within R (Figure 12.2) is given by equation 2.10,

$$U(P) = \frac{1}{4\pi} \int_S \left(\frac{1}{r} \frac{\partial U}{\partial n} - U \frac{\partial}{\partial n} \frac{1}{r} \right) dS, \qquad (12.1)$$

where S denotes the boundary of R, n the outward normal direction, and r the distance from P to the point of integration on S. Equation 12.1

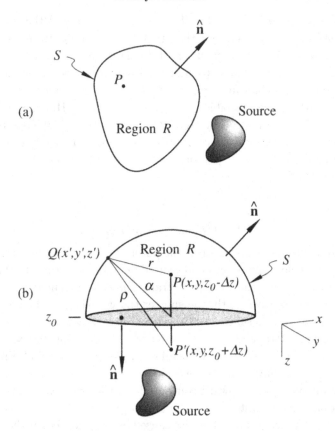

Fig. 12.2. (a) A function harmonic throughout region R can be evaluated at any point within R from its behavior on boundary S. Unit vector $\hat{\mathbf{n}}$ is normal to surface S. (b) Upward continuation from a horizontal surface. Potential field is known on horizontal plane $z = z_0$ and desired at point $P(x, y, z_0 - \Delta z)$ ($\Delta z > 0$). Surface S consists of the horizontal plane plus a hemispherical cap of radius α. Point P' is the mirror image of P projected through the plane. The point of integration Q is on surface S, and r and ρ denote distance from Q to P and from Q to P', respectively.

illustrates the essential principle of upward continuation: A potential field can be calculated at any point within a region from the behavior of the field on a surface enclosing the region. No knowledge is required about the sources of the field, except that none may be located within the region.

12.1.1 Level Surface to Level Surface

The simplest form of continuation is for potential fields measured on a level surface, and here we follow the classical derivation described by Henderson [125, 124]. Using a cartesian coordinate system with z axis directed downward, we assume that the potential field was measured on a level surface at $z = z_0$ and that the field is desired at a single point $P(x, y, z_0 - \Delta z)$ above the level surface, where $\Delta z > 0$. Surface S is composed of both the level surface plus a hemisphere of radius α, as shown in Figure 12.2. All sources lie at $z > z_0$. As α becomes large, it is easy to show that integration of equation 12.1 over the hemisphere becomes small. Hence, as $\alpha \to \infty$,

$$U(x, y, z_0 - \Delta z) =$$

$$\frac{1}{4\pi} \int\limits_{-\infty}^{\infty} \int\limits_{-\infty}^{\infty} \left(\frac{1}{r} \frac{\partial U(x', y', z_0)}{\partial z'} - U(x', y', z_0) \frac{\partial}{\partial z'} \frac{1}{r} \right) dx' \, dy', \quad (12.2)$$

where $r = \sqrt{(x - x')^2 + (y - y')^2 + (z_0 - \Delta z - z')^2}$, and where $\Delta z > 0$.

Unfortunately, equation 12.2 requires not only values of U on the surface but also values of the vertical gradient of U, a combination that is unlikely to be available in most practical applications. Hence, we need a way to eliminate the derivative term in equation 12.2, and as shown in Section 2.3.2, Green's second identity provides a way to do so. If V is another function also harmonic throughout R, then Green's second identity yields

$$\frac{1}{4\pi} \int\limits_{S} \left(V \frac{\partial U}{\partial n} - U \frac{\partial V}{\partial n} \right) dS = 0,$$

and adding this result to equation 12.1 provides

$$U(P) = \frac{1}{4\pi} \int\limits_{S} \left[\left(V + \frac{1}{r} \right) \frac{\partial U}{\partial n} - U \frac{\partial}{\partial n} \left(V + \frac{1}{r} \right) \right] dS. \quad (12.3)$$

To eliminate the first term of the integrand, a harmonic V is needed such that $V + \frac{1}{r} = 0$ at each point of S. We construct P', the mirror image of P, at $(x, y, z_0 + \Delta z)$ and let $V = -\frac{1}{\rho}$ where

$$\rho = \sqrt{(x - x')^2 + (y - y')^2 + (z_0 + \Delta z - z')^2}.$$

Note that V defined in this way satisfies the necessary requirements: $V + \frac{1}{r} = 0$ on the horizontal surface, $V + \frac{1}{r}$ will vanish on the hemisphere as α becomes large, and V is always harmonic because ρ never vanishes. Hence, equation 12.3 becomes

$$U(P) = \frac{1}{4\pi} \int_S \left[\left(\frac{1}{r} - \frac{1}{\rho} \right) \frac{\partial U}{\partial n} - U \frac{\partial}{\partial n} \left(\frac{1}{r} - \frac{1}{\rho} \right) \right] dS .$$

As the hemisphere becomes large, the first term vanishes at each point of S, and the second term vanishes except on the horizontal surface,

$$U(x, y, z_0 - \Delta z) = -\frac{1}{4\pi} \int_{-\infty}^{\infty} \int_{-\infty}^{\infty} U(x', y', z_0) \frac{\partial}{\partial z'} \left[\frac{1}{r} - \frac{1}{\rho} \right] dx' \, dy' .$$

Carrying out the derivative and letting z' move to the horizontal surface leads to

$$U(x, y, z_0 - \Delta z) = \frac{\Delta z}{2\pi} \int_{-\infty}^{\infty} \int_{-\infty}^{\infty} \frac{U(x', y', z_0)}{[(x - x')^2 + (y - y')^2 + \Delta z^2]^{3/2}} \, dx' \, dy' ,$$

$$\Delta z > 0 . \tag{12.4}$$

Equation 12.4 is the *upward-continuation integral.* It shows how to calculate the value of a potential field at any point above a level, horizontal surface from complete knowledge of the field on the surface. Some compromises will be required in practical applications, of course; we never will know the potential field precisely at each point of an infinite plane. It is particularly important to know the field well beyond the lateral extent of all sources, a recommendation that is difficult to implement in practice.

Fourier-Domain Representation

Equation 12.4 can be used to continue data measured on a level surface to another surface, level or not. For each point of the new surface, the two-dimensional integral must be evaluated, a computationally intensive task. The procedure can be made more efficient and insightful, however, if it is recast in the Fourier domain. In doing so, we will have to settle for level-surface-to-level-surface applications.

Note that equation 12.4 is simply a two-dimensional convolution,

$$U(x, y, z_0 - \Delta z) = \int_{-\infty}^{\infty} \int_{-\infty}^{\infty} U(x', y', z_0) \, \psi_u(x - x', y - y', \Delta z) \, dx' \, dy' ,$$

where

$$\psi_{\mathrm{u}}(x, y, \Delta z) = \frac{\Delta z}{2\pi} \frac{1}{(x^2 + y^2 + \Delta z^2)^{3/2}} \, . \tag{12.5}$$

If the potential field U measured on surface $z = z_0$ satisfies inequality 11.3, then it has a Fourier transform $\mathcal{F}[U]$. The Fourier-domain representation of equation 12.4 is found by transforming both sides of equation 12.4 to the Fourier domain and applying the Fourier-convolution theorem (Section 11.1.5),

$$\mathcal{F}[U_{\mathrm{u}}] = \mathcal{F}[U]\,\mathcal{F}[\psi_{\mathrm{u}}] \, , \tag{12.6}$$

where $\mathcal{F}[U_{\mathrm{u}}]$ is the Fourier transform of the upward-continued field. All that is needed is an analytical expression for $\mathcal{F}[\psi_{\mathrm{u}}]$, which can be found from the Fourier transform of equation 12.5. First note that

$$\psi_{\mathrm{u}}(x, y, \Delta z) = -\frac{1}{2\pi} \frac{\partial}{\partial \Delta z} \frac{1}{r} \, , \tag{12.7}$$

where $r = \sqrt{x^2 + y^2 + \Delta z^2}$. With the help of equation 11.21, therefore, the Fourier transform of equation 12.7 is given by

$$\mathcal{F}[\psi_{\mathrm{u}}] = -\frac{1}{2\pi} \frac{\partial}{\partial \Delta z} \mathcal{F}\left[\frac{1}{r}\right]$$

$$= -\frac{\partial}{\partial \Delta z} \frac{e^{-|k|\Delta z}}{|k|}$$

$$= e^{-\Delta z|k|}, \qquad \Delta z > 0 \, . \tag{12.8}$$

A level-to-level continuation can be achieved, therefore, by Fourier transforming the measured data, multiplying by the exponential term of equation 12.8, and inverse Fourier transforming the product.

It is clear from equation 12.8 and shown by Figure 12.3 that (1) the process of upward continuation attenuates all wavenumbers except $|k| = 0$, (2) each wavenumber is attenuated to a greater degree than all lower wavenumbers, and (3) the degree of attenuation increases with increasing Δz. Equation 12.8 is a real function, that is, it has no phase component, and consequently imparts no phase changes to the upward-continued field. As an example, Figure 12.4 shows the total-field anomaly of central Nevada (Figure 12.1(a)) continued upward 5 km. Notice that the shortest wavelengths of the original anomalies are essentially eliminated in Figure 12.4, whereas the fundamental anomalies remain in a smoother form.

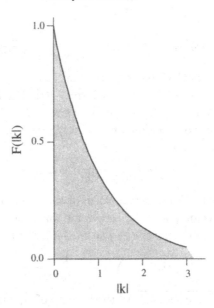

Fig. 12.3. Fourier-domain representation of level-to-level upward continuation. Each wavenumber is attenuated with respect to all lower wavenumbers. The filter is a real function, that is, it imparts no phase changes to the original anomaly.

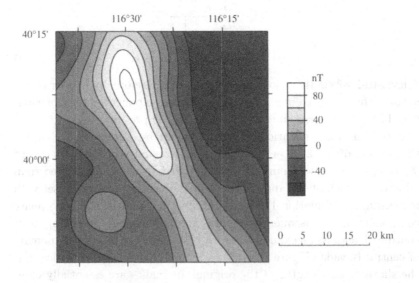

Fig. 12.4. The total-field anomaly of Figure 12.1(a) continued upward 5 km.

The function U in the previous discussion can be any potential field; that is, equations 12.4 and 12.6 apply to any component of a gravity or magnetic field measured on a horizontal surface. It also applies to the total-field magnetic anomaly, under the assumptions described in Section 8.3. Subroutine B.24 in Appendix B is a Fortran implementation of equations 12.6 and 12.8, appropriate for potential fields specified on a rectangular grid.

It should be clear that equation 12.6 is strictly valid only for level-to-level continuation, because U is specified on a level surface and Δz is constant. Cordell [67], however, adapted equation 12.6 for continuation to uneven surfaces. In his method, equations 12.6 and 12.8 are used to find the field on a series of level surfaces at successively higher elevations, some above and some below the desired uneven surface. The field at any point of the uneven surface can be found approximately by interpolating in the vertical direction using corresponding values from the level-to-level continuations. The method is called "chessboard" (because the stack of gridded data brings to mind a three-dimensional chess game) and is available as a Fortran algorithm (Cordell et al. [73]).

Downward Continuation

The foregoing discussion has assumed that all sources are located below the observation surface and that all points of the continuation are above the observation surface, that is, continuation is away from all sources. On first consideration, it may seem legitimate to try continuing measured data into regions *closer* to sources, so long as we are certain that no sources actually exist in the region of continuation. This calculation, called *downward continuation*, would be very useful in an interpretation of gravity or magnetic data because it would tend to accentuate the details of the source distribution, especially the shallowest components.

Downward continuation is, however, a risky proposition. Upward continuation is a smoothing operation. This is easily seen in equation 12.4 where $U(x, y, z_0 - \Delta z)$ at any point is simply the weighted average of all values of $U(x, y, z_0)$. Downward continuation, on the other hand, is the calculation of $U(x, y, z_0)$ from $U(x, y, z_0 - \Delta z)$, the inverse of equation 12.4. It is an "unsmoothing" operation, and as discussed in Section 10.1, such calculations are unstable. Small changes to $U(x, y, z_0 - \Delta z)$ can cause large and unrealistic variations in the calculated $U(x, y, z_0)$.

This problem is demonstrated by writing the inverse of equation 12.6,

$$\mathcal{F}[U] = \mathcal{F}[U_u]\,\mathcal{F}^{-1}[\psi_u]$$

$$= \mathcal{F}[U_u]\,e^{+|k|\Delta z}\,.$$

In this case, $\mathcal{F}[U_u]$ is the Fourier transform of the observed field, and $\mathcal{F}[U]$ is the desired field continued downward a distance Δz. Clearly the shortest wavelengths of the measured data will be greatly amplified by this procedure to a degree that depends on the value of Δz and the sample interval of the data. Any errors present and perhaps undetected in the measured data may appear in the calculated field as large and unrealistic variations. These complexities have obvious similarities with the inverse problem, and the subject of downward continuation has been treated formally as such (e.g., Huestis and Parker [133]; Courtillot, Duncruix, and Le Mouël [77]). Downward continuation is often used in spite of the potential peril. Indeed each of the upward-continuation methods discussed subsequently has been formulated for downward continuation.

12.1.2 Uneven Surfaces

As demonstrated by equations 12.6 and 12.8, continuing potential fields upward from one level surface to another level surface is a straightforward calculation. Continuing from a level surface to an uneven surface can be achieved with equation 12.4, which requires the evaluation of a two-dimensional integral at each desired point of the new surface, or it might be estimated more efficiently with the chessboard technique of Cordell [67].

The problem is more challenging, however, when the data are measured on an uneven surface. A variety of techniques have been described in recent years. Several authors (Courtillot, Ducruix, and Le Mouël [77]; Ducruix, Le Mouël, and Courtillot [80]; Huestis and Parker [133]) have treated the calculation as a formal inverse problem using the formalism of Backus and Gilbert [8]. Parker and Klitgord [209] used the Schwarz–Christoffel transformation to map magnetic measurements along an uneven profile onto a horizontal line, a method applicable only to profile data measured over assumed two-dimensional sources. Two other methods are outlined in the following sections.

Equivalent Source

If a source distribution could be found that generates an observed potential field, that distribution could be used to calculate the field anywhere

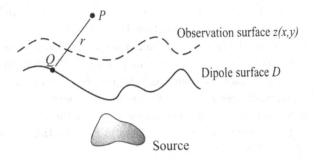

Fig. 12.5. Equivalent source.

above the original measurements. This two-step procedure, an inverse problem followed by a forward calculation, would provide a way to continue potential fields from surface to surface. The calculated source in all likelihood will not resemble the true source distribution in any way, but this is of no importance in this application; Green's third identity (Chapter 2) assures us that alternate sources can cause the same potential field in restricted regions. The source distribution must produce a potential field that is harmonic in the area of interest, vanishes at infinity, and reproduces the observed field. This is the essential logic behind the *equivalent-source* technique (Dampney [79], Bhattacharyya and Chan [17], Pedersen [218], Emilia [83], Nakatsuka [193], Hansen and Miyazaki [111], Arkani-Hamed [5]).

As before, we desire the field at point P above the surface on which the field has been measured (Figure 12.5). The observation surface is now uneven, and its elevation can be represented as a function of horizontal position $z(x, y)$. In this example, we following Hansen and Miyazaki [111] and assume that the potential at P is caused by a double distribution $\mathbf{m}$ spread over a surface D,

$$U(P) = \int_D \mathbf{m} \cdot \nabla_Q \frac{1}{r} \, dS$$

$$= \int_D m(x', y') \frac{\hat{\mathbf{m}} \cdot \hat{\mathbf{r}}}{r^2} \, dS, \qquad (12.9)$$

where $Q(x', y')$ is the point of integration and $\mathbf{r}$ is the vector of length r directed from Q to P (Figure 12.5). The double distribution has the same form as a spread of magnetic dipoles, but this does not limit its application to magnetic fields; U can represent a potential field of any kind. The shape and location of surface D is yet to be determined, but it should lie at or below the observation surface and above all true sources. This source distribution will achieve our goals if $U(P)$ is harmonic above the observation surface, vanishes as z approaches $-\infty$, and equals the observed field whenever P lies on the observation surface. The first two conditions are satisfied because equation 12.9 is essentially the potential field of a dipole distribution (see Section 5.1). The third condition is satisfied if we select $m(x,y)$ so that

$$U(x, y, z(x,y)) = \int\limits_{D} m(x', y') \frac{\hat{\mathbf{m}} \cdot \hat{\mathbf{r}}}{r^2} \, dS \, . \qquad (12.10)$$

Because the left side of equation 12.10 is known, the problem reduces to (1) selecting models for the unit vector $\hat{\mathbf{m}}$ (which is not necessarily uniform) and surface D, (2) finding a solution for $m(x, y)$, and (3) using $m(x, y)$ in equation 12.9 to find $U(P)$, the upward-continued field.

If P is always above (or on) the observation surface, then D can be placed *at* the observation surface, in which case equation 12.10 becomes

$$U(x, y, z(x,y)) = 2\pi m(x, y)\, \hat{\mathbf{m}} \cdot \hat{\mathbf{n}} + \int\limits_{D=z(x,y)} m(x', y') \frac{\hat{\mathbf{m}} \cdot \hat{\mathbf{r}}}{r^2} \, dS \, , \quad (12.11)$$

where $\hat{\mathbf{n}}$ is normal to the observation surface (Hansen and Miyazaki [111]). Equation 12.11 can be solved for $m(x, y)$ by the methods of successive approximations once a model is selected for $\hat{\mathbf{m}}$. Bhattacharyya and Chan [17] oriented $\hat{\mathbf{m}}$ normal to the observation surface, whereas Nakatsuka [193] let $\hat{\mathbf{m}}$ be vertical. In practice, the observed potential field usually is interpolated to a grid, so the surface integrals in equations 12.9 and 12.11 can be replaced with a double summation. The spread of dipoles can be approximated by discrete dipoles at grid intersections (e.g., Bhattacharyya and Chan [17]) or by rectangular facets with each facet centered about a grid intersection and possessing uniform dipole moment (Hansen and Miyazaki [111]).

Taylor's-Series Approximation

A Taylor's series uses derivatives of a function evaluated at one point to extrapolate the function to nearby points. Taylor's series was discussed in Section 10.3.1 as a way to modify in an optimal way the shape of causative bodies in order to solve the nonlinear inverse problem. Taylor's series also can be used to predict the value of a potential field at points away from the observation surface (Cordell and Grauch [70], Pilkington and Roest [224]).

First consider the continuation of a potential field $U(x, y, z_0)$ measured on a level surface ($z_0 = $ constant) and desired on an uneven surface $z(x, y)$. The value of the potential field at one point (x, y, z) of the new surface is given by

$$U(x, y, z) = U(x, y, z_0) + (z - z_0)\frac{\partial}{\partial z}U(x, y, z_0)$$

$$+ \frac{(z - z_0)^2}{2}\frac{\partial^2}{\partial z^2}U(x, y, z_0) + \cdots$$

$$= \sum_{n=0}^{\infty} \frac{(z - z_0)^n}{n!}\frac{\partial^n}{\partial z^n}U(x, y, z_0). \tag{12.12}$$

Cordell and Grauch [70] found empirically that convergence of equation 12.12 is most rapid if z_0 is placed at the mean of $z(x, y)$, and this can be achieved with a level-to-level continuation using equations 12.6 and 12.8. A solution to equation 12.12 requires vertical derivatives of the measured field, and these can be found using the Fourier domain. We will show in Section 12.2 that the Fourier transform of the nth vertical derivative of a potential field is given by the Fourier transform of the potential field times $|k|^n$, that is,

$$\mathcal{F}\left[\frac{\partial^n}{\partial z^n}U\right] = |k|^n \mathcal{F}[U]. \tag{12.13}$$

Using equation 12.13, the various vertical derivatives of the observed field can be found, and these can be used in summation 12.12 to find the field on surface $z(x, y)$. Cordell and Grauch [70] found that the first three terms of the summation are generally sufficient to achieve satisfactory results.

Exercise 12.1 Show that if $z(x, y)$ is a constant, equation 12.12 reduces to equations 12.6 and 12.8. Hint: Equation 12.13 may be helpful.

Equation 12.12 is appropriate for continuation from a level surface to an uneven surface. The more difficult case of continuation from an uneven surface can be achieved by rearranging the terms of equation 12.12. Isolating the first term of the summation on the left side of the equation yields

$$U(x, y, z_0) = U(x, y, z) - \sum_{n=1}^{\infty} \frac{(z - z_0)^n}{n!}\frac{\partial^n}{\partial z^n}U(x, y, z_0). \tag{12.14}$$

The desired quantity $U(x, y, z_0)$ can be estimated by successive approximations; that is, $U(x, y, z_0)$ determined at the ith iteration can be used to find $U(x, y, z_0)$ at the $(i + 1)$th iteration,

$$U(x, y, z_0)^{(i+1)} = U(x, y, z) - \sum_{n=1}^{\infty} \frac{(z - z_0)^n}{n!} \frac{\partial^n}{\partial z^n} U(x, y, z_0)^{(i)} .$$

An initial guess at $U(x, y, z_0)$ is needed. Cordell and Grauch [70] suggested on empirical grounds that this initial guess can be found by assuming (erroneously) that $U(x, y, z)$ is actually measured on a horizontal surface and continuing it to the uneven surface $z(x, y)$ using summation 12.12. As before, the vertical derivatives in equation 12.14 can be estimated with equation 12.13, although this relationship does not strictly apply to data measured on uneven surfaces.

Grauch [100] described a Fortran program implementing the Taylor's series method, and this program is available on diskette (Cordell et al. [73]).

12.2 Directional Derivatives

Consider a smoothly varying scalar quantity $\phi(x, y)$ measured on a horizontal surface. The horizontal derivatives of $\phi(x, y)$ are easily estimated using simple finite-difference methods and discrete measurements of $\phi(x, y)$. For example, if the values ϕ_{ij}, $i = 1, 2, \ldots, j = 1, 2, \ldots$, represent discrete measurements of $\phi(x, y)$ at uniform sample intervals Δx and Δy, then the horizontal derivatives of $\phi(x, y)$ at point i, j are given approximately by

$$\frac{d\phi(x, y)}{dx} \approx \frac{\phi_{i+1,j} - \phi_{i-1,j}}{2\Delta x} ,$$

$$\frac{d\phi(x, y)}{dy} \approx \frac{\phi_{i,j+1} - \phi_{i,j-1}}{2\Delta y} .$$

Horizontal derivatives are easily done in the Fourier domain as well (e.g., Pedersen [217]). According to the differentiation theorem (Section 11.1.2), the horizontal derivatives of $\phi(x, y)$ are given by

$$\mathcal{F}\left[\frac{d^n \phi}{dx^n}\right] = (ik_x)^n \mathcal{F}[\phi] , \qquad (12.15)$$

$$\mathcal{F}\left[\frac{d^n \phi}{dy^n}\right] = (ik_y)^n \mathcal{F}[\phi] . \qquad (12.16)$$

Hence, $(ik_x)^n$ and $(ik_y)^n$ are filters that transform a function measured on a horizontal surface into nth-order derivatives with respect to x or y, respectively.

If ϕ is a potential, we also have the ability to calculate vertical gradients. Indeed, the second vertical derivative is a direct consequence of Laplace's equation, for if ϕ is a potential, then $\nabla^2 \phi = 0$ and

$$\frac{\partial^2 \phi}{\partial z^2} = -\frac{\partial^2 \phi}{\partial x^2} - \frac{\partial^2 \phi}{\partial y^2}. \tag{12.17}$$

If ϕ is measured on a horizontal surface, then Laplace's equation can be transformed to the Fourier domain with the help of equations 12.15 and 12.16, that is,

$$\mathcal{F}\left[\frac{\partial^2 \phi}{\partial z^2}\right] = k_x^2 \, \mathcal{F}[\phi] + k_y^2 \, \mathcal{F}[\phi]$$

$$= |k|^2 \, \mathcal{F}[\phi]. \tag{12.18}$$

Hence, the second vertical derivative of a potential field measured on a horizontal surface is framed as a three-step filtering operation: Fourier transform the potential field, multiply by $|k|^2$, and inverse Fourier transform the product.

The second vertical derivative was an early mainstay of interpretation techniques (Evjen [86], Henderson and Zietz [126]) because it helps to resolve and accentuate shallow sources. To see why this should be the case, consider two monopoles observed at point P, one at shallow depth d_1 and the other at greater depth d_2. The field of each monopole is inversely proportional to the squared distance to P. Hence, as P moves toward the monopoles, the field due to the shallow monopole will increase more rapidly than the field of the deep monopole. It stands to reason that the second vertical derivative will have the same effect. Similarly, the second vertical derivative accentuates and helps to resolve edges of magnetic or gravity sources. These characteristics are illustrated by Figure 12.6, which shows the second vertical derivative of the total-field anomaly of central Nevada.

These properties of the second vertical derivative also follow from equation 12.18: Multiplying the potential field by $|k|^2$ clearly amplifies short-wavelength components of the field at the expense of long-wavelength components. Needless to say, all of the cautionary comments made earlier concerning downward continuation are applicable to the second vertical derivative as well.

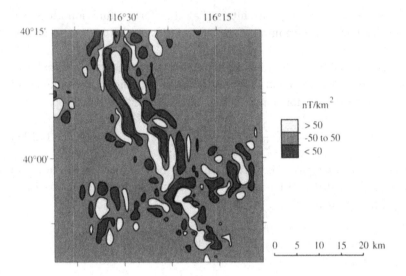

Fig. 12.6. The second vertical derivative of the total-field anomaly of Figure 12.1.

The second vertical derivative is a direct consequence of Laplace's equation. Indeed, vertical derivatives of any order are obtainable from a potential field. This follows from the earlier discussion on upward continuation. Using the usual conventions that z increases downward and that $\Delta z > 0$, the vertical derivative of first order is given by

$$\frac{\partial}{\partial z}\phi(x,y,z) = \lim_{\Delta z \to 0} \frac{\phi(x,y,z) - \phi(x,y,z - \Delta z)}{\Delta z},$$

and transforming to the Fourier domain yields

$$\mathcal{F}\left[\frac{\partial \phi}{\partial z}\right] = \lim_{\Delta z \to 0} \frac{\mathcal{F}[\phi] - \mathcal{F}[\phi]e^{-|k|\Delta z}}{\Delta z}$$

$$= \lim_{\Delta z \to 0} \frac{1 - e^{-|k|\Delta z}}{\Delta z}\mathcal{F}[\phi]$$

$$= |k|\,\mathcal{F}[\phi].$$

In a similar fashion, we could show that the nth-order vertical gradient is equal to the Fourier transform of the potential times $|k|^n$, or in general,

$$\mathcal{F}\left[\frac{\partial^n \phi}{\partial z^n}\right] = |k|^n \mathcal{F}[\phi]. \tag{12.19}$$

Subroutine B.25 in Appendix B is an implementation of the three steps implied by equation 12.19; it Fourier transforms gridded potential-field data, multiplies by $|k|^n$, and inverse Fourier transforms the product.

Transformations from Field to Potential

Equations 12.15, 12.16, and 12.19 provide ways to calculate derivatives in all three orthogonal directions from a potential field measured on a horizontal surface. We can write these relationships more compactly as a single equation,

$$\mathcal{F}\left[\nabla\phi\right] = \mathbf{k}\,\mathcal{F}\left[\phi\right], \tag{12.20}$$

where $\mathbf{k} = (ik_x, ik_y, |k|)$, as shown by Pedersen [216] using a somewhat different derivation.

A force field $\mathbf{F}$ is related to its potential ϕ according to $\mathbf{F} = \nabla\phi$, and equation 12.20 expresses this relationship in the Fourier domain. The component of the force field in any direction $\hat{\mathbf{f}}$, therefore, is simply a directional derivative of ϕ, given in the Fourier domain by

$$\mathcal{F}\left[\hat{\mathbf{f}} \cdot \nabla\phi\right] = (\hat{\mathbf{f}} \cdot \mathbf{k})\,\mathcal{F}\left[\phi\right]. \tag{12.21}$$

The vertical attraction of gravity, for example, is given by $g_z = \frac{\partial}{\partial z}U$, or in the Fourier domain by

$$\mathcal{F}\left[g_z\right] = |k|\,\mathcal{F}\left[U\right]. \tag{12.22}$$

Similarly, the total-field anomaly measured in an ambient field $\mathbf{f}$ is, to a good approximation, given by $\Delta T = -\hat{\mathbf{f}} \cdot \nabla V$, where V is the magnetic potential of the anomalous body. In the Fourier domain this is expressed by

$$\mathcal{F}\left[\Delta T\right] = -(\hat{\mathbf{f}} \cdot \mathbf{k})\,\mathcal{F}\left[V\right]$$

$$= -\Theta_{\mathrm{f}}|k|\,\mathcal{F}\left[V\right], \tag{12.23}$$

where Θ_{f} was defined in Section 11.2.1 as

$$\Theta_{\mathrm{f}} = \hat{f}_z + i\frac{\hat{f}_x k_x + \hat{f}_y k_y}{|k|}.$$

Equations 12.22 and 12.23 express the vertical attraction of gravity and the total-field anomaly in terms of their respective potentials. Rearranging terms provides

$$\mathcal{F}\left[U\right] = \frac{1}{|k|}\mathcal{F}\left[g_z\right], \qquad |k| \neq 0, \tag{12.24}$$

$$\mathcal{F}[V] = -\frac{1}{\Theta_f |k|} \mathcal{F}[\Delta T], \qquad |k| \neq 0, \qquad (12.25)$$

and these equations describe vertical gravity anomalies and total-field magnetic anomalies measured on a horizontal surface in terms of their respective potentials on that same surface.

12.3 Phase Transformations

The shape of a gravity anomaly naturally depends on the shape and distribution of mass, as described by the density distribution $\rho(x, y, z)$. Magnetic anomalies have an added complexity: The anomaly depends not only on the distribution of magnetization $M(x, y, z)$, but also on the direction of magnetization and on the direction in which the field is measured.† For a total-field anomaly, of course, the component of measurement is parallel to the ambient field.

This added complexity is easily described in the Fourier domain. Consider a three-dimensional distribution of magnetization $M(x, y, z)$ located entirely below the plane of observation at z_0. The distribution will be defined as zero outside a finite region with horizontal dimensions smaller than the dimensions of the survey, and we assume that the direction of magnetization (but not the intensity of magnetization) is uniform throughout the body. The Fourier transform of the total-field anomaly caused by $M(x, y, z)$ is given by equation 11.47,

$$\mathcal{F}[\Delta T] = 2\pi C_m \Theta_m \Theta_f |k| e^{|k| z_0} \int_{z_0}^{\infty} e^{-|k| z'} \mathcal{F}[M(z')] \, dz', \qquad (12.26)$$

where $\mathcal{F}[M(z')]$ denotes the Fourier transform of the magnetization on one horizontal slice through the body at depth z'. The parameters Θ_m and Θ_f in equation 12.26 are given by

$$\Theta_m = \hat{m}_z + i \frac{\hat{m}_x k_x + \hat{m}_y k_y}{|k|},$$

$$\Theta_f = \hat{f}_z + i \frac{\hat{f}_x k_x + \hat{f}_y k_y}{|k|},$$

where $\hat{\mathbf{m}} = (\hat{m}_x, \hat{m}_y, \hat{m}_z)$ and $\hat{\mathbf{f}} = (\hat{f}_x, \hat{f}_y, \hat{f}_z)$ are unit vectors in the

---

† Of course, the shape of a gravity anomaly also depends on the direction in which the field is measured, but conventional gravity meters measure the total gravity field of the earth, which is always vertical.

direction of the magnetization and in the direction of the ambient field, respectively. The integral term in equation 12.26 carries all of the information concerning $M(x, y, z)$, including the shape of the magnetic body. This term certainly will contribute to the phase of the anomaly, depending on how the magnetization is distributed, and this contribution will be difficult to assess without some a priori knowledge about $M(x, y, z)$. The functions Θ_m and Θ_f also contribute to the phase of the anomaly and contain all information regarding the directions of the magnetization and of the component being measured.

Suppose that a different component of the magnetic field or the effect of a different direction of magnetization is of interest. In either case, only $\hat{f}$ and $\hat{m}$ are to be altered in equation 12.26; $M(x, y, z)$ remains unchanged. Denoting the new directions of magnetization and ambient field as $\hat{m}' = (\hat{m}'_x, \hat{m}'_y, \hat{m}'_z)$ and $\hat{f}' = (\hat{f}'_x, \hat{f}'_y, \hat{f}'_z)$, respectively, the transformed anomaly will be

$$\mathcal{F}\left[\Delta T_t\right] = 2\pi C_m \Theta'_m \Theta'_f |k| \, e^{|k|z_0} \int_{z_0}^{\infty} e^{-|k|z'} \mathcal{F}\left[M(z')\right] dz' , \qquad (12.27)$$

where

$$\Theta'_m = \hat{m}'_z + i\frac{\hat{m}'_x k_x + \hat{m}'_y k_y}{|k|}$$

$$\Theta'_f = \hat{f}'_z + i\frac{\hat{f}'_x k_x + \hat{f}'_y k_y}{|k|} .$$

Combining equations 12.26 and 12.27 in order to eliminate the common factors provides

$$\mathcal{F}\left[\Delta T_t\right] = \mathcal{F}\left[\Delta T\right] \mathcal{F}\left[\psi_t\right] , \qquad (12.28)$$

where

$$\mathcal{F}\left[\psi_t\right] = \frac{\Theta'_m \Theta'_f}{\Theta_m \Theta_f} . \qquad (12.29)$$

Equations 12.28 and 12.29 describe a filtering operation that transforms a total-field anomaly with given directions of magnetization and ambient field into a new anomaly caused by the same distribution of magnetization but with new vector directions. No assumptions are necessary concerning the shape of the body or the distribution of magnetization, except that it should be sufficiently localized for the Fourier transform to exist.

The denominator in equation 12.29 can be zero in parts of the k_x, k_y plane for certain orientations of $\hat{\mathbf{m}}$ and $\hat{\mathbf{f}}$, in which case $\mathcal{F}\left[\psi_t\right]$ is not defined. If $\hat{\mathbf{m}}$ and $\hat{\mathbf{f}}$ are nearly parallel to these singular directions, $\mathcal{F}\left[\psi_t\right]$ can reach high amplitudes in parts of the k_x, k_y plane, and the filtering operation of equation 12.28 may be unstable. Some of these situations will be discussed in Section 12.3.1. Moreover, the imaginary part of equation 12.29 is discontinuous through the origin of the k_x, k_y plane, and this discontinuity can affect the long-wavelength parts of ΔT (Kis [147]).

The steps involved in applying equation 12.28 are (1) Fourier transform the measured ΔT, (2) multiply by $\mathcal{F}\left[\psi_t\right]$, and (3) inverse Fourier transform the product. Subroutine B.26 in Appendix B shows the implementation of this three-step procedure.

12.3.1 Reduction to the Pole

Positive gravity anomalies tend to be located over mass concentrations, but the same is not necessarily true for magnetic anomalies when the magnetization and ambient field are not both directed vertically. Unless $\hat{\mathbf{m}}$ and $\hat{\mathbf{f}}$ are both vertical, Θ_m and Θ_f will contribute a phase to the magnetic anomaly, which can shift the anomaly laterally, distort its shape, and even change its sign (Figure 12.7).

Exercise 12.2 What directions for $\hat{\mathbf{m}}$ and $\hat{\mathbf{f}}$ cause the phase contribution of $\Theta_m \Theta_f$ to be -1?

In general terms, if the magnetization and ambient field are not vertical, a symmetrical distribution of magnetization (such as a uniformly magnetized sphere) will produce a "skewed" rather than a symmetrical magnetic anomaly (see Figure 4.9 for several extreme examples).

This added complexity can be eliminated from a magnetic survey by using equations 12.28 and 12.29. If we require $\hat{\mathbf{m}}' = \hat{\mathbf{f}}' = (0,0,1)$ in

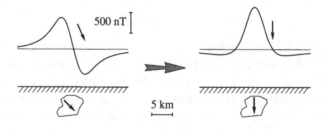

Fig. 12.7. A magnetic anomaly before and after being reduced to the pole.

equation 12.29, then equation 12.28 will transform a measured total-field anomaly into the vertical component of the field caused by the same source distribution magnetized in the vertical direction. The transformed anomaly in the Fourier domain is given by

$$\mathcal{F}\left[\Delta T_{\mathrm{r}}\right] = \mathcal{F}\left[\psi_{\mathrm{r}}\right]\mathcal{F}\left[\Delta T\right], \qquad (12.30)$$

where

$$\mathcal{F}\left[\psi_{\mathrm{r}}\right] = \frac{1}{\Theta_{\mathrm{m}}\Theta_{\mathrm{f}}}$$

$$= \frac{|k|^2}{a_1 k_x^2 + a_2 k_y^2 + a_3 k_x k_y + i|k|(b_1 k_x + b_2 k_y)}, \quad |k| \neq 0, \quad (12.31)$$

$$a_1 = \hat{m}_z \hat{f}_z - \hat{m}_x \hat{f}_x,$$

$$a_2 = \hat{m}_z \hat{f}_z - \hat{m}_y \hat{f}_y,$$

$$a_3 = -\hat{m}_y \hat{f}_x - \hat{m}_x \hat{f}_y,$$

$$b_1 = \hat{m}_x \hat{f}_z + \hat{m}_z \hat{f}_x,$$

$$b_2 = \hat{m}_y \hat{f}_z + \hat{m}_z \hat{f}_y.$$

The application of $\mathcal{F}\left[\psi_{\mathrm{r}}\right]$ is called *reduction to the pole* (Baranov and Naudy [10]) because ΔT_{r} is the anomaly that would be measured at the north magnetic pole, where induced magnetization and ambient field both would be directed vertically down (Figure 12.7). Reduction to the pole removes one level of complexity from the interpretive process: It shifts anomalies laterally to be located over their respective sources and alters their shape so that symmetrical sources cause symmetrical anomalies. The direction of magnetization and ambient field are required in equation 12.31, but no other assumptions about the distribution of magnetization are necessary, except those concerning the lateral dimensions of the sources described earlier.

Many of the comments concerning Θ_{m} and Θ_{f} in Section 11.2.1 apply to $\mathcal{F}\left[\psi_{\mathrm{r}}\right]$. The filter attains a constant value along any ray projected from the origin of the k_x, k_y plane. Two rays directed in opposite directions are complex conjugates of one another, so the imaginary part of $\mathcal{F}\left[\psi_{\mathrm{r}}\right]$ is discontinuous at the origin. Each ray, in general, differs in value from neighboring rays, but the average of $\mathcal{F}\left[\psi_{\mathrm{r}}\right]$ around any circle concentric about the origin is independent of the radius of the circle. Thus reduction to the pole has no effect on the shape of the radially averaged spectrum.

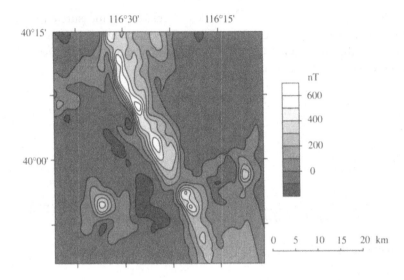

Fig. 12.8. The total-field anomaly of Figure 12.1 reduced to the pole.

Exercise 12.3 Demonstrate the validity of each of these points using equation 12.31.

Figure 12.8 shows the total-field anomaly from central Nevada (Figure 12.1(a)) reduced to the pole. Notice that the transformed anomalies are generally more symmetric than their original counterparts. In particular, the isolated anomaly in the southwestern part of the map is more centrally located over its apparent source, a Tertiary granitic pluton (Figure 12.1(b)). Subroutine B.26 in Appendix B can be used for reduction to the pole by specifying vertical inclinations for magnetization and ambient field.

Problems at Low Latitudes

Figure 12.9 shows two examples of the amplitude and phase of $\mathcal{F}[\psi_r]$: Figure 12.9(a) assumes an inclination of 60° for both the ambient field and magnetization, typical of a mid-latitude survey, whereas Figure 12.9(b) assumes an inclination of 10°, as might be found near the magnetic equator. Both amplitude spectra in Figure 12.9 vary smoothly throughout the k_x, k_y plane (except at the origin), but the amplitude spectrum in Figure 12.9(b) attains large values within narrow pie-shaped segments. Moreover, these high-amplitude segments extend to the longest

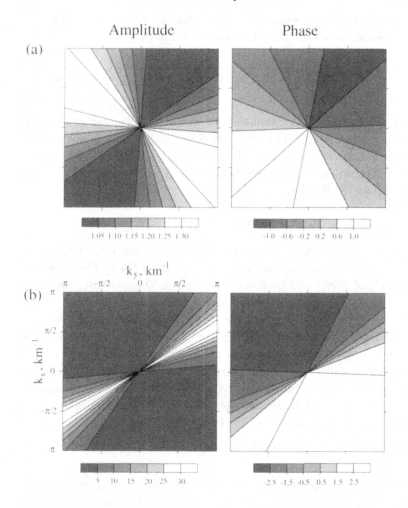

Fig. 12.9. Amplitude and phase spectra for two reduction-to-pole filters. (a) Mid-latitude case, with inclination 60°, declination 30°; (b) low-latitude case, with inclination 10°, declination −30°.

wavenumbers (shortest wavelengths) of the spectrum. This is a characteristic of $\mathcal{F}\left[\psi_r\right]$ in low-latitude situations. In applying equation 12.30 in such situations, the measured total-field anomaly, and any noise included within the measurements, will experience this high, directionally selective amplification. The result can appear as short-wavelength artifacts elongated in the direction of the declination.

We can better understand the reason for this selective amplification by converting equation 12.31 to polar coordinates,

$$\mathcal{F}[\psi_r] = \frac{1}{a_1 \cos^2 \lambda + a_2 \sin^2 \lambda + a_3 \cos \lambda \sin \lambda + i(b_1 \cos \lambda + b_2 \sin \lambda)},$$
$$(12.32)$$

where $\lambda = \arctan(k_y/k_x)$. Equation 12.32 can be calculated at all points of the k_x, k_y plane (except at $|k| = 0$) if the magnetization and ambient field have nonzero inclinations, but $\mathcal{F}[\psi_r]$ is undefined for certain λ if the magnetization and ambient field are both horizontal. For example, if $\hat{\mathbf{m}} = \hat{\mathbf{f}} = (1, 0, 0)$, then

$$\mathcal{F}[\psi_r] = \frac{-1}{\cos^2 \lambda},$$

and in this case $\mathcal{F}[\psi_r]$ is not defined along the k_y axis. If the inclinations are shallow but nonzero, as they are in Figure 12.9(b), the amplitude spectrum of $\mathcal{F}[\psi_r]$ will be defined at each point away from the origin but will attain large values within narrow pie-shaped segments.

Several techniques have been developed to improve reduction to the pole in low-latitude situations. One straightforward approach is to reduce low-latitude magnetic data to the magnetic equator rather than to the pole (Leu [164], Gibert and Galdeano [96]). Although this procedure will tend to center anomalies over their respective sources, the anomalies will be stretched in the east–west direction relative to the horizontal dimensions of the source (e.g., see Figure 4.9(b)).

Exercise 12.4 Derive an expression like equation 12.31 for *reduction to the equator*.

Reduction to the pole can be achieved using an equivalent-source scheme (Silva [258], Emilia [83], Bott and Ingles [40]), much in the same way that upward continuation was treated in Section 12.1.2. Such methods consist of two steps: a solution to the inverse problem followed by a forward calculation. First, the observed anomaly is used to find a hypothetical source distribution that can produce the observed anomaly, assuming an appropriate direction of magnetization for the survey site. Second, the vertical component of the magnetic field is found assuming that the equivalent source is vertically magnetized. The equivalent source most likely will have no resemblance to the true source, but this is of no importance so long as it produces the observed magnetic field. Equivalent sources typically are modeled as a layer of discrete magnetic sources, such as pairs of monopoles of opposite sign (Silva [258]), lines

of dipoles (Emilia [83]), or uniformly magnetized blocks (Bott and Ingles [40]). According to Silva [258], the equivalent-source technique is
largely free of the instabilities often associated with reduction to the
pole of low-latitude data. But this improvement comes at the expense of
significantly more computational effort as compared to the application
of equation 12.30.

Hansen and Pawlowski [112] and Pearson and Skinner [214] applied
Wiener filtering to the problem of low-latitude reduction to the pole.
Equation 12.30 can be considered to be the inverse of

$$\mathcal{F}[\Delta T] = \mathcal{F}^{-1}[\psi_r]\,\mathcal{F}[\Delta T_r]\,.$$

This operation is always stable, as can be demonstrated with the recipro-
cal of equation 12.32. In real situations, an observed total-field anomaly
ΔT_0 will include a noise component N so that $\Delta T_0 = \Delta T + N$. The
Fourier transform of the observed anomaly is

$$\mathcal{F}[\Delta T_0] = \mathcal{F}^{-1}[\psi_r]\,\mathcal{F}[\Delta T_r] + \mathcal{F}[N]\,.$$

An operator α is then required such that $|\alpha\mathcal{F}[\Delta T_0] - \mathcal{F}[\Delta T_r]|^2$ is a
minimum when averaged over all values. The filter α is a regulated ver-
sion of $\mathcal{F}[\psi_r]$ suitable for noisy data. According to Wiener theory, α can
be written in terms of $\mathcal{F}[\psi_r]$, which can easily be calculated, and the
power spectra of N and ΔT. The spectra can be found by assuming the
statistical nature of the noise and the ideal spectral characteristics of
reduced-to-pole anomalies. Hansen and Pawlowski [112] reported significant improvement using this technique in low-latitude applications of
reduction to the pole.

Nonuniform Direction of Magnetization and Ambient Field

The foregoing has assumed that the magnetization and regional field are
uniform throughout the study area, often an appropriate assumption for
small-scale surveys where the direction of the geomagnetic field varies by
only small amounts over the limits of the survey. Over continental-scale
areas, however, the geomagnetic field varies to a considerable degree, and
algorithms that assume uniformly directed magnetization and regional
field may produce errors when applied to similarly sized compilations of
aeromagnetic data.

Arkani-Hamed [5] discussed a reduction-to-pole technique that permits variable directions of magnetization and regional field while main-
taining the efficiency of the Fourier-domain methodology. His technique
employs an equivalent layer of magnetization; variations in the direction

336   *Transformations*

of magnetization and in the direction of the ambient field are described
by perturbations about uniform directions.

The magnetic potential observed at point P is given by

$$V(P) = C_m \int \mathbf{M}(Q) \cdot \nabla_Q \frac{1}{r} \, dv, \qquad (12.33)$$

where Q is the point of integration and r is the distance from Q to P.
Both the intensity and direction of magnetization are permitted to vary
in this case. We restrict P to a horizontal plane at $z = z_0$ and model the
magnetic source as a vanishingly thin layer at depth d $(d > z_0)$, that is,
we let

$$\mathbf{M}(Q) = \mathbf{M}(x', y') \, \delta(z' - d).$$

Substituting this magnetization into equation 12.33 and integrating over
z' reduces the triple integral to a two-dimensional convolution over
x' and y'. Fourier transforming both sides and applying the Fourier-
convolution theorem leads to

$$\mathcal{F}[V] = -2\pi C_m \frac{e^{|k|(z_0 - d)}}{|k|} \mathbf{k} \cdot \mathcal{F}[\mathbf{M}],$$

where $\mathbf{k} = (ik_x, ik_y, |k|)$. Now we let the equivalent layer be located at
the plane of observation so that $d \to z_0$. We also resolve the direction of
magnetization into two components, a uniform vector $\mathbf{m}_0$ and a small
perturbation $\Delta\mathbf{m}$ about this uniform direction, that is, we let $\mathbf{M}(x, y) =
M(x, y)(\mathbf{m}_o + \Delta\mathbf{m}(x, y))$. Making these substitutions provides

$$\mathcal{F}[V] = -\frac{2\pi C_m}{|k|} \{\mathbf{k} \cdot \mathbf{m}_0 \mathcal{F}[M] + \mathbf{k} \cdot \mathcal{F}[M\Delta\mathbf{m}]\},$$

and isolating $\mathcal{F}[M]$ on one side of the equation yields

$$\mathcal{F}[M] = -\left\{\frac{\mathcal{F}[V]|k|}{2\pi C_m} + \mathbf{k} \cdot \mathcal{F}[M\Delta\mathbf{m}]\right\} / (\mathbf{k} \cdot \mathbf{m}_0). \qquad (12.34)$$

Equation 12.34 can be solved recursively in order to derive $M(x, y)$ from
$V(x, y)$. An initial guess is made for $M(x, y)$ $(M(x, y) = 0$, for example),
and equation 12.34 is used to obtain a second approximation for $M(x, y)$.
This new solution is substituted back into equation 12.34 to get a third
approximation, and the procedure continues until the solution converges
appropriately.

Equation 12.34 requires the magnetic potential, and this can be obtained from the observed total-field anomalies. Like the magnetization,
we describe the direction of the ambient field as the sum of a uniform

vector $\mathbf{f}_0$ plus a perturbation $\Delta\mathbf{f}$ about $\mathbf{f}_0$, and the total-field anomaly becomes

$$\mathcal{F}\left[\Delta T\right] = -\mathcal{F}\left[\hat{\mathbf{f}} \cdot \nabla_P V\right]$$

$$= -\mathbf{f}_0 \cdot \mathcal{F}\left[\nabla_P V\right] - \mathcal{F}\left[\Delta\mathbf{f} \cdot \nabla_P V\right].$$

As shown in Section 12.2, the gradient of a potential observed on a horizontal plane has a Fourier transform given by

$$\mathcal{F}\left[\nabla_P V\right] = \mathbf{k}\mathcal{F}\left[V\right],$$

and making this substitution in the previous equation leads to

$$\mathcal{F}\left[V\right] = -\{\mathcal{F}\left[\Delta T\right] + \mathcal{F}\left[\Delta\mathbf{f} \cdot \nabla_P V\right]\}/(\mathbf{k} \cdot \mathbf{f}_0). \qquad (12.35)$$

Like equation 12.34, equation 12.35 can be solved iteratively in order to derive the magnetic potential from the observed anomaly, which in turn can be used in equation 12.34 to find the magnetization of the equivalent layer. It is then a simple matter to transform the magnetization into the reduced-to-pole anomaly; that is, we let $\Delta\mathbf{m} = \Delta\mathbf{f} = 0$ and let $\mathbf{m}_0 = \mathbf{f}_0 = (0, 0, 1)$ in equations 12.34 and 12.35 to get

$$\mathcal{F}\left[\Delta T_P\right] = 2\pi C_{\mathrm{m}}|k|\,\mathcal{F}\left[M(x, y)\right]. \qquad (12.36)$$

Thus reduction to the pole is a three-step procedure: A total-field anomaly can be transformed to an equivalent layer at zero depth using the recursive relation 12.34, the potential of the equivalent layer can be found using the recursive relation 12.35, and the potential can be converted to a reduced-to-pole anomaly using equation 12.36 (Arkani-Hamed [5]). The method requires knowledge over the entire survey of the ambient field direction and the direction of magnetization. The first of these can be obtained from a suitable field model, such as the IGRF, but the magnetization direction is more problematic unless purely induced magnetization is assumed.

Two-Dimensional Sources

Consider the total-field anomaly $\Delta T(x)$ measured along a line over and perpendicular to a two-dimensional body. We use the convention shown in Figure 9.6, that is, the long axis of the body is parallel to the y axis, and the profile is in the direction of the x axis. As discussed in previous chapters, the y component of magnetization for a two-dimensional body contributes nothing to the magnetic field, and the y component of the magnetic field is zero. The x and z components of magnetization and ambient field are given by

$$\hat{m}_x = \cos I_\mathrm{m} \cos(D_\mathrm{m} - \alpha),$$

$$\hat{m}_z = \sin I_\mathrm{m},$$

$$\hat{f}_x = \cos I_\mathrm{f} \cos(D_\mathrm{m} - \alpha),$$

$$\hat{f}_z = \sin I_\mathrm{f},$$

where I_m and D_m are the inclination and declination of the magnetization, I_f and D_f are the inclination and declination of the ambient field, and α is the azimuth of the x axis (i.e., the azimuth of the profile). Because the y components of $\hat{\mathbf{m}}$ and $\hat{\mathbf{f}}$ are irrelevant in the two-dimensional case, it is useful to consider the projections of $\hat{\mathbf{m}}$ and $\hat{\mathbf{f}}$ onto the plane normal to the two-dimensional body. The inclinations of these projected vectors are called the *effective inclinations* (Schouten[252]) and are given by

$$I'_\mathrm{m} = \arctan \frac{\hat{m}_z}{\hat{m}_x}$$

$$= \arctan \frac{\tan I_\mathrm{m}}{\cos(D - \alpha)} \tag{12.37}$$

$$I'_\mathrm{f} = \arctan \frac{\hat{f}_z}{\hat{f}_x},$$

$$= \arctan \frac{\tan I_\mathrm{m}}{\cos(D - \alpha)}. \tag{12.38}$$

The effective inclinations are always equal to or steeper than the true inclinations, and they are vertical when the long axis of the body lies parallel to the declination.

First we need the Fourier transform of an anomaly due to a two-dimensional body of arbitrary cross section. One element of such a body is a line of dipoles, and the Fourier transform of the anomaly caused by a line of dipoles is given by equation 11.29, where $\mathbf{m}' = m'\hat{\mathbf{m}}$ is dipole moment per unit length. Fourier transforms are one dimensional in this case, that is, they are functions of a single variable, namely, wavenumber k. We replace m' in equation 11.29 with a single element $M(x, z')\, dx\, dz'$ of the two-dimensional body. As before, $M(x, z)$ will be understood to be nonzero only for $z > 0$ and within a finite part of the x, z plane.

Integrating over x and z provides the Fourier transform of the anomaly caused by an arbitrary two-dimensional body

$$\mathcal{F}[\Delta T] = 2\pi C_\mathrm{m}\vartheta_\mathrm{m}\vartheta_\mathrm{f}|k| \int_{z'} e^{-|k|z'} \mathcal{F}[M(x,z')]\,dz'\,, \qquad (12.39)$$

where

$$\vartheta_\mathrm{m} = \hat{m}_z + i\hat{m}_x\,\mathrm{sgn}\,k\,,$$

$$\vartheta_\mathrm{f} = \hat{f}_z + i\hat{f}_x\,\mathrm{sgn}\,k\,.$$

Rearranging terms, we can show that the amplitude and phase of $\vartheta_\mathrm{m}\vartheta_\mathrm{f}$ are

$$|\vartheta_\mathrm{m}\vartheta_\mathrm{f}| = \frac{\sin I_\mathrm{m}\sin I_\mathrm{f}}{\sin I'_\mathrm{m}\sin I'_\mathrm{f}}\,,$$

$$\mathrm{phase}\,\vartheta_\mathrm{m}\vartheta_\mathrm{f} = (\pi - I'_\mathrm{m} - I'_\mathrm{f})\,\mathrm{sgn}\,k\,. \qquad (12.40)$$

According to equation 12.39, the phase of $\mathcal{F}[\Delta T]$ originates from two aspects of the source: The integral term, depending on the distribution of magnetization $M(x,z)$, and the factors ϑ_m and ϑ_f, which depend exclusively on the directions of magnetization and ambient field. Equation 12.40 shows that the phase originating from ϑ_m and ϑ_f is related very simply to the effective inclinations I'_m and I'_f. Figure 12.10 shows various total-field anomalies over the same distribution of magnetization, each with a different value of $\vartheta_\mathrm{m}\vartheta_\mathrm{f}$.

At the north magnetic pole, $\vartheta_\mathrm{m} = \vartheta_\mathrm{f} = 1$, and equation 12.39 becomes

$$\mathcal{F}[\Delta T_\mathrm{r}] = 2\pi C_\mathrm{m}|k| \int_{z'} e^{-|k|z'} \mathcal{F}[M(x,z')]\,dz'\,.$$

Combining this equation with equation 12.39 in order to eliminate the common terms yields

$$\mathcal{F}[\Delta T_\mathrm{r}] = \mathcal{F}[\Delta T]\,\mathcal{F}[\psi_\mathrm{r}]\,, \qquad (12.41)$$

where

$$\mathcal{F}[\psi_\mathrm{r}] = \frac{1}{\vartheta_\mathrm{m}\vartheta_\mathrm{f}}$$

$$= \frac{\sin I'_\mathrm{m}\sin I'_\mathrm{f}}{\sin I_\mathrm{m}\sin I_\mathrm{f}}\,e^{i(I'_\mathrm{m}+I'_\mathrm{f}-\pi)\,\mathrm{sgn}\,k}\,. \qquad (12.42)$$

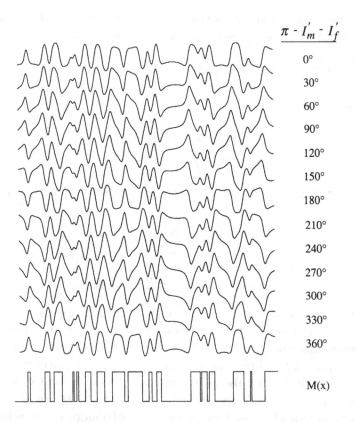

Fig. 12.10. Various total-field anomaly profiles caused by the same distribution of magnetization but with various effective inclinations. The source is a horizontal layer, 1km thick and 3 km below the level of measurement. The layer is uniformly magnetized in the vertical direction so $M(x, z) = M(x)$. At the north magnetic pole, both the magnetization and regional field are directed vertically down so $\pi - I'_m - I'_f = 0$. At the magnetic equator, both vectors are horizontal, and $\pi - I'_m - I'_f = 180°$. At the south pole, both vectors are directed vertically up, and $\pi - I'_m - I'_f = 360°$.

The function $\mathcal{F}[\psi_r]$ is the reduction-to-pole filter for the two-dimensional case. It has constant amplitude for all k given by

$$|\mathcal{F}[\psi_r]| = \frac{\sin I'_m}{\sin I_m}\frac{\sin I'_f}{\sin I_f}.$$

Hence, the instabilities that accompany reduction to the pole of anomalies caused by three-dimensional sources at low latitudes are not a concern for the two-dimensional case. It should be clear, however, that the

long axis of the two-dimensional body should not be parallel to the declination for bodies near the equator.

The reduction-to-pole filter has constant phase for all $k > 0$ and all
$k < 0$,

$$\text{phase}\, \mathcal{F}\,[\psi_r] = \text{sgn}\, k\, (\pi - I'_m - I'_f)\,.$$

Notice that the phase of $\mathcal{F}\,[\psi_r]$ is simply the negative of the phase of
$\vartheta_m \vartheta_f$; that is, when $\mathcal{F}\,[\Delta T]$ is multiplied by $\mathcal{F}\,[\psi_r]$ in equation 12.41, the
phases add to zero. Hence, just as for the three-dimensional case, reduc-
tion to the pole of anomalies caused by two-dimensional sources removes
one element of complexity from the anomalies. It shifts anomalies laterally, tending to place them over their respective sources, and it distorts
the shapes of anomalies so that symmetric sources produce symmetric
anomalies. For example, it could be argued that the topmost profile in
Figure 12.10, the profile that would be observed at the north magnetic
pole, is more easily interpreted in terms of $M(x)$ than its phase-shifted
counterparts; in particular, its positive anomalies lie over positive parts
of $M(x)$.

Two applications related to reduction to the pole are worth mention.
First, reduction to the pole can help in comparing anomalies from widely
separated areas, in the special case where the distribution of magnetization is expected to be similar at the two locations. Consider, for example,
two sets of magnetic anomalies from widely separated ocean basins but
measured over ocean crust of the same geologic age. The two regions
of ocean crust may have recorded the same sequence of geomagnetic
reversals so the distributions of magnetization at the two sites are similar. But comparison of the anomalies may be hampered because the
two sites have different directions of magnetization and ambient field.
Such comparisons are more easily achieved if both sets of anomalies have
been reduced to the pole (Schouten [252], Schouten and McCamy [253],
Blakely and Cox [27, 26]).

Second, phase shifting of anomalies may help in estimating the true
direction of magnetization, in the special case where the distribution of
magnetization is known in advance (Schouten and Cande [254]; Cande
[51]). The distribution of magnetization recorded by seafloor spreading,
for example, can be modeled to a first approximation as a square wave
(e.g., $M(x)$ at the bottom of Figure 12.10). Such magnetizations produce anomalies at the magnetic pole with a characteristic form (e.g., the
top profile in Figure 12.10). A measured magnetic profile can be phase
shifted by varying amounts until it assumes this characteristic form, and

the total phase shift will be an estimate of $I'_m + I'_f - \pi$. Because I'_f presumably is known at the survey site, the total phase shifting provides a direct measure of the effective inclination of magnetization I'_m.

The effective inclination in turn provides a means of estimating the virtual geomagnetic pole for the oceanic plate, under the assumption that the direction of magnetization is parallel to the paleomagnetic field. Although a virtual geomagnetic pole can be calculated uniquely from the paleomagnetic inclination and declination observed at a single site (e.g., Butler [47, pp. 157–60]), here we only have an effective inclination, and according to equation 12.37, an effective inclination has an infinite variety of compatible inclination–declination pairs. Because of this ambiguity, a single effective inclination is compatible with an infinite set of permissible paleomagnetic poles all lying along a semi-great circle, which is not a particularly instructive result. However, a second set of anomalies of the same age and on the same rigid plate could be analyzed in the same way in order to produce a second great circle of permissible pole positions. The intersection of the two great circles provides an estimate of the single paleomagnetic pole compatible with both sites and for the rigid plate over which both sets of anomalies are measured (Schouten and Cande [252], Cande [51]). Systematic discrepancies in the intersection of more than two great circles, called *anomalous skewness*, indicate that one or more of the underlying assumptions are in error; such discrepancies may reflect complexities in the behavior of the paleomagnetic field or in the way reversals are recorded at spreading centers (Cande [51], Cande and Kent [52]).

12.3.2 Calculation of Vector Components

Equations 12.28 and 12.29 also can be used to transform a total-field anomaly into some other component of the magnetic field. For example, the vertical component B_z of the magnetic field can be calculated from a total-field anomaly by letting $\Theta'_m = \Theta_m$ and $\Theta'_f = 1$ in equation 12.29, that is,

$$\mathcal{F}\left[B_z\right] = \mathcal{F}\left[\Delta T\right] \mathcal{F}\left[\psi_z\right],$$

where

$$\mathcal{F}\left[\psi_z\right] = \frac{1}{\Theta_f}$$

$$= \frac{|k|}{|k|\,\hat{f}_z + i(k_x \hat{f}_x + k_y \hat{f}_y)}, \qquad |k| \neq 0.$$

Knowing the direction of the ambient field, therefore, is sufficient to transform a total-field anomaly into the vertical component of the magnetic field.

It should be clear that such transformations will be limited for certain ambient-field directions in the same way that reduction to the pole is limited at low latitudes. Consider, for example, a total-field anomaly measured near the magnetic equator and caused by a body with purely induced magnetization. Both $\hat{\mathbf{f}}$ and $\hat{\mathbf{m}}$ will have shallow inclinations in this case, and transforming the total-field anomaly into the vertical component of the magnetic field can be expected to be an unstable operation. Any noise within the measurements will generate artifacts, typically short in wavelength and elongated parallel to the declination.

Exercise 12.5 Derive Fourier-domain filters that will transform a total-field anomaly into the horizontal-north component and the horizontal-east component of the magnetic field. Discuss the conditions under which these filters are expected to be unstable.

12.4 Pseudogravity Transformation

Poisson's relation was discussed at some length in Section 5.4, where it was shown that the magnetic potential V and gravitational potential U caused by a uniformly dense and uniformly magnetized body are related by a directional derivative, that is,

$$V = -\frac{C_{\mathrm{m}}}{\gamma}\frac{M}{\rho}\hat{\mathbf{m}} \cdot \nabla_P U$$

$$= -\frac{C_{\mathrm{m}}}{\gamma}\frac{M}{\rho}g_{\mathrm{m}}\,, \tag{12.43}$$

where ρ is the density, M is the intensity of magnetization, $\hat{\mathbf{m}}$ is the direction of magnetization, and g_{m} is the component of the gravity field in the direction of magnetization $\hat{\mathbf{m}}$. In deriving Poisson's relation, we assumed that M and ρ are constant. However, we can consider a variable distribution of magnetization or density to be composed of arbitrarily small regions of uniform magnetization or density; equation 12.43 is appropriate for each of these small regions and, invoking the superposition principle, must be appropriate for variable distributions of density and magnetization.

Baranov [9] described an application of Poisson's relation in which the total-field magnetic anomaly is converted into the gravity anomaly that

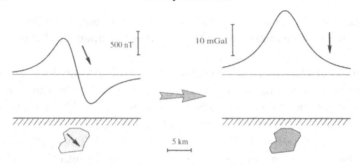

Fig. 12.11. A magnetic anomaly and its pseudogravity transform.

would be observed if the magnetization distribution were to be replaced with an identical density distribution (i.e., $\frac{M}{\rho}$ is a constant throughout the source). He called the resulting quantity a *pseudogravity anomaly*, and the transformation itself is generally referred to as a *pseudogravity transformation* (Figure 12.11). These are perhaps unfortunate names since mass is not involved in any way. As we shall see shortly, the transformation may more appropriately be considered to be a conversion from magnetic field to magnetic potential. Nevertheless, we will use the conventional term, pseudogravity, in the following discussion.

The pseudogravity transformation has several important applications. Some geologic units may be both highly magnetic and anomalously dense. A mafic pluton surrounded by sedimentary rocks, for example, may produce both a gravity and magnetic anomaly. A pseudogravity anomaly, calculated from the measured magnetic field, can be compared directly with measurements of the gravity field. Such comparisons might help to build an interpretation of the shape and size of the source, or at least permit an investigation of the ratio M/ρ and how it varies within the source (e.g., Kanasewich and Agarwal [145], Bott and Ingles [40], Cordell and Taylor [74], Chandler and Malek [55]).

A pseudogravity transformation might be a useful strategy in interpreting magnetic anomalies, not because we believe that a mass distribution actually corresponds to the magnetic distribution beneath the magnetic survey, but because gravity anomalies are in some ways more instructive and easier to interpret and quantify than magnetic anomalies. Gravity anomalies over tabular bodies have steepest horizontal gradients approximately over the edges of the bodies, and this property can be exploited in a magnetic interpretation by transforming the magnetic anomaly to a pseudogravity anomaly and searching the pseudogravity anomaly for maximum horizontal gradients. This application of the pseudogravity transform will be discussed in more detail in the next section.

The pseudogravity transform is more easily understood and more easily undertaken in the Fourier domain. Assuming that the ratio ρ/M is a constant at each point, the Fourier transform of equation 12.43 is given by

$$\mathcal{F}\left[g_{\mathrm{m}}\right] = -\frac{\gamma}{C_{\mathrm{m}}}\frac{\rho}{M}\,\mathcal{F}\left[V\right], \qquad (12.44)$$

and combining with equation 12.25 provides

$$\mathcal{F}\left[g_{\mathrm{m}}\right] = \frac{\gamma}{C_{\mathrm{m}}|k|\Theta_{\mathrm{f}}}\frac{\rho}{M}\,\mathcal{F}\left[\Delta T\right].$$

This equation relates the total-field anomaly to one component of the gravity field, the component parallel to the magnetization. We are more interested in the vertical component of the gravity anomaly, however, and this can be found by dividing both sides by Θ_{m}. Hence, denoting the pseudogravity anomaly as ΔT_{psg}, we get

$$\mathcal{F}\left[\Delta T_{\mathrm{psg}}\right] = \mathcal{F}\left[\Delta T\right]\mathcal{F}\left[\psi_{\mathrm{psg}}\right], \qquad (12.45)$$

where

$$\mathcal{F}\left[\psi_{\mathrm{psg}}\right] = \frac{\gamma}{C_{\mathrm{m}}|k|\Theta_{\mathrm{m}}\Theta_{\mathrm{f}}}\frac{\rho}{M}, \qquad |k| \neq 0, \qquad (12.46)$$

and ρ/M is a constant. The function $\mathcal{F}\left[\psi_{\mathrm{psg}}\right]$ is a filter that transforms a total-field anomaly measured on a horizontal surface into the pseudogravity anomaly. As we have seen in previous sections of this chapter, the transformation amounts to a three-step procedure: Fourier transform the total-field anomaly, multiply by $\mathcal{F}\left[\psi_{\mathrm{psg}}\right]$, and inverse Fourier transform the product. Subroutine B.27 in Appendix B is an implementation of this three-step transformation.

Notice the similarities between $\mathcal{F}\left[\psi_{\mathrm{psg}}\right]$ and the reduction-to-pole filter $\mathcal{F}\left[\psi_{\mathrm{r}}\right]$. In particular, the two filters are related by

$$\mathcal{F}\left[\psi_{\mathrm{psg}}\right] = \frac{A}{|k|}\,\mathcal{F}\left[\psi_{\mathrm{r}}\right],$$

where A is a constant. Thus the two filters have certain spectral properties in common. Indeed, the phase spectrum of $\mathcal{F}\left[\psi_{\mathrm{psg}}\right]$ is identical to that of $\mathcal{F}\left[\psi_{\mathrm{r}}\right]$ (Figure 12.9), and we can expect the pseudogravity transformation to have limitations when the magnetization and ambient field have shallow inclinations, as might be expected at low latitudes. The amplitude spectrum of $\mathcal{F}\left[\psi_{\mathrm{psg}}\right]$ is proportional to the amplitude spectrum of $\mathcal{F}[\psi_r]$ (Figure 12.9) weighted by $1/|k|$. Hence, the radial amplitude spectrum is proportional to $1/|k|$; that is, the pseudogravity transformation amplifies low wavenumbers (long wavelengths) and attenuates high wavenumbers (short wavelengths). The low-wavenumber amplification is cause for some concern; any long-wavelength noise contained in the measured total-field data will be amplified along with authentic anomalies.

Also note the relationship between pseudogravity and magnetic potential,

$$\mathcal{F}[g_{\mathrm{psg}}] = \frac{B}{\Theta_{\mathrm{m}}} \mathcal{F}[V],$$

where B is a constant. In particular, the pseudogravity anomaly of a magnetic source is proportional to the magnetic potential of the same source with vertical magnetization.

Figure 12.12 shows the pseudogravity transform of the total-field anomaly from central Nevada. Note its similarities with the upward-continued field (Figure 12.4). In particular, long-wavelength features of the original map have been amplified at the expense of short-wavelength anomalies.

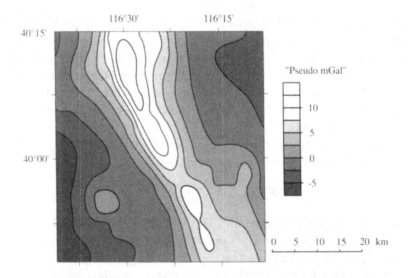

Fig. 12.12. The pseudogravity anomaly transformed from the total-field anomaly of Figure 12.1.

12.4.1 Pseudomagnetic Calculation

The inverse of equation 12.45 can be used to transform a measured gravity anomaly into the magnetic anomaly that would be observed if the density distribution were replaced by a magnetic distribution in one-to-one proportion. Rearranging equation 12.45 yields

$$\mathcal{F}\left[g_{\text{psm}}\right] = \mathcal{F}\left[g\right]\mathcal{F}\left[\psi_{\text{psm}}\right], \tag{12.47}$$

where g_{psm} denotes the transformed anomaly and

$$\mathcal{F}\left[\psi_{\text{psm}}\right] = \frac{C_{\text{m}}|k|\Theta_{\text{m}}\Theta_{\text{f}}}{\gamma}\frac{M}{\rho}. \tag{12.48}$$

This operation, called a *pseudomagnetic transformation*, does not suffer from the low-latitude limitations of its pseudogravity counterpart, but it clearly can suffer from instabilities. In particular, short-wavelength components of g_z are amplified; the shorter the wavelength, the greater will be the amplification. Any noise present at these wavelengths will be similarly amplified, and this can lead to high-amplitude, short-wavelength artifacts in the transformed result.

12.5 Horizontal Gradients and Boundary Analysis

The steepest horizontal gradient of a gravity anomaly $g_z(x, y)$ (or of a pseudogravity anomaly) caused by a tabular body tends to overlie the edges of the body. Indeed, the steepest gradient will be located directly over the edge of the body if the edge is vertical and far removed from all other edges or sources.

Exercise 12.6 Consider a uniform, horizontal, semi-infinite slab with vertical face (i.e., the slab occupies the region $0 \leq x < \infty$, $-\infty < y < \infty$, $z_1 < z < z_2$), like that shown in Figures 9.15(a) and 9.15(b). Show from equation 9.2.2 that a gravity profile over the edge of the slab measured in the x direction will have its maximum gradient over the edge of the slab.

We can exploit this characteristic of gravity anomalies in order to locate abrupt lateral changes in density directly from gravity measurements (Cordell [66]). Moreover, the same technique could be applied to magnetic measurements by first transforming them to pseudogravity anomalies, in which case the steepest horizontal gradients would reflect abrupt lateral changes in magnetization (Cordell and Grauch [68, 70]).

The *magnitude of the horizontal gradient* of the gravity or pseudogravity anomaly, loosely referred to here as the *horizontal gradient*, is

348   *Transformations*

given by

$$h(x,y) = \left[\left(\frac{\partial g_z(x,y)}{\partial x}\right)^2 + \left(\frac{\partial g_z(x,y)}{\partial y}\right)^2\right]^{\frac{1}{2}} \qquad (12.49)$$

and is easily calculated using simple finite-difference relationships, as shown in Subroutine B.28 in Appendix B. The horizontal gradient tends to have maxima located over edges of gravity (or pseudogravity) sources, as shown for a magnetic profile in Figure 12.13. When applied to two-dimensional surveys, the horizontal gradient tends to place narrow ridges over abrupt changes in magnetization or density. Locating maxima in the horizontal gradient can be done by simple inspection, but Blakely and Simpson [33] automated the procedure with an algorithm that scans the rows and columns of gridded data and records the locations of maxima in a file for later analysis and plotting.

Figure 12.14 shows an application to the total-field anomaly of central Nevada. The gridded magnetic data of Figure 12.1 first were transformed to pseudogravity anomalies (Figure 12.12) and then converted to horizontal gradients. The two parallel ridges in Figure 12.14, trending north-northwest through the center of the map, indicate the edges of a roughly two-dimensional source causing the north-northwest trending anomaly in Figure 12.1. The positions of these two parallel ridges presumably mark the lateral extent of a mid-Miocene rift zone in this area (Zoback and Thompson [295]).

Interpreting the horizontal gradient in terms of density or magnetization contrasts, and ultimately in terms of geology, requires several

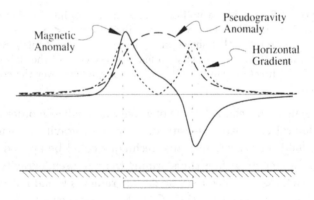

Fig. 12.13. The magnetic anomaly, pseudogravity anomaly, and magnitude of the horizontal gradient over a tabular body.

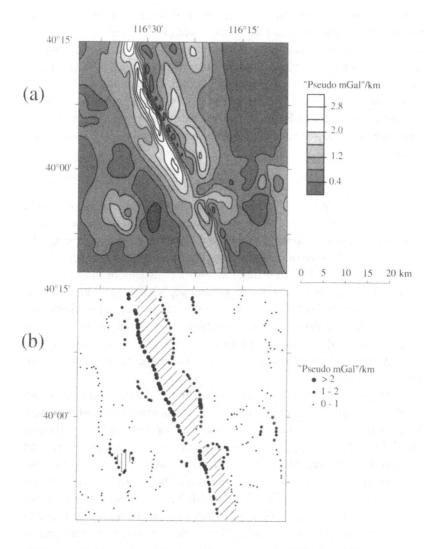

Fig. 12.14. Magnetic boundaries of central Nevada. (a) Total-field anomalies of Figure 12.1 were transformed to pseudogravity anomalies, then converted to horizontal gradients. (b) Dots show maxima in the horizontal gradient automatically located by the method of Blakely and Simpson [33]. Diagonal hatching represents interpreted basaltic rocks associated with a mid-Miocene rift event; vertical hatching shows the location of a granitic pluton of Tertiary age.

underlying assumptions. In particular, we have assumed that contrasts in physical properties occur across vertical and abrupt boundaries isolated from other sources. Geology, unfortunately, is not so simple. Magnetization and density can vary in all directions within a geologic unit, and contacts between units are not generally vertical. Grauch and Cordell [103] have quantified some of the errors that can occur when these underlying assumptions are violated, but in spite of these potential pitfalls, the technique has proven effective in interpretation of regional-scale data bases.

12.5.1 Terracing

Cordell and McCafferty [72] described a related technique to produce a kind of equivalent source from measured gravity or magnetic anomalies. The guiding principle behind their technique is that density or magnetization within the equivalent layer should remain constant except across abrupt boundaries. The resulting patchwork of density or magnetization values brings to mind a terraced landscape, where the "hills and valleys" of the potential field are transformed into horizontal surfaces with abrupt edges. Consequently, the technique is referred to as *terracing*. Maps of terraced physical properties have the general appearance of a geology map, where lithologies are depicted as uniform bodies except across contacts.

Terracing is not an inverse scheme in the usual sense. No inverse calculations are made, but rather the values of the gravity or pseudogravity field are iteratively increased or decreased in order to approach the terraced form. Specifically, the value at each point of the grid is revised upward or downward based on the algebraic sign of the curvature (i.e., the second vertical derivative) at that point. Repeated adjustments are made until the terraced effect is achieved. The resulting terraces do not directly reflect density or magnetization but can be scaled in order to do so. For example, the magnetization or density of each terrace can be treated as one unknown in an over-determined system of linear equations. Assuming top and bottom surfaces for the layer, least-squares techniques can be used to solve for the best single value of magnetization or density for each terrace.

12.6 Analytic Signal

The *analytic signal* is formed through a combination of the horizontal and vertical gradients of a magnetic anomaly. The analytic signal has a form over causative bodies that depends on the locations of the bodies but not on their directions of magnetization. The application of analytic signals to magnetic interpretation was pioneered by Nabighian [190, 191] for the two-dimensional case, primarily as a tool to estimate depth and position of sources. More recently, the method has been expanded to three-dimensional problems (Nabighian [192]; Roest, Verhoef, and Pilkington [247]) as a mapping and depth-to-source technique and as a way to learn about the nature of the causative magnetization (Roest, Arkani-Hamed, and Verhoef [245]; Roest and Pilkington [246]).

12.6.1 Hilbert Transforms

The Hilbert transform plays an important role in the analytic signal, and we begin the discussion with a very brief review thereof. The interested reader is referred to texts by Bracewell [42] and Papoulis [202] for additional information.

The *Hilbert transform* of $f(x)$ is given by

$$F_I(x) = -\frac{1}{\pi} \int_{-\infty}^{\infty} \frac{f(x')}{x - x'} \, dx' , \qquad (12.50)$$

and its inverse by

$$f(x') = \frac{1}{\pi} \int_{-\infty}^{\infty} \frac{F_I(x)}{x' - x} \, dx . \qquad (12.51)$$

Equation 12.50 amounts to the convolution of $f(x)$ with $-1/\pi x$. It has, therefore, a one-dimensional Fourier transform given by the Fourier transform of $f(x)$ times the Fourier transform of $-1/\pi x$, namely,

$$\mathcal{F}[F_I] = i \operatorname{sgn} k \, \mathcal{F}[f] . \qquad (12.52)$$

The Hilbert transform has a curious effect on $f(x)$, as demonstrated by equation 12.52: It leaves the amplitude of $f(x)$ unchanged but shifts the phase of $f(x)$ by $\pi/2$ for $k > 0$ and by $-\pi/2$ for $k < 0$.

The *analytic signal* of $f(x)$ is defined as

$$a(x) = f(x) - i \, F_I(x) \qquad (12.53)$$

(Bracewell [42]). With equation 12.52 it is easy to derive the Fourier transform of the analytic signal,

$$\mathcal{F}[a] = \mathcal{F}[f] (1 + \operatorname{sgn} k) . \qquad (12.54)$$

Hence, the spectrum of the analytic signal of $f(x)$ is twice the Fourier transform of $f(x)$ at $k > 0$ and zero for $k < 0$. It follows that the analytic signal of $f(x)$ could be found in either of two ways: (1) by directly calculating the Hilbert transform of $f(x)$, as in equation 12.50, and then adding this transform to $f(x)$, as in equation 12.53; or (2) by Fourier transforming $f(x)$, setting to zero all values at $k < 0$, doubling all values at $k > 0$, and inverse Fourier transforming the result.

Exercise 12.7 Find the analytic signal for $f(x) = \cos ax$.

12.6.2 Application to Potential Fields

In Section 12.2, we found simple relationships, expressed by equation 12.20, between a potential field measured on a horizontal surface and its derivatives with respect to any direction. Now consider a potential field $\phi(x, z)$ measured along the x axis and caused by a two-dimensional source aligned parallel to the y axis, as in Figure 9.6. In the two-dimensional case, equation 12.20 can be expressed as

$$\mathcal{F}\left[\frac{\partial \phi}{\partial x}\right] = ik\,\mathcal{F}\left[\phi\right], \tag{12.55}$$

$$\mathcal{F}\left[\frac{\partial \phi}{\partial z}\right] = |k|\,\mathcal{F}\left[\phi\right]. \tag{12.56}$$

From these relationships, it is easily shown that $\frac{\partial \phi}{\partial x}$ and $\frac{\partial \phi}{\partial z}$ are a Hilbert transform pair. In particular, equation 12.52 is satisfied if we let $f(x) = \frac{\partial \phi}{\partial x}$ and $F_I(x) = -\frac{\partial \phi}{\partial z}$, and we form the analytic signal from equation 12.53 as

$$a(x, z) = \frac{\partial \phi}{\partial x} + i\,\frac{\partial \phi}{\partial z}. \tag{12.57}$$

This version of $a(x, z)$ satisfies the Cauchy–Riemann conditions (Section 1.3.3), the proof of which is left to the problem set at the end of this chapter, hence, $a(x, z)$ is analytic at each point of the x axis. In Section 1.3.3 it was shown that if $a(x, z)$ is analytic in a domain, then the real part of $a(x, z)$ must be harmonic; that is, in the present case,

$$\nabla^2 \frac{\partial \phi}{\partial x} = \frac{\partial}{\partial x} \nabla^2 \phi$$

$$= 0,$$

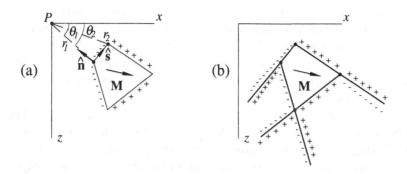

Fig. 12.15. Uniformly magnetized, two-dimensional polygon. The n-sided polygon can be replaced with $2n$ semi-infinite sheets, two sheets per corner, without affecting the magnetic anomaly.

which is consistent with our starting assumption that ϕ is a potential.

Nabighian [190, 191, 192] showed how the analytical signal can be applied to the interpretation of total-field magnetic anomalies. Consider a two-dimensional body with uniform magnetization $\mathbf{M}$ and with polygonal cross section, as shown in Figure 12.15. The ambient field is in the direction $\hat{\mathbf{f}}$ and the total-field anomaly is measured along the x axis. All corners of the body have z coordinates greater than zero. The anomaly caused by one side of the n-sided prism, extending from (x_1, z_1) to (x_2, z_2), is given by equations 9.27 and 9.28,

$$\Delta T(x, z) = \alpha \log \frac{r_2}{r_1} + \beta (\theta_2 - \theta_1), \qquad (12.58)$$

where r_1, r_2, θ_1, and θ_2 are defined as shown in Figure 12.15, and where

$$\alpha = -2C_{\mathrm{m}}(\mathbf{M} \cdot \hat{\mathbf{n}})(\hat{\mathbf{f}} \cdot \hat{\mathbf{s}}),$$

$$\beta = -2C_{\mathrm{m}}(\mathbf{M} \cdot \hat{\mathbf{n}})(\hat{\mathbf{f}} \times \hat{\mathbf{s}}).$$

The horizontal and vertical derivatives of the anomaly caused by this one side are

$$\frac{\partial}{\partial x} \Delta T(x, z) = \frac{\alpha(x - x_2) + \beta(z - z_2)}{r_2^2} - \frac{\alpha(x - x_1) + \beta(z - z_1)}{r_1^2},$$

$$\frac{\partial}{\partial z} \Delta T(x, z) = \frac{\beta(x - x_2) - \alpha(z - z_2)}{r_2^2} - \frac{\beta(x - x_1) - \alpha(z - z_1)}{r_1^2}.$$

We now replace each side of the polygon with two semi-infinite sheets arranged so that the sheets do not cross the x axis, as shown in Figure 12.15. It should be clear from Figure 12.15 that this trick will not affect the magnetic anomaly in any way. Now the polygon side extending from (x_1, z_1) to (x_2, z_2) is replaced with two semi-infinite sheets, with opposite sign but identical α and β, one extending away from (x_1, z_1) and the other from (x_2, z_2).

Consider just the semi-infinite sheet extending from (x_2, z_2). The derivative of the total-field anomaly caused by this sheet can be found by letting $r_1 \to \infty$ in the previous equations. Parameters α and β do not change, but the second term of each equation vanishes. The analytic signal of this one sheet then is given by

$$a(x, z) = \frac{\alpha + i\beta}{x - x_2 - i(z - z_2)}. \qquad (12.59)$$

It can be shown that this equation satisfies the Cauchy–Riemann conditions for the complex variable $\omega = x - iz$, so $a(x, z)$ is analytic. Moreover, it has a Fourier transform given by

$$\mathcal{F}[a] = -\pi i(\alpha + i\beta)e^{-|k|(z_2 - z)}e^{-ikx_2}(1 + \operatorname{sgn} k), \quad z < z_2,$$

which satisfies the requirement that $\mathcal{F}[a] = 0$ for $k < 0$.

The analytic signal expressed by equation 12.59 is of special interest. It is a function of a complex variable $\omega = x - iz$ and can be written

$$a(\omega) = \frac{\alpha + i\beta}{\omega - \omega_2},$$

where $\omega_2 = x_2 - iz_2$. The numerator depends strictly on the orientation of the sheet (as defined by $\hat{n}$ and $\hat{s}$) with respect to $\mathbf{M}$ and $\hat{\mathbf{f}}$. The denominator, on the other hand, depends exclusively on the location at which the sheet terminates, that is, (x_2, z_2). The point ω_2 is an isolated singular point of the complex plane, meaning that $a(\omega)$ has a simple pole at that location.

Nabighian [190, 191] showed how $a(x, z)$ can be used to locate the end of the sheet. The amplitude of $a(x, z)$ is given by

$$|a(x, z)| = \left[\frac{\alpha^2 + \beta^2}{(x - x_2)^2 + (z - z_2)^2}\right]^{\frac{1}{2}}.$$

Thus $|a(x, z)|$ as measured along the x axis has the form of a symmetric, bell-shaped curve centered about $x = x_2$, and the width of the curve is related to z_2. Specifically, the width of the curve at half its maximum height is $2|z - z_2|$. Hence, the horizontal position and depth of the end of the sheet (x_2, z_2) can be estimated through simple examination of $a(x, z)$ along the x axis.

Now we return to the n-sided polygon. The polygon is equivalent to $2n$ sheets, each sheet terminating at a polygon corner. Hence, the analytic signal constructed from the total-field anomaly of the prism will have a peak centered over each corner of the polygon, and the width of each peak will be related to the depth of its respective corner. An analysis of $\Delta T(x, z)$ as measured along the x axis might involve the following steps:

1. Differentiate $\Delta T(x, z)$ with respect to x.
2. Find the analytic signal by

    (a) Fourier transforming $\frac{\partial}{\partial x} \Delta T(x, z)$,
    (b) doubling values at $k > 0$ and canceling values at $k < 0$, and
    (c) inverse Fourier transforming.

3. Calculate and interpret $|a(x, z)|$.

The difficulty comes in resolving the various bell-shaped curves of $a(x, z)$. The various curves superimpose and coalesce when polygon corners are close together or lie above one another.

For the three-dimensional case, the analytic signal is given by

$$\mathbf{A}(x, y, z) = \frac{\partial \Delta T}{\partial x}\hat{\mathbf{i}} + \frac{\partial \Delta T}{\partial y}\hat{\mathbf{j}} + i\,\frac{\partial \Delta T}{\partial z}\hat{\mathbf{k}}, \qquad (12.60)$$

(Roest et al. [247]), where $\hat{\mathbf{i}}$, $\hat{\mathbf{j}}$, and $\hat{\mathbf{k}}$ are unit vectors in the x, y, and z directions, respectively, and where now $\Delta T(x, y, z)$ is measured on the x, y plane. This function possesses the necessary property of analytic signals, that is, its real and imaginary parts form a Hilbert transform pair. Roest et al. [247] described how this function could be used in the interpretation of gridded data in terms of three-dimensional sources. One strategy would be to (1) form $|\mathbf{A}(x, y, z)|$ from the partial derivatives of the magnetic anomaly; (2) locate the maxima of this function, perhaps with the technique of Blakely and Simpson [33]; (3) automatically calculate the half-width at each maximum in the direction perpendicular to the strike of the maximum; and (4) plot the maxima in a map projection with a symbol that indicates depth to source.

Analytic Signal and Direction of Magnetization

It is clear from equation 12.59 that the magnitude of the analytic signal is independent of the direction of magnetization for the two-dimensional case, and this can similarly be shown to be true for the three-dimensional

356 Transformations

case. This fact can be used to investigate the nature of the magnetization of causative bodies. Roest and Pilkington [246] described a method that involves computing from measured total-field anomalies both $|\mathbf{A}(x,y,z)|$ and the horizontal gradient of the pseudogravity anomaly. The pseudogravity calculation implicitly requires an assumption about the direction of magnetization. If this direction is selected properly, the horizontal gradient of the pseudogravity anomaly will have a form similar to $|\mathbf{A}(x,y,z)|$. Hence, the appropriate magnetization can be estimated by repeated adjustment of the direction of magnetization, calculation of the horizontal gradient of the pseudogravity anomaly, and comparison with $|\mathbf{A}(x,y,z)|$. Roest et al. [245] quantified this comparison using cross-correlation methods and thereby applied the method to investigate anomalous skewness in marine magnetic anomalies (see Section 12.3.1).

12.7 Problem Set

1. Consider a marine magnetic profile $\Delta T(x)$ measured over an ocean basin of given age. Assume that the source of $\Delta T(x)$ is a flat-lying crustal layer and approximate the crust as a horizontal slab with top and bottom at depths z_1 and z_2, respectively. Let the magnetization of the slab be uniform in both the y and z directions so $M(x,y,z) = M(x)$. Assume that $M(x)$ is a precise recording of the paleomagnetic field as the crust evolved at an ancient spreading center with spreading rate V.

   (a) Design a filter that will transform $\Delta T(x)$ into the profile that would be observed over a different ocean basin of the same age but created with a faster spreading rate V_2 $(V_2 > V)$. Describe the stability of the filter.

   (b) What happens to the stability of the filter if $V_2 < V$?

2. Consider a potential field measured on a horizontal surface. Discuss the amplification that would occur at Nyquist wavenumbers if the potential field were downward continued a distance Δz. Express the amplification in terms of Δz and the sample intervals Δx and Δy. Do the same for the pseudogravity transform and the second vertical derivative.

3. Consider the total-field anomaly over a two-dimensional body located at the magnetic equator and lying parallel to the magnetic declination. If the magnetization is entirely induced, what happens to

the effective inclinations? Discuss the implications for the total-field
anomaly and the reduction-to-the-pole filter.

4. A measured total-field anomaly has the unusual form

$$\Delta T(x, y) = A_1 \sin \frac{2\pi x}{\lambda_1} + A_2 \sin \frac{2\pi y}{\lambda_2} .$$

Assume that the anomaly continues in this periodic way outside of the
limits of the survey. Derive an expression for the total-field anomaly
measured on a surface that is Δz higher than the original survey.
Express the result in terms of A_1, A_2, λ_1, λ_2, and Δz. Hints:

$$\mathcal{F}[\sin ax] = i\pi[\delta(k + a) - \delta(k - a)],$$

$$f(x)\,\delta(x - a) = f(a)\,\delta(x - a) .$$

5. Prove the following theorems:

(a) The Fourier transform of the gradient of a scalar is the Fourier
transform of the scalar times a vector; that is,

$$\mathcal{F}[\nabla \phi] = \left(ik_x, ik_y, \frac{\partial}{\partial z} \right) \mathcal{F}[\phi] .$$

(b) If ϕ represents any potential measured on a horizontal surface,
then

$$\mathcal{F}\left[\frac{\partial}{\partial z} \phi \right] = \frac{\partial}{\partial z}\mathcal{F}[\phi] = |k|\mathcal{F}[\phi] .$$

6. Consider a vertical gravity anomaly $g_z(x, y)$ measured on a horizontal surface and caused by a density distribution entirely below the
surface.

(a) Use the Fourier domain to find an equivalent density distribution
that causes $g_z(x, y)$ but that is spread over a vanishingly thin
layer at depth d below the level of observation, that is, the density
is described by $\rho(x, y, z) = \rho(x, y)\,\delta(z - d)$.

(b) Show that $\rho(x, y)$ is proportional to $g_z(x, y)$ if $d \to 0$.

(c) Repeat (a) for a total-field anomaly; that is, given a total-field
anomaly $\Delta T(x, y)$, find an equivalent magnetization $M(x, y)$ on
a vanishingly thin layer at depth d.

(d) Show that the distribution of magnetization $M(x, y)$ is proportional to the pseudogravity anomaly if $d \to 0$. In other words,
prove that an equivalent layer at the elevation of a total-field
anomaly is proportional to the pseudogravity anomaly.

7. Let $\phi(x, z)$ be a potential field caused by two-dimensional sources and observed along the horizontal x axis over and perpendicular to the sources. Show that

$$a(x, z) = \frac{\partial}{\partial x}\phi + i\frac{\partial}{\partial z}\phi$$

is analytic along the line. In particular, show that this equation satisfies the Cauchy–Riemann conditions of Section 1.3.3. Caution: The coordinate system in Section 1.3.3 was different than in the present chapter.

8. Find the analytic signal $a(x, z)$ for the total field anomaly over and perpendicular to a single line of dipoles. Express $a(x, z)$ in terms of the depth and position of the line of dipoles.

   (a) Show that this $a(x, z)$ satisfies the Cauchy–Riemann conditions.
   (b) Show that $\mathcal{F}[a]$ is zero for $k < 0$.
   (c) Describe how $a(x, z)$ could be used to estimate the depth and lateral position of the line source.

Appendix A
Review of Vector Calculus

For everything works through innate forces shown by lines, angles and figures.

(Roger Bacon)

Many standard texts on advanced calculus and vector analysis provide thorough treatments of vector calculus. The purpose here is not to reproduce those treatments in detail, but rather to introduce the vector notation used in this text and to provide a reference for the important differential operations (gradient, divergence, and so forth) used herein. Readers wanting a more extensive review are referred to Marion [173, pp. 1–55].

A.1 Vector Representations

The geometric vector is a mathematical quantity with both direction and magnitude. Vectors in this text are represented by boldface characters (e.g., $\mathbf{A}$), whereas scalar quantities are not so emphasized (e.g., ψ). The magnitude of a vector is written as either A or $|\mathbf{A}|$. Vectors of equal magnitude and having the same direction are said to be equal. Two vectors pointing in opposite directions are negatives of one another, that is, $-\mathbf{A}$ is directed opposite to $\mathbf{A}$. Multiplying a vector with a scalar leaves the direction unchanged but increases the magnitude by a factor equal to the scalar, that is, the vector $\mathbf{B} = \psi\mathbf{A}$ has the same direction as $\mathbf{A}$ but magnitude $\psi|\mathbf{A}|$. Unit vectors are vectors with unit magnitude and are represented by boldface characters topped with a circumflex (e.g., $\hat{\mathbf{A}}$). The unit vector directed parallel to vector $\mathbf{A}$ is given by

$$\hat{\mathbf{A}} = \frac{\mathbf{A}}{|\mathbf{A}|}.$$

Vectors can be combined by addition, for example, $\mathbf{C} = \mathbf{A} + \mathbf{B}$. Vector addition is commutative,

$$\mathbf{A} + \mathbf{B} = \mathbf{B} + \mathbf{A},$$

associative,

$$(\mathbf{A} + \mathbf{B}) + \mathbf{C} = \mathbf{A} + (\mathbf{B} + \mathbf{C}),$$

and distributive,

$$\psi (\mathbf{A} + \mathbf{B}) = \psi\mathbf{A} + \psi\mathbf{B}.$$

A geometric vector can be expanded along a complete set of linearly independent basis vectors. A set of basis vectors is complete if any arbitrary vector can be expanded in terms of its projections along the basis vectors. Here we are concerned either with three-dimensional or two-dimensional orthogonal spaces. In such cases, vectors can be expanded in terms of two (in two-dimensional cases) or three (in three-dimensional cases) orthogonal unit vectors, with each unit vector weighted by the projection of the vector parallel to the direction of the basis vector. For example, if $\hat{\mathbf{x}}_1$, $\hat{\mathbf{x}}_2$, and $\hat{\mathbf{x}}_3$ represent three orthogonal unit vectors, then vector $\mathbf{A}$ can be expanded into

$$\mathbf{A} = A_1\hat{\mathbf{x}}_1 + A_2\hat{\mathbf{x}}_2 + A_3\hat{\mathbf{x}}_3,$$

where A_1, A_2, and A_3 are the projections of $\mathbf{A}$ parallel to $\hat{\mathbf{x}}_1$, $\hat{\mathbf{x}}_2$, and $\hat{\mathbf{x}}_3$, respectively.

This text deals with only three coordinate systems: the cartesian, cylindrical, and spherical systems. The notation for each of these is described briefly in the following sections.

Cartesian Coordinates

The cartesian coordinate system is represented in this text with orthogonal axes x, y, and z, as shown in Figure A.1. Unit vectors $\hat{\mathbf{i}}$, $\hat{\mathbf{j}}$, and $\hat{\mathbf{k}}$ are directed parallel to the x, y, and z axes, respectively, so vector $\mathbf{A}$ can be represented as

$$\mathbf{A} = A_x\hat{\mathbf{i}} + A_y\hat{\mathbf{j}} + A_z\hat{\mathbf{k}},$$

where A_x, A_y, and A_z are components of $\mathbf{A}$ found by projecting $\mathbf{A}$ onto the x, y, and z axes, respectively. The magnitude of $\mathbf{A}$ is given by

$$|\mathbf{A}| = \left(A_x^2 + A_y^2 + A_z^2\right)^{\frac{1}{2}}.$$

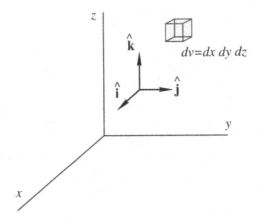

Fig. A.1. Cartesian coordinate system.

The unit vector in the direction of **A** is given by

$$\hat{\mathbf{A}} = \hat{A}_x\hat{\mathbf{i}} + \hat{A}_y\hat{\mathbf{j}} + \hat{A}_z\hat{\mathbf{k}},$$

where

$$\hat{A}_x = \frac{A_x}{A},$$

$$\hat{A}_y = \frac{A_y}{A},$$

$$\hat{A}_z = \frac{A_z}{A}.$$

It is obvious from this definition that $|\hat{\mathbf{A}}| = 1$. An element of volume in the cartesian system is $dv = dx\,dy\,dz$.

Cylindrical Coordinates

In cylindrical coordinates, vector **A** is denoted by

$$\mathbf{A} = A_r\hat{\mathbf{r}} + A_\theta\hat{\mathbf{\Theta}} + A_z\hat{\mathbf{z}},$$

where $\hat{\mathbf{r}}$, $\hat{\mathbf{\Theta}}$, and $\hat{\mathbf{z}}$ are unit vectors in the direction of increasing r, θ, and z, respectively, as shown in Figure A.2. Note that the three unit vectors so defined are always orthogonal to each other. Magnitudes and unit vectors in the cylindrical system are written analogously to those in

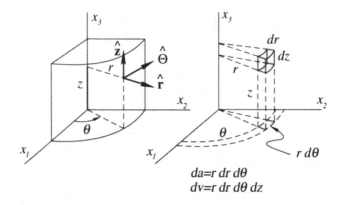

$da = r\, dr\, d\theta$
$dv = r\, dr\, d\theta\, dz$

Fig. A.2. Cylindrical coordinate system.

the cartesian system. An element of volume in the cylindrical coordinate system is given by $dv = r\, dr\, d\theta\, dz$.

Spherical Coordinates

Vector **A** is represented in the spherical coordinate system by

$$\mathbf{A} = A_r \hat{\mathbf{r}} + A_\theta \hat{\mathbf{\Theta}} + A_\phi \hat{\mathbf{\Phi}}.$$

Unit vectors $\hat{\mathbf{r}}$, $\hat{\mathbf{\Theta}}$, and $\hat{\mathbf{\Phi}}$ are always orthogonal and directed in the direction of increasing r, θ, and ϕ, respectively. Vector magnitudes and unit vectors are analogous to those in the cartesian system. An element of volume is given by $dv = r^2 \sin\theta\, dr\, d\theta\, d\phi$. Figure A.3 shows these relationships and an element of volume in spherical coordinates.

A.2 Vector Multiplication

A.2.1 Scalar Product

The *scalar product*, or *dot product*, of two vectors equals the product of the magnitudes of the two vectors scaled by the cosine of the angle formed by their unit vectors, that is,

$$\mathbf{A} \cdot \mathbf{B} = AB \cos\theta.$$

It follows that the scalar product is largest when **A** and **B** are parallel and vanishes if **A** and **B** are perpendicular. The scalar product of a

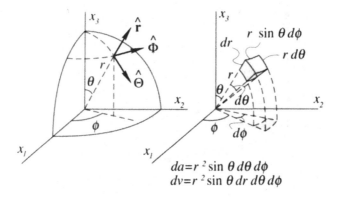

$$da = r^2 \sin \theta \, d\theta \, d\phi$$
$$dv = r^2 \sin \theta \, dr \, d\theta \, d\phi$$

Fig. A.3. Spherical coordinate system.

vector **A** and a unit vector $\hat{\mathbf{B}}$ is equal to the component of **A** in the direction of $\hat{\mathbf{B}}$. Scalar products in the three coordinate systems are written as follows:

Cartesian coordinates:

$$\mathbf{A} \cdot \mathbf{B} = A_x B_x + A_y B_y + A_z B_z.$$

Cylindrical coordinates:

$$\mathbf{A} \cdot \mathbf{B} = A_r B_r + A_\theta B_\theta + A_z B_z.$$

Spherical coordinates:

$$\mathbf{A} \cdot \mathbf{B} = A_r B_r + A_\theta B_\theta + A_\phi B_\phi.$$

A.2.2 Vector Product

The *vector product*, or *cross product*, of two vectors is represented by $\mathbf{A} \times \mathbf{B}$. The vector product produces a new vector with direction normal to both original vectors. For example, vector $\mathbf{C} = \mathbf{A} \times \mathbf{B}$ has a direction perpendicular to both **A** and **B** and is in the direction of advance of a right-handed screw rotated from **A** to **B**. The magnitude of **C** is the area of a parallelogram defined by vectors **A** and **B**, as in Figure A.4. Note that $\mathbf{A} \times \mathbf{B} = -\mathbf{B} \times \mathbf{A}$.

The following equations show the vector product in the three coordinate systems, where determinants are used as a shorthand notation:

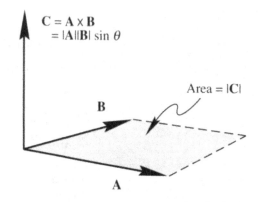

Fig. A.4. Cross product in cartesian coordinate system.

Cartesian coordinates:

$$\mathbf{A} \times \mathbf{B} = \begin{vmatrix} \hat{\mathbf{i}} & \hat{\mathbf{j}} & \hat{\mathbf{z}} \\ A_x & A_y & A_z \\ B_x & B_y & B_z \end{vmatrix}.$$

Cylindrical coordinates:

$$\mathbf{A} \times \mathbf{B} = \begin{vmatrix} \hat{\mathbf{r}} & \hat{\mathbf{\Theta}} & \hat{\mathbf{k}} \\ A_r & A_\theta & A_z \\ B_r & B_\theta & B_z \end{vmatrix}.$$

Spherical coordinates:

$$\mathbf{A} \times \mathbf{B} = \begin{vmatrix} \hat{\mathbf{r}} & \hat{\mathbf{\Theta}} & \hat{\mathbf{\Phi}} \\ A_r & A_\theta & A_\phi \\ B_r & B_\theta & B_\phi \end{vmatrix}.$$

A.3 Differential Operations

Scalar functions of position, such as the temperature distribution within a volume of material, are called *scalar fields*. Vector quantities likewise can be functions of spatial coordinates, with both direction and magnitude varying from point to point. Heat flow $\mathbf{J}(x, y, z)$ within a material, for example, is a function of coordinates x, y, z. Such quantities are called *vector fields*.

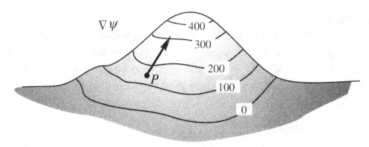

Fig. A.5. Contour map of scalar quantity ψ. The gradient at any point is normal to the contours and in the direction of maximum increase in ψ at that point.

A.3.1 Gradient of Scalar

The gradient of a scalar field ψ, designated by $\nabla\psi$, represents the direction and rate of change of the field. The gradient of ψ is itself a vector directed normal to lines of constant ψ and, consequently, in the direction of the maximum change in ψ. On a contour map showing equal values of ψ, for example, the gradient $\nabla\psi$ will be a vector that points "uphill" and is everywhere perpendicular to the contours (Figure A.5). Note that $\nabla\psi\cdot\hat{\mathbf{A}}$ is the derivative of ψ in the direction of $\hat{\mathbf{A}}$. The gradient is represented in the three coordinate systems as follows:

Cartesian coordinates:

$$\nabla\psi = \hat{\mathbf{i}}\frac{\partial\psi}{\partial x} + \hat{\mathbf{j}}\frac{\partial\psi}{\partial y} + \hat{\mathbf{k}}\frac{\partial\psi}{\partial z}.$$

Cylindrical coordinates:

$$\nabla\psi = \hat{\mathbf{r}}\frac{\partial\psi}{\partial r} + \hat{\boldsymbol{\Theta}}\frac{1}{r}\frac{\partial\psi}{\partial\theta} + \hat{\mathbf{z}}\frac{\partial\psi}{\partial z}.$$

Spherical coordinates:

$$\nabla\psi = \hat{\mathbf{r}}\frac{\partial\psi}{\partial r} + \hat{\boldsymbol{\Theta}}\frac{1}{r}\frac{\partial\psi}{\partial\theta} + \hat{\boldsymbol{\Phi}}\frac{1}{r\sin\theta}\frac{\partial\psi}{\partial\phi}.$$

A.3.2 Divergence of a Vector

The divergence of a vector $\mathbf{A}$, designated by $\nabla \cdot \mathbf{A}$, is a scalar that represents the three-dimensional spatial derivative of a vector field. For example, if $\mathbf{A}(x, y, z)$ describes fluid flow at point (x, y, z), then $\nabla \cdot \mathbf{A}$ is the *net* outward flow per unit volume from a small volume surrounding point (x, y, z). If the flux of $\mathbf{A}$ entering the volume equals the flux leaving the volume, then $\nabla \cdot \mathbf{A} = 0$, and $\mathbf{A}$ is said to be a *divergenceless* or *solenoidal* vector. In the three coordinate systems we have the following representations for divergence:

Cartesian coordinates:

$$\nabla \cdot \mathbf{A} = \frac{\partial A_x}{\partial x} + \frac{\partial A_y}{\partial y} + \frac{\partial A_z}{\partial z}.$$

Cylindrical coordinates:

$$\nabla \cdot \mathbf{A} = \frac{1}{r} \frac{\partial}{\partial r} (r A_r) + \frac{1}{r} \frac{\partial A_\theta}{\partial \theta} + \frac{\partial A_z}{\partial z}.$$

Spherical coordinates:

$$\nabla \cdot \mathbf{A} = \frac{1}{r^2} \frac{\partial}{\partial r} (r^2 A_r) + \frac{1}{r \sin \theta} \frac{\partial}{\partial \theta} (\sin \theta \, A_\theta) + \frac{1}{r \sin \theta} \frac{\partial A_\phi}{\partial \phi}.$$

A.3.3 Curl of a Vector

The curl of a vector $\mathbf{A}$ is represented by $\mathbf{B} = \nabla \times \mathbf{A}$. Like the divergence, the curl also represents the spatial derivative of a vector field but in a different way: It is a vector (rather than a scalar) that measures the circulation of the field. One way to visualize the curl of a vector is to imagine a paddle wheel placed in a fluid that flows in accordance with the vector. The paddle wheel will rotate if the curl of the vector is nonzero. The curl of a vector is sometimes called the *vorticity* of the vector. A vector field with nonzero curl is called a *rotational* or *vortex field*. If the curl is zero, the vector is said to be *irrotational* or *conservative*. The curl is represented in the three coordinate systems as follows:

Cartesian coordinates:

$$\nabla \times \mathbf{A} = \begin{vmatrix} \hat{\mathbf{i}} & \hat{\mathbf{j}} & \hat{\mathbf{k}} \\ \frac{\partial}{\partial x} & \frac{\partial}{\partial y} & \frac{\partial}{\partial z} \\ A_x & A_y & A_z \end{vmatrix}.$$

Cylindrical coordinates:

$$\nabla \times \mathbf{A} = \hat{\mathbf{r}} \left(\frac{1}{r} \frac{\partial A_z}{\partial \theta} - \frac{\partial A_\theta}{\partial z} \right) + \hat{\boldsymbol{\Theta}} \left(\frac{\partial A_r}{\partial z} - \frac{\partial A_z}{\partial r} \right) + \hat{\mathbf{z}} \left(\frac{1}{r} \frac{\partial}{\partial r} (r A_\theta) - \frac{1}{r} \frac{\partial A_r}{\partial \theta} \right).$$

Spherical coordinates:

$$\nabla \times \mathbf{A} = \hat{\mathbf{r}} \frac{1}{r \sin \theta} \left(\frac{\partial}{\partial \theta} (\sin \theta A_\phi) - \frac{\partial A_\theta}{\partial \phi} \right)$$

$$+ \hat{\mathbf{\Theta}} \frac{1}{r \sin \theta} \left(\frac{\partial A_r}{\partial \phi} - \sin \theta \frac{\partial}{\partial r} (r A_\phi) \right)$$

$$+ \hat{\mathbf{\Phi}} \frac{1}{r} \left(\frac{\partial}{\partial r} (r A_\theta) - \frac{\partial A_r}{\partial \theta} \right).$$

A.3.4 Laplacian of a Scalar

Laplace's equation $\nabla^2 \psi = 0$ is a second-order differential equation of particular importance to this text. The ∇^2 operator, sometimes referred to as the *Laplacian*, is equivalent to the divergence of a gradient, that is, $\nabla^2 \psi = \nabla \cdot (\nabla \psi)$. It has the following forms in the three orthogonal coordinate systems.

Cartesian coordinates:

$$\nabla^2 \psi = \frac{\partial^2 \psi}{\partial x^2} + \frac{\partial^2 \psi}{\partial y^2} + \frac{\partial^2 \psi}{\partial z^2}.$$

Cylindrical coordinates:

$$\nabla^2 \psi = \frac{1}{r} \frac{\partial}{\partial r} \left(r \frac{\partial \psi}{\partial r} \right) + \frac{1}{r^2} \frac{\partial^2 \psi}{\partial \theta^2} + \frac{\partial^2 \psi}{\partial z^2}.$$

Spherical coordinates:

$$\nabla^2 \psi = \frac{1}{r^2} \frac{\partial}{\partial r} \left(r^2 \frac{\partial \psi}{\partial r} \right) + \frac{1}{r^2 \sin \theta} \frac{\partial}{\partial \theta} \left(\sin \theta \frac{\partial \psi}{\partial \theta} \right) + \frac{1}{r^2 \sin^2 \theta} \frac{\partial^2 \psi}{\partial \phi^2}.$$

A.4 Vector Identities

In the following, U and V will be scalar quantities and $\mathbf{A}$ and $\mathbf{B}$ will be vectors. The following relations hold at points where the functions are defined.

1. $\nabla(UV) = U\nabla V + V\nabla U.$
2. $\nabla(\mathbf{A} \cdot \mathbf{B}) = (\mathbf{A} \cdot \nabla)\mathbf{B} + (\mathbf{B} \cdot \nabla)\mathbf{A} + \mathbf{A} \times (\nabla \times \mathbf{B}) + \mathbf{B} \times (\nabla \times \mathbf{A}).$
3. $\nabla \times (\nabla U) = 0.$
4. $\nabla \cdot (\nabla \times \mathbf{A}) = 0.$

5. $\nabla \times \nabla \times \mathbf{A} = \nabla(\nabla \cdot \mathbf{A}) - \nabla^2 \mathbf{A}$.

6. $\nabla \cdot (U\mathbf{A}) = U\nabla \cdot \mathbf{A} + (\mathbf{A} \cdot \nabla)U$.

7. $\nabla \cdot (\mathbf{A} \times \mathbf{B}) = \mathbf{B} \cdot (\nabla \times \mathbf{A}) - \mathbf{A} \cdot (\nabla \times \mathbf{B})$.

8. $\nabla \times (U\mathbf{A}) = U(\nabla \times \mathbf{A}) + (\nabla U) \times \mathbf{A}$.

9. $\nabla \times (\mathbf{A} \times \mathbf{B}) = \mathbf{A}(\nabla \cdot \mathbf{B}) - \mathbf{B}(\nabla \cdot \mathbf{A}) + (\mathbf{B} \cdot \nabla)\mathbf{A} - (\mathbf{A} \cdot \nabla)\mathbf{B}$.

10. $\int_S \mathbf{A} \cdot \hat{\mathbf{n}} \, dS = \int_v \nabla \cdot \mathbf{A} \, dv$ (Divergence theorem).

11. $\int_S \mathbf{A} \cdot \mathbf{dl} = \int_S (\nabla \times \mathbf{A}) \cdot \hat{\mathbf{n}} \, dS$ (Stokes's theorem).

12. $\int_S U\hat{\mathbf{n}} \, dS = \int_v \nabla U \, dv$.

13. $\int_S (\hat{\mathbf{n}} \times \mathbf{A}) \, dS = \int_v (\nabla \times \mathbf{A}) \, dv$.

14. $\oint (\mathbf{A} \cdot \mathbf{B}) \, \mathbf{dl} = \int_S \left(\nabla(\mathbf{A} \cdot \mathbf{B}) \right) \times \hat{\mathbf{n}} \, dS$.

Appendix B
Subroutines

Computers are useless. They can only give you answers.

(Pablo Picasso)

The following pages provide listings of all subroutines referred to in various parts of the text. They appear in the order in which they were referenced. Table B.1 lists them alphabetically and summarizes their functions.

All subroutines are written in ANSI-standard Fortran 77. They have been compiled on a Sun SPARCstation running SunOS 4.13 and Sun Fortran version 1.3, and therefore should compile under any standard Fortran 77 compiler. Each subroutine has been tested in simple applications, but users should be prepared for unknown bugs. All input and output parameters are passed as subroutine arguments rather than through common blocks. Single precision is used throughout (except in subroutine FOURN); users may find it advisable to switch to double precision in some applications.

These subroutines are designed to instruct rather than to be efficient. For example, several of the algorithms require the computation of radial wavenumber

$$|k| = \sqrt{k_x^2 + k_y^2}$$

at grid intersections throughout the k_x, k_y plane. This is done here by calculating $|k|$ explicitly at each grid intersection, so that the reader can easily see the logic involved. A clever programmer, however, could apply some simple tricks in order to exploit the symmetry of $|k|$ in the k_x, k_y plane and thereby greatly speed these computations.

Several of the subroutines (GLAYER, MLAYER, MTOPO, CONTIN, VERDER, NEWVEC, and PSEUDO) make use of the two-dimensional Fourier transform and its inverse. As discussed in Appendix C, the

369

Table B.1. *List of subroutines by function.*[a]

Name	Function	Number
contin	Continue gridded potential fields from one level to another	B.24
cross *	Calculate vector products	B.13
cylind	Calculate gravitational attraction of a cylinder	B.2
dipole	Calculate magnetic induction of a dipole	B.3
dircos *	Calculate direction cosines	B.9
expand	Add tapered rows and columns to a grid	B.29
fac *	Calculate factorials	B.5
facmag	Calculate magnetic induction of a polygonal face	B.10
fork	Calculate one-dimensional Fourier transform and inverse	B.16
fourn *	Calculate n-dimensional Fourier transform and inverse	B.17
gbox	Calculate gravitational attraction of rectangular prism	B.6
gfilt *	Calculate earth filter (gravity case) for horizontal layer	B.21
glayer	Calculate gravitational attraction of flat, horizontal layer	B.18
gpoly	Calculate gravitational attraction of a polygonal prism	B.7
hgrad	Calculate maximum horizontal gradient from gridded data	B.28
kvalue *	Calculate wavenumber coordinates	B.20
line *	Calculate intersection of two lines	B.12
mbox	Calculate the total-field anomaly of a rectangular prism	B.8
mfilt *	Calculate earth filter (magnetic case) for horizontal layer	B.22
mlayer	Calculate total-field anomaly of flat, horizontal layer	B.19
mtopo	Calculate total-field anomaly of layer with uneven topography	B.23
newvec	Transform vector directions in gridded magnetic anomalies	B.26
plane *	Calculate intersection of a plane and a perpendicular line	B.11
pseudo	Transform gridded magnetic anomaly into pseudogravity anomaly	B.27
ribbon	Calculate magnetic induction of a ribbon of magnetic charge	B.15
rot *	Find sense of rotation of one vector with respect to another	B.14
schmit	Calculate normalized associated Legendre polynomials	B.4
sphere	Calculate gravitational attraction of a sphere	B.1
verder	Calculate vertical derivatives of gridded potential fields	B.25

[a] *Note:* Asterisks indicate those subroutines required by other subroutines.

discrete Fourier transform inherently assumes that a rectangular grid is periodic, in effect repeating itself infinitely many times in all horizontal directions, like a vast checkerboard. This can cause undesirable "edge effects" if the edges of the data grid do not meet smoothly with their repetitive neighbors. It is strongly recommended, therefore, that grids be adjusted in some way prior to calling these subroutines in order to eliminate abrupt discontinuities at the edges. One way to do this is to add artificial rows and columns to the data grid in such a way as to produce smooth transitions to neighboring grids. These extra rows and columns can be eliminated from the output grid after the subroutine has executed. Subroutine EXPAND provides a simple algorithm to expand grids in this way.

```
      subroutine sphere(xq,yq,zq,a,rho,xp,yp,zp,gx,gy,gz)
c
c  Subroutine SPHERE calculates the three components of gravitational
c  attraction at a single point due to a uniform sphere.
c
c  Input parameters:
c    Observation point is (xp,yp,zp), and center of sphere is at
c    (xq,yq,zq).  Radius of sphere is a and density is rho.  Density
c    in units of kg/(m**3).  All distance parameters in units of km.
c
c  Output parameters:
c    Gravitational components (gx,gy,gz) in units of mGal.
c
      real km2m
      data gamma/6.67e-11/,si2mg/1.e5/,pi/3.14159265/,km2m/1.e3/
      ierror=0
      rx=xp-xq
      ry=yp-yq
      rz=zp-zq
      r=sqrt(rx**2+ry**2+rz**2)
      if(r.eq.0.)pause 'SPHERE:  Bad argument detected.'
      r3=r**3
      tmass=4.*pi*rho*(a**3)/3.
      gx=-gamma*tmass*rx/r3
      gy=-gamma*tmass*ry/r3
      gz=-gamma*tmass*rz/r3
      gx=gx*si2mg*km2m
      gy=gy*si2mg*km2m
      gz=gz*si2mg*km2m
      return
      end
```

Subroutine B.1.   Subroutine to calculate the three components of gravitational attraction due to a sphere of homogeneous density.

```
      subroutine cylind(xq,zq,a,rho,xp,zp,gx,gz)
c
c Subroutine CYLINDer calculates x and z components of gravitational
c attraction due to cylinder lying parallel to y axis.
c
c Input parameters:
c   Point of observation is (xp,zp).  Axis of cylinder penetrates
c   x,z plane at (xq,zq).  Radius of cylinder is a and density is
c   rho.  Density in kg/(m**3).  All distance parameters in km.
c
c Output parameters:
c   Components of gravitational attraction (gx,gz) in mGal.
c
      real km2m
      data gamma/6.67e-11/,si2mg/1.e5/,pi/3.14159265/,km2m/1.e3/
      rx=xp-xq
      rz=zp-zq
      r2=rx**2+rz**2
      if(r2.eq.0.)pause 'CYLIND:  Bad argument detected.'
      tmass=pi*(a**2)*rho
      gx=-2.*gamma*tmass*rx/r2
      gz=-2.*gamma*tmass*rz/r2
      gx=gx*si2mg*km2m
      gz=gz*si2mg*km2m
      return
      end
```

Subroutine B.2. Subroutine to calculate the gravitational attraction perpendicular to an infinitely extended cylinder. The cylinder lies parallel to the y axis, and x and z components of gravitational attraction are returned.

```
      subroutine dipole(xq,yq,zq,a,mi,md,m,xp,yp,zp,bx,by,bz)
c
c  Subroutine DIPOLE computes the three components of magnetic
c  induction caused by a uniformly magnetized sphere.  x axis
c  is north, z axis is down.
c
c  Input parameters:
c    Observation point located at (xp,yp,zp).  Sphere centered
c    at (xq,yq,zq).  Magnetization of sphere defined by
c    intensity m, inclination mi, and declination md.  Units
c    of distance irrelevant but must be consistent. All angles
c    in degrees.  Intensity of magnetization in A/m.  Requires
c    subroutine DIRCOS.
c
c  Output parameters:
c    The three components of magnetic induction (bx,by,bz) in
c    units of nT.
c
      real mi,md,m,mx,my,mz,moment
      data pi/3.14159265/,t2nt/1.e9/,cm/1.e-7/
      call dircos(mi,md,0.,mx,my,mz)
      rx=xp-xq
      ry=yp-yq
      rz=zp-zq
      r2=rx**2+ry**2+rz**2
      r=sqrt(r2)
      if(r.eq.0.)pause 'DIPOLE:  Bad argument detected.'
      r5=r**5
      dot=rx*mx+ry*my+rz*mz
      moment=4.*pi*(a**3)*m/3
      bx=cm*moment*(3.*dot*rx-r2*mx)/r5
      by=cm*moment*(3.*dot*ry-r2*my)/r5
      bz=cm*moment*(3.*dot*rz-r2*mz)/r5
      bx=bx*t2nt
      by=by*t2nt
      bz=bz*t2nt
      return
      end
```

Subroutine B.3. Subroutine to calculate the three components of magnetic
induction due to a uniformly magnetized sphere (equivalent to the magnetic
induction of a dipole). Coordinate system is arranged so that z is down and
x is north. Requires subroutine DIRCOS.

```
      function schmit(n,m,theta)
c
c  Returns Schmidt normalized associated Legendre polynomial.
c  Requires function fac. Modified from Press et al. (1986).
c
c  Input parameters:
c    Argument of polynomial is theta, in degrees. Degree and
c    order of polynomial are n and m, respectively. Parameter n
c    must be greater than zero, and m must be greater than or
c    equal to n.
c
      data d2rad/.017453293/
      x=cos(theta*d2rad)
      if(m.lt.0.or.m.gt.n)pause ' SCHMIT:   Bad argument detected'
      pmm=1.
      if(m.gt.0)then
         somx2=sqrt((1.-x)*(1.+x))
         fact=1.
         do 10 i=1,m
            pmm=-pmm*fact*somx2
            fact=fact+2.
10       continue
      end if
      if(n.eq.m)then
         schmit=pmm
      else
         pmmp1=x*(2*m+1)*pmm
         if(n.eq.m+1)then
            schmit=pmmp1
         else
            do 11 nn=m+2,n
               pnn=(x*(2*nn-1)*pmmp1-(nn+m-1)*pmm)/(nn-m)
               pmm=pmmp1
               pmmp1=pnn
11          continue
            schmit=pnn
         end if
      end if
      if(m.ne.0)then
         xnorm=sqrt(2*fac(n-m)/fac(n+m))
         schmit=xnorm*schmit
      end if
      return
      end
```

Subroutine B.4. Function to calculate the value of a Schmidt normalized associated Legendre polynomial $P_n^m(\theta)$. Modified from Press et al. [233].

```
      function fac(n)
c
c  Function FAC calculates n!
c
      if(n.lt.0)pause ' FAC:  Bad argument detected'
      if(n.eq.0.or.n.eq.1)then
        fac=1
        else
          fac=n
          fac2=fac
 30       fac2=fac2-1.
          fac=fac*fac2
          if(fac2.gt.2)go to 30
          end if
      return
      end
```

Subroutine B.5. Function to calculate the factorial of n.

Subroutines

```
      subroutine gbox(x0,y0,z0,x1,y1,z1,x2,y2,z2,rho,g)
c
c Subroutine GBOX computes the vertical attraction of a
c rectangular prism. Sides of prism are parallel to x,y,z axes,
c and z axis is vertical down.
c
c Input parameters:
c   Observation point is (x0,y0,z0). The prism extends from x1
c   to x2, from y1 to y2, and from z1 to z2 in the x, y, and z
c   directions, respectively. Density of prism is rho. All
c   distance parameters in units of km; rho in units of
c   kg/(m**3).
c
c Output parameters:
c   Vertical attraction of gravity g, in mGal.
c
      real km2m
      dimension x(2),y(2),z(2),isign(2)
      data isign/-1,1/,gamma/6.670e-11/,twopi/6.2831853/,
     &     si2mg/1.e5/,km2m/1.e3/
      x(1)=x0-x1
      y(1)=y0-y1
      z(1)=z0-z1
      x(2)=x0-x2
      y(2)=y0-y2
      z(2)=z0-z2
      sum=0.
      do 1 i=1,2
        do 1 j=1,2
          do 1 k=1,2
            rijk=sqrt(x(i)**2+y(j)**2+z(k)**2)
            ijk=isign(i)*isign(j)*isign(k)
            arg1=atan2((x(i)*y(j)),(z(k)*rijk))
            if(arg1.lt.0.)arg1=arg1+twopi
            arg2=rijk+y(j)
            arg3=rijk+x(i)
            if(arg2.le.0.)pause 'GBOX:  Bad field point'
            if(arg3.le.0.)pause 'GBOX:  Bad field point'
            arg2=alog(arg2)
            arg3=alog(arg3)
            sum=sum+ijk*(z(k)*arg1-x(i)*arg2-y(j)*arg3)
1         continue
      g=rho*gamma*sum*si2mg*km2m
      return
      end
```

Subroutine B.6.   Subroutine to calculate the vertical attraction of gravity due to a rectangular prism.

```
      subroutine gpoly(x0,z0,xcorn,zcorn,ncorn,rho,g)
c
c  Subroutine GPOLY computes the vertical attraction of a two-
c  dimensional body with polygonal cross section.  Axes are
c  right-handed system with y axis parallel to long direction
c  of body and z axis vertical down.
c
c  Input parameters:
c     Observation point is (x0,z0).  Arrays xcorn and zcorn (each
c     of length ncorn) contain the coordinates of the polygon
c     corners, arranged in clockwise order when viewed with x axis
c     to right.  Density of body is rho.  All distance parameters
c     in units of km; rho in units of kg/(m**3).
c
c  Output parameters:
c     Vertical attraction of gravity g, in mGal.
c
      real km2m
      dimension xcorn(ncorn),zcorn(ncorn)
      data gamma/6.670e-11/,si2mg/1.e5/,km2m/1.e3/
      sum=0.
      do 1 n=1,ncorn
         if(n.eq.ncorn)then
            n2=1
         else
             n2=n+1
             end if
         x1=xcorn(n)-x0
         z1=zcorn(n)-z0
         x2=xcorn(n2)-x0
         z2=zcorn(n2)-z0
         r1sq=x1**2+z1**2
         r2sq=x2**2+z2**2
         if(r1sq.eq.0.)pause 'GPOLY:  Field point on corner'
         if(r2sq.eq.0.)pause 'GPOLY:  Field point on corner'
         denom=z2-z1
         if(denom.eq.0.)denom=1.e-6
         alpha=(x2-x1)/denom
         beta=(x1*z2-x2*z1)/denom
         factor=beta/(1.+alpha**2)
         term1=0.5*(alog(r2sq)-alog(r1sq))
         term2=atan2(z2,x2)-atan2(z1,x1)
         sum=sum+factor*(term1-alpha*term2)
    1    continue
      g=2.*rho*gamma*sum*si2mg*km2m
      return
      end
```

Subroutine B.7.   Subroutine to calculate the vertical gravitational attraction
due to a two-dimensional prism of polygonal cross section. Axis of prism lies
parallel to y axis; z axis is down. Polygon corners are contained within arrays
xcorn and **zcorn** in clockwise order as viewed with the x axis to the right.

```
subroutine mbox(x0,y0,z0,x1,y1,z1,x2,y2,mi,md,fi,fd,m,theta,t)
c
c  Subroutine MBOX computes the total-field anomaly of an infinitely
c  extended rectangular prism.  Sides of prism are parallel to x,y,z
c  axes, and z is vertical down.  Bottom of prism extends to infinity.
c  Two calls to mbox can provide the anomaly of a prism with finite
c  thickness; e.g.,
c
c      call mbox(x0,y0,z0,x1,y1,z1,x2,y2,mi,md,fi,fd,m,theta,t1)
c      call mbox(x0,y0,z0,x1,y1,z1,x2,y2,mi,md,fi,fd,m,theta,t2)
c      t=t1-t2
c
c  Requires subroutine DIRCOS.  Method from Bhattacharyya (1964).
c
c  Input parameters:
c    Observation point is (x0,y0,z0).  Prism extends from x1 to
c    x2, y1 to y2, and z1 to infinity in x, y, and z directions,
c    respectively.  Magnetization defined by inclination mi,
c    declination md, intensity m.  Ambient field defined by
c    inclination fi and declination fd.  x axis has declination
c    theta. Distance units are irrelevant but must be consistent.
c    Angles are in degrees, with inclinations positive below
c    horizontal and declinations positive east of true north.
c    Magnetization in A/m.
c
c  Output parameters:
c    Total-field anomaly t, in nT.
c
       real alpha(2),beta(2),mi,md,m,ma,mb,mc
       data cm/1.e-7/,t2nt/1.e9/
       call dircos(mi,md,theta,ma,mb,mc)
       call dircos(fi,fd,theta,fa,fb,fc)
       fm1=ma*fb+mb*fa
       fm2=ma*fc+mc*fa
       fm3=mb*fc+mc*fb
       fm4=ma*fa
       fm5=mb*fb
       fm6=mc*fc
       alpha(1)=x1-x0
       alpha(2)=x2-x0
       beta(1)=y1-y0
       beta(2)=y2-y0
       h=z1-z0
       t=0.
       hsq=h**2
```

Subroutine B.8.   Subroutine to calculate the total-field anomaly due to a rectangular prism with infinite depth extent. The anomaly of a prism with finite thickness can be found by calling MBOX twice, once with z1 equal to the top of the prism and once with z1 equal to the bottom, and subtracting the second result from the first. Subroutine DIRCOS also required.

```
      do 1 i=1,2
        alphasq=alpha(i)**2
        do 1 j=1,2
          sign=1.
          if(i.ne.j)sign=-1.
          r0sq=alphasq+beta(j)**2+hsq
          r0=sqrt(r0sq)
          r0h=r0*h
          alphabeta=alpha(i)*beta(j)
          arg1=(r0-alpha(i))/(r0+alpha(i))
          arg2=(r0-beta(j))/(r0+beta(j))
          arg3=alphasq+r0h+hsq
          arg4=r0sq+r0h-alphasq
          tlog=fm3*alog(arg1)/2.+fm2*alog(arg2)/2.
     &           -fm1*alog(r0+h)
          tatan=-fm4*atan2(alphabeta,arg3)
     &           -fm5*atan2(alphabeta,arg4)
     &           +fm6*atan2(alphabeta,r0h)
    1 t=t+sign*(tlog+tatan)
      t=t*m*cm*t2nt
      return
      end
```

Continuation of Subroutine B.8.

```
      subroutine dircos(incl,decl,azim,a,b,c)
c
c  Subroutine DIRCOS computes direction cosines from inclination
c  and declination.
c
c  Input parameters:
c     incl:  inclination in degrees positive below horizontal.
c     decl:  declination in degrees positive east of true north.
c     azim:  azimuth of x axis in degrees positive east of north.
c
c  Output parameters:
c     a,b,c:  the three direction cosines.
c
      real incl
      data d2rad/.017453293/
      xincl=incl*d2rad
      xdecl=decl*d2rad
      xazim=azim*d2rad
      a=cos(xincl)*cos(xdecl-xazim)
      b=cos(xincl)*sin(xdecl-xazim)
      c=sin(xincl)
      return
      end
```

Subroutine B.9. Subroutine to calculate the three direction cosines of a vector from its inclination and declination.

```
      subroutine facmag(mx,my,mz,x0,y0,z0,x,y,z,n,fx,fy,fz)
c
c  Subroutine FACMAG computes the magnetic field due to surface
c  charge  on a polygonal face.  Repeated calls can build the
c  field of an arbitrary polyhedron.  x axis is directed north,
c  z axis vertical down.  Requires subroutines CROSS, ROT, LINE,
c  and PLANE.  Algorithm from Bott (1963).
c
c  Input parameters:
c    Observation point is (x0,y0,z0).  Polygon corners defined
c    by arrays x, y, and z of length n.  Magnetization given by
c    mx,my,mz.  Polygon limited to 10 corners.  Distance units
c    are irrelevant but must be consistent; magnetization in A/m.
c
c  Output parameters:
c    Three components of magnetic field (fx,fy,fz), in nT.
c
      real mx,my,mz,nx,ny,nz
      dimension u(10),v2(10),v1(10),s(10),xk(10),yk(10),
     &        zk(10),xl(10),yl(10),zl(10),x(10),y(10),z(10)
      data cm/1.e-7/,t2nt/1.e9/,epsilon/1.e-20/
      fx=0.
      fy=0.
      fz=0.
      x(n+1)=x(1)
      y(n+1)=y(1)
      z(n+1)=z(1)
      do 1 i=1,n
        xl(i)=x(i+1)-x(i)
        yl(i)=y(i+1)-y(i)
        zl(i)=z(i+1)-z(i)
        rl=sqrt(xl(i)**2+yl(i)**2+zl(i)**2)
        xl(i)=xl(i)/rl
        yl(i)=yl(i)/rl
    1   zl(i)=zl(i)/rl
      call cross(xl(2),yl(2),zl(2),xl(1),yl(1),zl(1),nx,ny,nz,rn)
      nx=nx/rn
      ny=ny/rn
      nz=nz/rn
      dot=mx*nx+my*ny+mz*nz
      if(dot.eq.0.)return
      call plane(x0,y0,z0,x(1),y(1),z(1),x(2),y(2),z(2),x(3),
     &          y(3),z(3),px,py,pz,w)
      do 2 i=1,n
        call rot(x(i),y(i),z(i),x(i+1),y(i+1),z(i+1),nx,ny,nz,
     &          px,py,pz,s(i))
        if(s(i).eq.0.)go to 2
        call line(px,py,pz,x(i),y(i),z(i),x(i+1),y(i+1),z(i+1),
```

Subroutine B.10. Subroutine to calculate the magnetic field due to magnetic charge on a flat polygonal face. By repeated calls, the magnetic field of a uniformly magnetized polyhedron can be computed. Subroutines PLANE, LINE, CROSS, and ROT are required. Algorithm from Bott [36].

```
     &              u1,v,w1,v1(i),v2(i),u(i))
                rk=sqrt((u1-px)**2+(v-py)**2+(w1-pz)**2)
                xk(i)=(u1-px)/rk
                yk(i)=(v-py)/rk
     2          zk(i)=(w1-pz)/rk
          do 3 j=1,n
                if(s(j).eq.0.)go to 3
                us=u(j)**2
                v2s=v2(j)**2
                v1s=v1(j)**2
                a2=v2(j)/u(j)
                a1=v1(j)/u(j)
                f2=sqrt(1.+a2*a2)
                f1=sqrt(1.+a1*a1)
                rho2=sqrt(us+v2s)
                rho1=sqrt(us+v1s)
                r2=sqrt(us+v2s+w**2)
                r1=sqrt(us+v1s+w**2)
                if(w.ne.0.)then
                     fu2=(a2/f2)*alog((r2+rho2)/abs(w))
     &                    -.5*alog((r2+v2(j))/(r2-v2(j)))
                     fu1=(a1/f1)*alog((r1+rho1)/abs(w))-
     &                    .5*alog((r1+v1(j))/(r1-v1(j)))
                     fv2=(1./f2)*alog((r2+rho2)/abs(w))
                     fv1=(1./f1)*alog((r1+rho1)/abs(w))
                     fw2=atan2((a2*(r2-abs(w))),(r2+a2*a2*abs(w)))
                     fw1=atan2((a1*(r1-abs(w))),(r1+a1*a1*abs(w)))
                     fu=dot*(fu2-fu1)
                     fv=-dot*(fv2-fv1)
                     fw=(-w*dot/abs(w))*(fw2-fw1)
                else
                     fu2=(a2/f2)*(1.+alog((r2+rho2)/epsilon))-
     &                    .5*alog((r2+v2(j))/(r2-v2(j)))
                     fu1=(a1/f1)*(1.+alog((r1+rho1)/epsilon))-
     &                    .5*alog((r1+v1(j))/(r1-v1(j)))
                     fv2=(1./f2)*(1.+alog((r2+rho2)/epsilon))
                     fv1=(1./f1)*(1.+alog((r1+rho1)/epsilon))
                     fu=dot*(fu2-fu1)
                     fv=-dot*(fv2-fv1)
                     fw=0.
                end if
          fx=fx-s(j)*(fu*xk(j)+fv*xl(j)+fw*nx)
          fy=fy-s(j)*(fu*yk(j)+fv*yl(j)+fw*ny)
          fz=fz-s(j)*(fu*zk(j)+fv*zl(j)+fw*nz)
     3 continue
          fx=fx*cm*t2nt
          fy=fy*cm*t2nt
          fz=fz*cm*t2nt
          return
          end
```

Continuation of Subroutine B.10.

```
subroutine plane(x0,y0,z0,x1,y1,z1,x2,y2,z2,x3,y3,z3,x,y,z,r)
c
c  Subroutine PLANE computes the intersection (x,y,z) of a plane
c  and a perpendicular line.  The plane is defined by three points
c  (x1,y1,z1), (x2,y2,z2), and (x3,y3,z3).  The line passes through
c  (x0,y0,z0).  Computation is done by a transformation and inverse
c  transformation of coordinates systems.
c
       x2n=x2-x1
       y2n=y2-y1
       z2n=z2-z1
       x0n=x0-x1
       y0n=y0-y1
       z0n=z0-z1
       x3n=x3-x1
       y3n=y3-y1
       z3n=z3-z1
       call cross(x3n,y3n,z3n,x2n,y2n,z2n,cx,cy,cz,c)
       call cross(x2n,y2n,z2n,cx,cy,cz,dx,dy,dz,d)
       a=sqrt(x2n**2+y2n**2+z2n**2)
       t11=x2n/a
       t12=y2n/a
       t13=z2n/a
       t21=cx/c
       t22=cy/c
       t23=cz/c
       t31=dx/d
       t32=dy/d
       t33=dz/d
       tx0=t11*x0n+t12*y0n+t13*z0n
       tz0=t31*x0n+t32*y0n+t33*z0n
       r=t21*x0n+t22*y0n+t23*z0n
       x=t11*tx0+t31*tz0
       y=t12*tx0+t32*tz0
       z=t13*tx0+t33*tz0
       x=x+x1
       y=y+y1
       z=z+z1
       return
       end
```

Subroutine B.11.   Subroutine to calculate the intersection of a plane and a
perpendicular line.

```
      subroutine line(x0,y0,z0,x1,y1,z1,x2,y2,z2,x,y,z,v1,v2,r)
c
c  Subroutine LINE determines the intersection (x,y,z) of two
c  lines.  First line is defined by points (x1,y1,z1) and
c  (x2,y2,z2).  Second line is perpendicular to the first and
c  passes through point (x0,y0,z0).  Distance between (x,y,z)
c  and (x0,y0,z0) is returned as r.  Computation is done by a
c  transformation of coordinate systems.
c
      tx0=x0-x1
      ty0=y0-y1
      tz0=z0-z1
      tx2=x2-x1
      ty2=y2-y1
      tz2=z2-z1
      a=sqrt(tx2**2+ty2**2+tz2**2)
      call cross(tx2,ty2,tz2,tx0,ty0,tz0,cx,cy,cz,c)
      call cross(cx,cy,cz,tx2,ty2,tz2,dx,dy,dz,d)
      tt11=tx2/a
      tt12=ty2/a
      tt13=tz2/a
      tt21=dx/d
      tt22=dy/d
      tt23=dz/d
      tt31=cx/c
      tt32=cy/c
      tt33=cz/c
      u0=tt11*tx0+tt12*ty0+tt13*tz0
      r=tt21*tx0+tt22*ty0+tt23*tz0
      x=tt11*u0+x1
      y=tt12*u0+y1
      z=tt13*u0+z1
      v1=-u0
      v2=a-u0
      return
      end
```

Subroutine B.12.   Subroutine to calculate the intersection of two lines.

```
      subroutine cross(ax,ay,az,bx,by,bz,cx,cy,cz,r)
c
c  Subroutine CROSS computes the vector product of two vectors; i.e.,
c
c               (cx,cy,cz) = (ax,ay,az) X (bx,by,bz)
c
      cx=ay*bz-az*by
      cy=az*bx-ax*bz
      cz=ax*by-ay*bx
      r=sqrt(cx**2+cy**2+cz**2)
      return
      end
```

Subroutine B.13.   Subroutine to calculate vector products.

```
      subroutine rot(ax,ay,az,bx,by,bz,nx,ny,nz,px,py,pz,s)
c
c  Subroutine ROT finds the sense of rotation of the vector
c  from (ax,ay,az) to (bx,by,bz) with respect to a second
c  vector through point (px,py,pz).  The second vector has
c  components given by (nx,ny,nz).  Returned parameter s is
c  1 if anticlockwise, -1 if clockwise, or 0 if collinear.
c
      real nx,ny,nz
      x=bx-ax
      y=by-ay
      z=bz-az
      call cross(nx,ny,nz,x,y,z,cx,cy,cz,c)
      u=px-ax
      v=py-ay
      w=pz-az
      d=u*cx+v*cy+w*cz
      if(d)2,3,4
    2 s=1.
      go to 1
    3 s=0.
      go to 1
    4 s=-1.
    1 continue
      return
      end
```

Subroutine B.14.   Subroutine to find the sense of rotation of one vector with respect to another.

```
      subroutine ribbon(x0,z0,x1,z1,x2,z2,mx,mz,fx,fz,ier)
c
c  Subroutine RIBBON computes the x and z components of magnetic
c  induction due to a single side of a two-dimensional prism with
c  polygonal cross section.  The prism is assumed infinitely
c  extended parallel to the y axis; z axis is vertical down.
c
c  Input parameters:
c    Observation point is (x0,z0).  Coordinates (x1,z1) and
c    (x2,z2) are two consecutive corners of the polygon taken in
c    clockwise order around the polygon as viewed with the x
c    axis to the right.  The x and z components of magnetization
c    are mx and mz.  Distance units are irrelevant but must be
c    consistent; magnetization in A/m.
c
c  Output parameters:
c    Components of magnetic field fx and fz, in nT.
c    Errors are recorded by ier:
c        ier=0, no errors;
c        ier=1, two corners are too close (no calculation);
c        ier=2, field point too close to corner (calculation
c               continues).
c
      real mx,mz
      data pi/3.14159265/,small/1.e-18/,cm/1.e-7/,t2nt/1.e9/
      ier=0
      sx=x2-x1
      sz=z2-z1
      s=sqrt(sx**2+sz**2)
c
c  --  If two corners are too close, return
c
      if (s.lt.small)then
        ier=1
        return
        end if
      sx=sx/s
      sz=sz/s
      qs=mx*sz-mz*sx
      rx1=x1-x0
      rz1=z1-z0
      rx2=x2-x0
      rz2=z2-z0
```

Subroutine B.15. Subroutine to calculate the magnetic induction of a flat ribbon of magnetic charge. The ribbon is infinitely extended and lies parallel to the y axis; z axis is down. With repeated calls, the magnetic field of a prism with polygonal cross section can be computed.

388   *Subroutines*

```
c
c  --  If field point is too near a corner, signal error
c
      if(abs(rx1).lt.small.and.abs(rz1).lt.small)ier=2
      if(abs(rx2).lt.small.and.abs(rz2).lt.small)ier=2
      if(ier.eq.2)then
         rx1=small
         rz1=small
         rx2=small
         rz2=small
         end if
      r1=sqrt(rx1**2+rz1**2)
      r2=sqrt(rx2**2+rz2**2)
      theta1=atan2(rz1,rx1)
      theta2=atan2(rz2,rx2)
      angle=theta1-theta2
      if (angle.gt.pi)angle=angle-2.*pi
      if (angle.lt.-pi)angle=angle+2.*pi
c
c  --  If field point is too close to side, signal error
c
      if (abs(angle).gt.(.995*pi))ier=2
      flog=alog(r2)-alog(r1)
      factor=-2.*cm*qs*t2nt
      fx=factor*(sx*flog-sz*angle)
      fz=factor*(sz*flog+sx*angle)
      return
      end
```

Continuation of Subroutine B.15.

```
      subroutine fork(lx,cx,signi)
c
c  Subroutine FORK calculates the Fourier transform of a one-
c  dimensional array. Algorithm from Claerbout (1976).
c
c  Input/output parameters:
c     Complex array cx of length lx is the input array. Upon
c     return, cx contains the transformed array. Length of
c     array must be a power of 2. If signi=-1., then the forward
c     calculation is performed; if signi=1., the inverse transform
c     is performed.
c
      complex cx(2050),carg,cexp,cw,ctemp
      j=1
      sc=sqrt(1./lx)
      do 5 i=1,lx
        if(i.gt.j)go to 2
          ctemp=cx(j)*sc
          cx(j)=cx(i)*sc
          cx(i)=ctemp
2       m=lx/2
3       if(j.le.m)go to 5
          j=j-m
          m=m/2
          if(m.ge.1)go to 3
5       j=j+m
      l=1
6     istep=2*l
      do 8 m=1,l
        carg=(0.,1.)*(3.14159265*signi*(m-1))/l
        cw=cexp(carg)
        do 8 i=m,lx,istep
          ipl=i+l
          ctemp=cw*cx(ipl)
          cx(ipl)=cx(i)-ctemp
8         cx(i)=cx(i)+ctemp
      l=istep
      if(l.lt.lx)go to 6
      return
      end
```

Subroutine B.16. Subroutine to calculate the one-dimensional discrete Fourier transform and its inverse. Input data is required to have a length equal to a power of two; zeroes can be added to the end of array cx in order to satisfy this requirement. This algorithm is a slight modification of a subroutine described by Claerbout [60].

```
      subroutine fourn(data,nn,ndim,isign)
c
c  Replaces DATA by its NDIM-dimensional discrete Fourier transform,
c  if ISIGN is input as 1.  NN is an integer array of length NDIM,
c  containing the lengths of each dimension (number of complex values),
c  which must all be powers of 2.  DATA is a real array of length twice
c  the product of these lengths, in which the data are stored as in a
c  multidimensional complex Fortran array.  If ISIGN is input as -1,
c  DATA is replaced by its inverse transform times the product of the
c  lengths of all dimensions.  From Press et al. (1986, pp. 451-3).
c
      real*8 wr,wi,wpr,wpi,wtemp,theta
      dimension nn(ndim),data(*)
      ntot=1
      do 11 iidim=1,ndim
        ntot=ntot*nn(iidim)
11    continue
      nprev=1
      do 18 iidim=1,ndim
        n=nn(iidim)
        nrem=ntot/(n*nprev)
        ip1=2*nprev
        ip2=ip1*n
        ip3=ip2*nrem
        i2rev=1
        do 14 i2=1,ip2,ip1
          if(i2.lt.i2rev)then
            do 13 i1=i2,i2+ip1-2,2
              do 12 i3=i1,ip3,ip2
                i3rev=i2rev+i3-i2
                tempr=data(i3)
                tempi=data(i3+1)
                data(i3)=data(i3rev)
                data(i3+1)=data(i3rev+1)
                data(i3rev)=tempr
                data(i3rev+1)=tempi
12            continue
13          continue
          endif
          ibit=ip2/2
1         if ((ibit.ge.ip1).and.(i2rev.gt.ibit)) then
            i2rev=i2rev-ibit
            ibit=ibit/2
          go to 1
          endif
          i2rev=i2rev+ibit
14      continue
        ifp1=ip1
```

Subroutine B.17. Subroutine to calculate the discrete Fourier transform and inverse Fourier transform of an n-dimensional array. Each dimension must be a power of two. From Press et al. [233].

```
2        if(ifp1.lt.ip2)then
           ifp2=2*ifp1
           theta=isign*6.28318530717959d0/(ifp2/ip1)
           wpr=-2.d0*dsin(0.5d0*theta)**2
           wpi=dsin(theta)
           wr=1.d0
           wi=0.d0
           do 17 i3=1,ifp1,ip1
             do 16 i1=i3,i3+ip1-2,2
               do 15 i2=i1,ip3,ifp2
                 k1=i2
                 k2=k1+ifp1
                 tempr=sngl(wr)*data(k2)-sngl(wi)*data(k2+1)
                 tempi=sngl(wr)*data(k2+1)+sngl(wi)*data(k2)
                 data(k2)=data(k1)-tempr
                 data(k2+1)=data(k1+1)-tempi
                 data(k1)=data(k1)+tempr
                 data(k1+1)=data(k1+1)+tempi
15               continue
16             continue
               wtemp=wr
               wr=wr*wpr-wi*wpi+wr
               wi=wi*wpr+wtemp*wpi+wi
17         continue
           ifp1=ifp2
           go to 2
         endif
         nprev=n*nprev
18       continue
         return
         end
```

Continuation of Subroutine B.17.

```
      subroutine glayer(rho,nx,ny,dx,dy,z1,z2,store)
c
c Subroutine GLAYER calculates the vertical gravitational
c attraction on a two-dimensional grid caused by a two-
c dimensional density confined to a horizontal layer.  The
c following steps are involved:  (1) Fourier transform the
c density, (2) multiply by the earth filter, and (3) inverse
c Fourier transform the product.  Density is specified on a
c rectangular grid with x and y axes directed north and east,
c respectively.  z axis is down.  Requires subroutines
c FOURN, KVALUE, and GFILT.
c
c Input parameters:
c    nx - number of elements in the south-to-north direction.
c    ny - number of elements in the west-to-east direction.
c         (NOTE: Both nx and ny must be powers of two.)
c    rho - a one-dimensional real array containing the
c          two-dimensional density, in kg/(m**3).  Elements
c          should be in order of west to east, then south to
c          north (i.e., element 1 is the southwest corner,
c          element ny is the southeast corner, element
c          (nx-1)*ny+1 is the northwest corner, and element ny*nx
c          is the northeast corner.
c    store - a one-dimensional real array used internally.
c          It should be dimensioned at least 2*nx*ny.
c    dx - sample interval in the x direction, in km.
c    dy - sample interval in the y direction, in km.
c    z1 - depth to top of layer, in km.  Must be > 0.
c    z2 - depth to bottom of layer, in km.  Must be > z1.
c
c Output parameters:
c    rho - upon output, rho will contain the gravity anomaly,
c          in mGal, with same orientation as before.
c
      complex crho,cmplx
      real kx,ky,km2m
      dimension rho(nx*ny),store(2*nx*ny),nn(2)
      data pi/3.14159265/,si2mg/1.e5/,km2m/1.e3/
      index(i,j,ncol)=(j-1)*ncol+i
      nn(1)=ny
      nn(2)=nx
      ndim=2
      dkx=2.*pi/(nx*dx)
      dky=2.*pi/(ny*dy)
      do 10 j=1,nx
         do 10 i=1,ny
            ij=index(i,j,ny)
            store(2*ij-1)=rho(ij)
10          store(2*ij)=0.
      call fourn(store,nn,ndim,-1)
```

Subroutine B.18. Subroutine to calculate the vertical attraction due to a horizontal layer with flat top, flat bottom, and a density distribution that varies in both horizontal directions. Input density and output anomaly are specified on the same rectangular grid; both dimensions must be a power of two. Requires subroutines FOURN, KVALUE, and GFILT.

```
      do 20 j=1,nx
        do 20 i=1,ny
          ij=index(i,j,ny)
          call kvalue(i,j,nx,ny,dkx,dky,kx,ky)
          crho=cmplx(store(2*ij-1),store(2*ij))
          crho=crho*gfilt(kx,ky,z1,z2)
          store(2*ij-1)=real(crho)
20        store(2*ij)=aimag(crho)
      call fourn(store,nn,ndim,+1)
      do 30 j=1,nx
        do 30 i=1,ny
          ij=index(i,j,ny)
30        rho(ij)=store(2*ij-1)*si2mg*km2m/(nx*ny)
      return
      end
```

Continuation of Subroutine B.18.

```
      subroutine mlayer(mag,nx,ny,dx,dy,z1,z2,mi,md,fi,fd,store)
c
c  Subroutine MLAYER calculates the total-field anomaly on a two-
c  dimensional grid due to a horizontal layer with two-
c  dimensional magnetization.  The following steps are involved:
c  (1) Fourier transform the magnetization, (2) multiply by the
c  earth filter, and (3) inverse Fourier transform the product.
c  Magnetization is specified on a rectangular grid with x and y
c  axes directed north and east, respectively.  z axis is down.
c  Distance units irrelevant but must be consistent.  Requires
c  subroutines FOURN, DIRCOS, KVALUE, and MFILT.
c
c  Input parameters:
c    nx - number of elements in the south-to-north direction.
c    ny - number of elements in the west-to-east direction.
c         (NOTE:  both nx and ny must be a power of two.)
c    mag - a one-dimensional real array containing the
c          two-dimensional magnetization (in A/m).  Elements should
c          be in order of west to east, then south to north (i.e.,
c          element 1 is southwest corner, element ny is
c          southeast corner, element (nx-1)*ny+1 is northwest
c          corner, and element ny*nx is northeast corner).
c    store - a one-dimensional real array used internally.
c            It should be dimensioned at least 2*nx*ny.
c    dx - sample interval in the x direction.
c    dy - sample interval in the y direction.
c    z1 - depth to top of layer.  Must be > 0.
c    z2 - depth to bottom of layer.  Must be > z1.
c    mi - inclination of magnetization, in degrees positive below
c         horizontal.
c    md - declination of magnetization, in degrees east of north.
c    fi - inclination of regional field.
c    fd - declination of regional field.
c
c  Output parameters:
c    mag - upon output, mag contains the total-field anomaly
c          (in nT) with same orientation as before.
c
      complex cmag,mfilt,cmplx
      real mag,mi,md,mx,my,mz,kx,ky
      dimension mag(nx*ny),store(2*nx*ny),nn(2)
      data pi/3.14159265/,t2nt/1.e9/
      index(i,j,ncol)=(j-1)*ncol+i
      nn(1)=ny
      nn(2)=nx
      ndim=2
      call dircos(mi,md,0.,mx,my,mz)
      call dircos(fi,fd,0.,fx,fy,fz)
      dkx=2.*pi/(nx*dx)
      dky=2.*pi/(ny*dy)
```

Subroutine B.19. Subroutine to calculate the total-field anomaly due to a horizontal layer with flat top, flat bottom, and magnetization that varies in both horizontal directions. Input magnetization and output total-field anomaly are specified on the same rectangular grid; both dimensions must be a power of two. Requires subroutines FOURN, DIRCOS, KVALUE, and MFILT.

```
     do 10 j=1,nx
        do 10 i=1,ny
        ij=index(i,j,ny)
        store(2*ij-1)=mag(ij)
10      store(2*ij)=0.
     call fourn(store,nn,ndim,-1)
     do 20 j=1,nx
        do 20 i=1,ny
        ij=index(i,j,ny)
        call kvalue(i,j,nx,ny,dkx,dky,kx,ky)
        cmag=cmplx(store(2*ij-1),store(2*ij))
        cmag=cmag*mfilt(kx,ky,mx,my,mz,fx,fy,fz,z1,z2)
        store(2*ij-1)=real(cmag)
20      store(2*ij)=aimag(cmag)
     call fourn(store,nn,ndim,+1)
     do 30 j=1,nx
        do 30 i=1,ny
        ij=index(i,j,ny)
30      mag(ij)=store(2*ij-1)*t2nt/(nx*ny)
     return
     end
```

Continuation of Subroutine B.19.

396   *Subroutines*

```
      subroutine kvalue(i,j,nx,ny,dkx,dky,kx,ky)
c  Subroutine KVALUE finds the wavenumber coordinates of one
c  element of a rectangular grid from subroutine FOURN.
c
c  Input parameters:
c    i  - index in the ky direction.
c    j  - index in the kx direction.
c    nx - dimension of grid in ky direction (a power of two).
c    ny - dimension of grid in kx direction (a power of two).
c    dkx - sample interval in the kx direction.
c    dky - sample interval in the ky direction.
c
c  Output parameters:
c    kx - the wavenumber coordinate in the kx direction.
c    ky - the wavenumber coordinate in the ky direction.
c
      real kx,ky
      nyqx=nx/2+1
      nyqy=ny/2+1
      if(j.le.nyqx)then
         kx=(j-1)*dkx
         else
            kx=(j-nx-1)*dkx
         end if
      if(i.le.nyqy)then
         ky=(i-1)*dky
         else
            ky=(i-ny-1)*dky
         end if
      return
      end
```

Subroutine B.20.   Subroutine to calculate the wavenumber coordinates of
elements of grids returned by subroutine FOURN.

```
      function gfilt(kx,ky,z1,z2)
c
c Function GFILT calculates the value of the gravitational
c earth filter at a single (kx,ky) location.
c
c Input parameters:
c   kx - the wavenumber coordinate in the kx direction, in
c        units of 1/km.
c   ky - the wavenumber coordinate in the ky direction, in
c        units of 1/km.
c   z1 - the depth to the top of the layer, in km.
c   z2 - the depth to the bottom of the layer, in km.
c
c Output parameters:
c   gfilt - the value of the earth filter.
c
      real kx,ky,k
      data pi/3.14159265/,gamma/6.67e-11/
      k=sqrt(kx**2+ky**2)
      if(k.eq.0.)then
         gfilt=2.*pi*gamma*(z2-z1)
         else
            gfilt=2.*pi*gamma*(exp(-k*z1)-exp(-k*z2))/k
            end if
      return
      end
```

Subroutine B.21. Function to calculate the earth filter (gravity case) for a horizontal layer.

```
      function mfilt(kx,ky,mx,my,mz,fx,fy,fz,z1,z2)
c
c  Function MFILT calculates the complex value of the earth
c  filter at a single (kx,ky) location.
c
c  Input parameters:
c     kx - the wavenumber coordinate in the kx direction.
c     ky - the wavenumber coordinate in the ky direction.
c     mx - the x direction cosine of the magnetization vector.
c     my - the y direction cosine of the magnetization vector.
c     mz - the z direction cosine of the magnetization vector.
c     fx - the x direction cosine of the regional field vector.
c     fy - the y direction cosine of the regional field vector.
c     fz - the z direction cosine of the regional field vector.
c     z1 - the depth to the top of the layer.
c     z2 - the depth to the bottom of the layer.
c
c  Output parameters:
c     mfilt - the complex value of the earth filter.
c
      complex mfilt,thetam,thetaf,cmplx
      real kx,ky,k,mx,my,mz
      data pi/3.14159265/,cm/1.e-7/
      k=sqrt(kx**2+ky**2)
      if(k.eq.0.)then
         mfilt=0.
         else
            thetam=cmplx(mz,(kx*mx+ky*my)/k)
            thetaf=cmplx(fz,(kx*fx+ky*fy)/k)
            mfilt=2.*pi*cm*thetam*thetaf*(exp(-k*z1)-exp(-k*z2))
            end if
      return
      end
```

Subroutine B.22. Subroutine to calculate the earth filter (magnetic case) for a horizontal layer.

```
      subroutine mtopo(mag,ztop,nx,ny,dx,dy,mi,md,fi,fd,nstop,err,
     &                 store,cstore)
c
c Subroutine MTOPO calculates the total-field anomaly on a two-
c dimensional grid due to an infinite half-space with uneven
c top surface and two-dimensional magnetization. Method
c according to Parker (1972, pp. 447-55).
c Magnetization is specified on a rectangular grid with x and y
c axes directed north and east, respectively. z axis is down
c and anomaly is calculated at z=0; topographic grid should be
c arranged accordingly. Units of distance irrelevant but must be
c consistent. Requires subroutines FOURN, DIRCOS, KVALUE, and FAC.
c
c Input parameters:
c   nx - number of elements in the south-to-north direction.
c   ny - number of elements in the west-to-east direction.
c        (NOTE: both nx and ny must be powers of two.)
c   mag - a one-dimensional real array containing the two-
c         dimensional magnetization (in A/m). Elements should
c         be in order of west to east, then south to north
c         (i.e., element 1 is southwest corner, element ny is
c         southeast corner, element (nx-1)*ny+1 is northwest
c         corner, and element ny*nx is northeast corner).
c   ztop - a one-dimensional real array containing the
c          topography of the upper surface, in same units as dx
c          and dy. Grid layout same as mag. Note: z axis is
c          positive down and anomaly is calculated at z=0. Array
c          ztop is modified by subroutine.
c   store - a one-dimensional real array used internally.
c           Should be dimensioned at least 2*nx*ny.
c   cstore - a one-dimensional complex array used internally.
c            Should be dimensioned at least nx*ny.
c   dx - sample interval in the x direction.
c   dy - sample interval in the y direction.
c   mi - inclination of magnetization, in degrees positive below
c        horizontal.
c   md - declination of magnetization, in degrees east of north.
c   fi - inclination of regional field.
c   fd - declination of regional field.
c   nstop - maximum number of iterations to try.
c   err - convergence criterion. Iterations stop when the
c         contribution of the last term to the summation is less
c         than err times the contribution of all previous terms.
c
```

Subroutine B.23. Subroutine to calculate the total-field anomaly due to a
semi-infinite half space with uneven top surface and infinite depth extent.
Two calls to subroutine MTOPO can provide the anomaly of a layer with
uneven top and bottom surfaces. Input arrays **mag** and **ztop** are specified on the
same rectangular grid. Edge discontinuities should be eliminated from input
grids (see subroutine EXPAND). Upon output, mag is replaced with the total-
field anomaly on the same rectangular grid. Requires subroutines FOURN,
DIRCOS, KVALUE, and FAC. Method according to Parker [204].

400 *Subroutines*

```
c  Output parameters:
c    mag - upon output, mag contains the total-field anomaly
c          (in nT) with same grid orientation as before.
c
       dimension mag(nx*ny),ztop(nx*ny),store(2*nx*ny),cstore(nx*ny),
     &            nn(2)
       real mag,mi,md,mx,my,mz,kx,ky,k
       complex cmplx,cstore,cstep,thetam,thetaf
       data pi/3.14159265/,t2nt/1.e9/,cm/1.e-7/
       index(i,j,ncol)=(j-1)*ncol+i
       nn(1)=ny
       nn(2)=nx
       ndim=2
       call dircos(mi,md,0.,mx,my,mz)
       call dircos(fi,fd,0.,fx,fy,fz)
       dkx=2.*pi/(nx*dx)
       dky=2.*pi/(ny*dy)
       ztpmax=-1.e20
       ztpmin= 1.e20
       do 1 j=1,nx
          do 1 i=1,ny
             ij=index(i,j,ny)
             ztpmax=amax1(ztpmax,ztop(ij))
    1        ztpmin=amin1(ztpmin,ztop(ij))
       ztpmed=ztpmin+(ztpmax-ztpmin)/2.
       do 2 j=1,nx
          do 2 i=1,ny
             ij=index(i,j,ny)
             ztop(ij)=ztop(ij)-ztpmed
    2        cstore(ij)=0.
       write(*,100)
  100  format(/,' Ratio = contribution of Nth term divided by',/,
     &         '              contribution of 0 through N-1 terms',//,
     &         '   N | Ratio',/,
     &         ' -----|----------')
       n=-1
    3  n=n+1
       do 4 j=1,nx
          do 4 i=1,ny
             ij=index(i,j,ny)
             store(2*ij-1)=mag(ij)*ztop(ij)**n
    4        store(2*ij)=0.
       call fourn(store,nn,ndim,-1)
       abnew=0.
       abold=0.
       do 5 j=1,nx
          do 5 i=1,ny
             ij=index(i,j,ny)
             call kvalue(i,j,nx,ny,dkx,dky,kx,ky)
             k=sqrt(kx**2+ky**2)
             arg=((-k)**n)*exp(-k*ztpmed)/fac(n)
             cstep=arg*cmplx(store(2*ij-1),store(2*ij))
             abnew=abnew+cabs(cstep)
             abold=abold+cabs(cstore(ij))
```

Continuation of Subroutine B.23.

```
 5       cstore(ij)=cstore(ij)+cstep
         if(n.eq.0.)go to 3
         ratio=abnew/abold
         write(*,101)n,ratio
101 format(1x,i5,g10.3)
         if(ratio.gt.err.and.n.lt.nstop)go to 3
         do 6 j=1,nx
            do 6 i=1,ny
               ij=index(i,j,ny)
               if(ij.eq.1)then
                  store(2*ij-1)=0.
                  store(2*ij)=0.
               else
                  call kvalue(i,j,nx,ny,dkx,dky,kx,ky)
                  k=sqrt(kx**2+ky**2)
                  thetam=cmplx(mz,(mx*kx+my*ky)/k)
                  thetaf=cmplx(fz,(fx*kx+fy*ky)/k)
                  cstore(ij)=2.*pi*cm*thetam*thetaf*cstore(ij)
                  store(2*ij-1)=real(cstore(ij))
                  store(2*ij)=aimag(cstore(ij))
               end if
 6       continue
         call fourn(store,nn,ndim,+1)
         do 7 j=1,nx
            do 7 i=1,ny
               ij=index(i,j,ny)
 7          mag(ij)=store(2*ij-1)*t2nt/(nx*ny)
         return
         end
```

Continuation of Subroutine B.23.

```
      subroutine contin(grid,nx,ny,dx,dy,dz,store)
c
c  Subroutine CONTIN upward continues gridded potential-field
c  data using the following steps:  (1) Fourier transform the field,
c  (2) multiply by the continuation filter, and (3) inverse Fourier
c  transform the product.  Field values are specified on a
c  rectangular grid with x and y axes directed north and east,
c  respectively.  z axis is down.  North is arbitrary.  Requires
c  subroutines FOURN and KVALUE.
c
c  Input parameters:
c     nx - number of elements in the south-to-north direction.
c     ny - number of elements in the west-to-east direction.
c        (NOTE:  both nx and ny must be a power of two.)
c     grid - a one-dimensional real array containing the
c        two-dimensional potential field.  Elements should
c        be in order of west to east, then south to north (i.e.,
c        element 1 is southwest corner, element ny is
c        southeast corner, element (nx-1)*ny+1 is northwest
c        corner, and element ny*nx is northeast corner).
c     store - a one-dimensional real array used internally.
c        It should be dimensioned at least 2*nx*ny in the
c        calling program.
c     dx - sample interval in the x direction.
c     dy - sample interval in the y direction.
c     dz - continuation distance,in same units as dx and dy.  Should
c        be greater than zero for upward continuation.
c
c  Output parameters:
c     grid - upon output, grid contains the upward-continued
c        potential field with same orientation as before.
c
      dimension grid(nx*ny),store(2*nx*ny),nn(2)
      real kx,ky,k
      complex cgrid,cmplx
      data pi/3.14159265/
      index(i,j,ncol)=(j-1)*ncol+i
      nn(1)=ny
      nn(2)=nx
      ndim=2
      dkx=2.*pi/(nx*dx)
      dky=2.*pi/(ny*dy)
      do 10 j=1,nx
         do 10 i=1,ny
            ij=index(i,j,ny)
            store(2*ij-1)=grid(ij)
10          store(2*ij)=0.
      call fourn(store,nn,ndim,-1)
      do 20 j=1,nx
         do 20 i=1,ny
```

Subroutine B.24. Subroutine to continue potential fields from one level to another. Field is specified on a rectangular grid, assumed to be horizontal. Both dimensions must be a power of two. Parameter dz should be greater than zero for upward continuation. Subroutines FOURN and KVALUE are also required. Edge discontinuities should be eliminated from input grids (see subroutine EXPAND).

```
         ij=index(i,j,ny)
         call kvalue(i,j,nx,ny,dkx,dky,kx,ky)
         k=sqrt(kx**2+ky**2)
         cont=exp(-k*dz)
         cgrid=cmplx(store(2*ij-1),store(2*ij))*cont
         store(2*ij-1)=real(cgrid)
20       store(2*ij)=aimag(cgrid)
      call fourn(store,nn,ndim,+1)
      do 30 j=1,nx
         do 30 i=1,ny
         ij=index(i,j,ny)
30       grid(ij)=store(2*ij-1)/(nx*ny)
      return
      end
```

Continuation of Subroutine B.24.

```
      subroutine verder(grid,nx,ny,dx,dy,norder,store)
c
c  Subroutine VERDER calculates the vertical derivative of
c  gridded potential-field data using the following steps:
c  (1) Fourier transform the field, (2) multiply by the vertical-
c  derivative filter, and (3) inverse Fourier transform the
c  product.  Field values are specified on a rectangular grid
c  with x and y axes directed north and east, respectively.
c  z axis is down.  North is arbitrary.  Requires subroutines
c  FOURN and KVALUE.
c
c  Input parameters:
c    nx - number of elements in the south-to-north direction.
c    ny - number of elements in the west-to-east direction.
c       (NOTE:  both nx and ny must be a power of two.)
c    grid - a one-dimensional real array containing the
c       two-dimensional potential field.  Elements should
c       be in order of west to east, then south to north (i.e.,
c       element 1 is southwest corner, element ny is
c       southeast corner, element (nx-1)*ny+1 is northwest
c       corner, and element ny*nx is northeast corner).
c    store - a one-dimensional real array used internally.
c       It should be dimensioned at least 2*nx*ny in the
c       calling program.
c    dx - sample interval in the x direction, units irrelevant.
c    dy - sample interval in the y direction, units irrelevant.
c    norder - the order of the vertical derivative.
c
c  Output parameters:
c    grid - upon output, grid contains the vertical derivative of
c       the potential field with same orientation as before.
c
      dimension grid(nx*ny),store(2*nx*ny),nn(2)
      complex cgrid,cmplx
      real kx,ky,k
      data pi/3.14159265/
      index(i,j,ncol)=(j-1)*ncol+i
      nn(1)=ny
      nn(2)=nx
      ndim=2
      dkx=2.*pi/(nx*dx)
      dky=2.*pi/(ny*dy)
      do 10 j=1,nx
         do 10 i=1,ny
            ij=index(i,j,ny)
            store(2*ij-1)=grid(ij)
10          store(2*ij)=0.
      call fourn(store,nn,ndim,-1)
      do 20 j=1,nx
         do 20 i=1,ny
```

Subroutine B.25.   Subroutine to calculate vertical derivatives of gridded potential-field data. Field is specified on a rectangular grid, assumed to be horizontal. Both dimensions must be a power of two. Requires subroutines FOURN and KVALUE. Edge discontinuities should be eliminated from input grids (see subroutine EXPAND).

```
      ij=index(i,j,ny)
      call kvalue(i,j,nx,ny,dkx,dky,kx,ky)
      k=sqrt(kx**2+ky**2)
      cgrid=cmplx(store(2*ij-1),store(2*ij))
      cgrid=cgrid*k**norder
      store(2*ij-1)=real(cgrid)
20    store(2*ij)=aimag(cgrid)
   call fourn(store,nn,ndim,+1)
   do 30 j=1,nx
      do 30 i=1,ny
      ij=index(i,j,ny)
30    grid(ij)=store(2*ij-1)/(nx*ny)
   return
   end
```

Continuation of Subroutine B.25.

406   *Subroutines*

```
      subroutine newvec(grid,nx,ny,dx,dy,fi1,fd1,mi1,md1,fi2,
     &                  fd2,mi2,md2,store)
c
c  Subroutine NEWVEC transforms a gridded total-field anomaly
c  into a new anomaly with new directions of magnetization and
c  ambient field.  NEWVEC uses the following steps:  (1) Fourier
c  transform the field, (2) multiply by the phase filter, and
c  (3) inverse Fourier transform the product.  Anomaly values
c  are specified on a rectangular grid with x and y axes
c  directed north and east, respectively.  z axis is down.
c  Requires subroutines FOURN, DIRCOS, and KVALUE.
c
c  Input parameters:
c    nx - number of elements in the south-to-north direction.
c    ny - number of elements in the west-to-east direction.
c         (NOTE:  both nx and ny must be a power of two.)
c    grid - a one-dimensional real array containing the
c           two-dimensional total-field anomaly.  Elements should
c           be in order of west to east, then south to north (i.e.,
c           element 1 is southwest corner, element ny is
c           southeast corner, element (nx-1)*ny+1 is northwest
c           corner, and element ny*nx is northeast corner).
c    store - a one-dimensional real array used internally.
c           It should be dimensioned at least 2*nx*ny in the
c           calling program.
c    dx - sample interval in the x direction, units irrelevant.
c    dy - sample interval in the y direction, units irrelevant.
c    mi1 - original inclination of magnetization, in degrees.
c    md1 - original declination of magnetization.
c    fi1 - original inclination of ambient field.
c    fd1 - original declination of ambient field.
c    mi2 - new inclination of magnetization, in degrees.
c    md2 - new declination of magnetization.
c    fi2 - new inclination of ambient field.
c    fd2 - new declination of ambient field.
c
c  Output parameters:
c    grid - upon output, grid contains the transformed total-
c           field anomaly with same orientation as before.
c
      dimension grid(nx*ny),store(2*nx*ny),nn(2)
      complex cgrid,cmplx,thetam1,thetam2,thetaf1,thetaf2,
     &        cphase
      real kx,ky,k,mi1,md1,mi2,md2,mx1,my1,mz1,mx2,my2,mz2
      data pi/3.14159265/
      index(i,j,ncol)=(j-1)*ncol+i
      nn(1)=ny
      nn(2)=nx
      ndim=2
      dkx=2.*pi/(nx*dx)
```

Subroutine B.26. Subroutine to transform the magnetization direction and ambient-field direction of a total-field anomaly. Anomalies can be reduced to the pole by letting mi2=fi2=90. Anomaly is specified on a rectangular grid, assumed to be horizontal. Both dimensions must be a power of two. Subroutines FOURN, KVALUE, and DIRCOS are required. Edge discontinuities should be eliminated from input grids (see 0subroutine EXPAND).

```
dky=2.*pi/(ny*dy)
call dircos(mi1,md1,0.,mx1,my1,mz1)
call dircos(fi1,fd1,0.,fx1,fy1,fz1)
call dircos(mi2,md2,0.,mx2,my2,mz2)
call dircos(fi2,fd2,0.,fx2,fy2,fz2)
do 10 j=1,nx
   do 10 i=1,ny
      ij=index(i,j,ny)
      store(2*ij-1)=grid(ij)
10    store(2*ij)=0.
call fourn(store,nn,ndim,-1)
do 20 j=1,nx
   do 20 i=1,ny
      ij=index(i,j,ny)
      if(ij.eq.1)then
         cphase=0.
      else
         call kvalue(i,j,nx,ny,dkx,dky,kx,ky)
         k=sqrt(kx**2+ky**2)
         thetam1=cmplx(mz1,(kx*mx1+ky*my1)/k)
         thetaf1=cmplx(fz1,(kx*fx1+ky*fy1)/k)
         thetam2=cmplx(mz2,(kx*mx2+ky*my2)/k)
         thetaf2=cmplx(fz2,(kx*fx2+ky*fy2)/k)
         cphase=thetam2*thetaf2/(thetam1*thetaf1)
      end if
      cgrid=cmplx(store(2*ij-1),store(2*ij))
      cgrid=cgrid*cphase
      store(2*ij-1)=real(cgrid)
20    store(2*ij)=aimag(cgrid)
call fourn(store,nn,ndim,+1)
do 30 j=1,nx
   do 30 i=1,ny
      ij=index(i,j,ny)
30    grid(ij)=store(2*ij-1)/(nx*ny)
return
end
```

Continuation of Subroutine B.26.

```
      subroutine pseudo(grid,nx,ny,dx,dy,fi,fd,mi,md,store)
c
c  Subroutine PSEUDO transforms a gridded total-field anomaly
c  into a pseudogravity anomaly using the following steps:
c  (1) Fourier transform the field, (2) multiply by the phase
c  filter, and (3) inverse Fourier transform the product. Anomaly
c  values are specified on a rectangular grid with x and y axes
c  directed north and east, respectively. z axis is down. Ratio
c  of density to magnetization assumed to be 100 kg/(m**3) per
c  1 A/m. Requires subroutines FOURN, DIRCOS, and KVALUE.
c
c  Input parameters:
c    nx - number of elements in the south-to-north direction.
c    ny - number of elements in the west-to-east direction.
c       (NOTE: both nx and ny must be a power of two.)
c    grid - a one-dimensional real array containing the
c         two-dimensional total-field anomaly in nT. Elements
c         should be in order of west to east, then south to north
c         (i.e., element 1 is southwest corner, element ny is
c         southeast corner, element (nx-1)*ny+1 is northwest
c         corner, and element ny*nx is northeast corner).
c    store - a one-dimensional real array used internally.
c         It should be dimensioned at least 2*nx*ny in the
c         calling program.
c    dx - sample interval in the x direction, in km.
c    dy - sample interval in the y direction, in km.
c    mi - inclination of magnetization, in degrees.
c    md - declination of magnetization.
c    fi - inclination of ambient field.
c    fd - declination of ambient field.
c
c  Output parameters:
c    grid - upon output, grid contains the pseudogravity anomaly
c         in mGal with same orientation as before.
c
      dimension grid(nx*ny),store(2*nx*ny),nn(2)
      complex cgrid,cmplx,thetam,thetaf,cpsgr
      real kx,ky,k,mi,md,mx,my,mz,mag,km2m
      data cm/1.e-7/,gamma/6.67e-11/,t2nt/1.e9/,si2mg/1.e5/
      data pi/3.14159265/,rho/100./,mag/1./,km2m/1.e3/
      index(i,j,ncol)=(j-1)*ncol+i
      const=gamma*rho*si2mg*km2m/(cm*mag*t2nt)
      nn(1)=ny
      nn(2)=nx
      ndim=2
      dkx=2.*pi/(nx*dx)
      dky=2.*pi/(ny*dy)
      call dircos(mi,md,0.,mx,my,mz)
      call dircos(fi,fd,0.,fx,fy,fz)
      do 10 j=1,nx
        do 10 i=1,ny
```

Subroutine B.27. Subroutine to transform a total-field anomaly into a pseudo-gravity anomaly. Anomalies are specified on a rectangular grid, assumed to be horizontal. Both dimensions must be a power of two. Requires subroutines FOURN, DIRCOS, and KVALUE. Edge discontinuities should be eliminated from input grids (see subroutine EXPAND).

```
         ij=index(i,j,ny)
         store(2*ij-1)=grid(ij)
10       store(2*ij)=0.
   call fourn(store,nn,ndim,-1)
   do 20 j=1,nx
      do 20 i=1,ny
         ij=index(i,j,ny)
         if(ij.eq.1)then
            cphase=0.
            else
               call kvalue(i,j,nx,ny,dkx,dky,kx,ky)
               k=sqrt(kx**2+ky**2)
               thetam=cmplx(mz,(kx*mx+ky*my)/k)
               thetaf=cmplx(fz,(kx*fx+ky*fy)/k)
               cpsgr=1./(thetam*thetaf*k)
            end if
         cgrid=cmplx(store(2*ij-1),store(2*ij))
         cgrid=cgrid*cpsgr
         store(2*ij-1)=real(cgrid)
20       store(2*ij)=aimag(cgrid)
   call fourn(store,nn,ndim,+1)
   do 30 j=1,nx
      do 30 i=1,ny
         ij=index(i,j,ny)
30       grid(ij)=store(2*ij-1)*const/(nx*ny)
   return
   end
```

Continuation of Subroutine B.27.

```
      subroutine hgrad(grid,nx,ny,dx,dy,store)
c
c  Subroutine HGRAD calculates the maximum horizontal
c  gradient of a two-dimensional function using simple finite-
c  difference relations.  Function is specified on a
c  rectangular grid with x and y axes directed north and east,
c  respectively.  North is arbitrary.
c
c  Input parameters:
c     nx - number of elements in the west-to-east direction.
c     ny - number of elements in south-to-north direction.
c     grid - a one-dimensional real array containing the
c         two-dimensional function.  Elements should
c         be in order of west to east, then south to north (i.e.,
c         element 1 is southwest corner, element ny is
c         southeast corner, element (nx-1)*ny+1 is northwest
c         corner, and element ny*nx is northeast corner).
c     dx - sample interval in the x direction, units irrelevant.
c     dy - sample interval in the y direction, units irrelevant.
c     store - one-dimensional real array used internally.
c         Should be dimensioned at least nx*ny in the calling
c         program.
c
c  Output parameters:
c     grid - upon output, grid contains the maximum horizontal
c         gradient with same orientation as before.
c
      dimension grid(nx*ny),store(nx*ny)
      index(i,j,ncol)=(j-1)*ncol+i
      dx2=2.*dx
      dy2=2.*dy
      do 10 j=1,nx
         jm1=j-1
         if(jm1.lt.1)jm1=1
         jp1=j+1
         if(jp1.gt.nx)jp1=nx
         do 10 i=1,ny
            im1=i-1
            if(im1.lt.1)im1=1
            ip1=i+1
            if(ip1.gt.ny)ip1=ny
            dfdx=(grid(index(ip1,j,ny))-grid(index(im1,j,ny)))/dx2
            dfdy=(grid(index(i,jp1,ny))-grid(index(i,jm1,ny)))/dy2
            store(index(i,j,ny))=sqrt(dfdx**2+dfdy**2)
 10      continue
      do 20 i=1,nx*ny
 20   grid(i)=store(i)
      return
      end
```

Subroutine B.28. Subroutine to calculate the maximum horizontal gradient of a two-dimensional function specified on a rectangular grid.

```
      subroutine expand(grid,ncol,nrow,grid2,ncol2,nrow2)
c
c  Subroutine EXPAND adds "tapered" rows and columns to a
c  grid.  Input grid(i,j), i=1,2,...,ncol, j=1,2,...,nrow, is
c  modified as follows:
c
c  (1) Each row is expanded in length from ncol to ncol2.
c      New elements of row j linearly change in value from
c      grid(ncol,j) to grid(1,j).
c  (2) Each column is expanded in length from nrow to nrow2.
c      New elements of column 1 linearly change in value
c      from grid(i,nrow) to grid(i,1).
c  (3) All elements at i < or = to ncol and j < or = nrow
c      are left unchanged.
c
c  Input parameters:
c     grid -        one-dimensional real array representing
c                   a two-dimensional grid.
c     ncol,nrow -   dimensions of input grid.
c     ncol2,nrow2 - dimensions of output grid (grid2).
c                   ncol2 must be > ncol; nrow2 must be > nrow.
c
c  Output parameters:
c     grid2 -       one-dimensional real array representing
c                   a two-dimension grid, as described earlier.
c
      dimension grid(ncol*nrow),grid2(ncol2*nrow2)
      index(i,j,ncol)=(j-1)*ncol+i
      do 10 j=1,nrow
        do 10 i=1,ncol
          ij=index(i,j,ncol)
          ij2=index(i,j,ncol2)
10        grid2(ij2)=grid(ij)
      if(ncol2.gt.ncol)then
        do 20 j=1,nrow
          i1=index(1,j,ncol2)
          i2=index(ncol,j,ncol2)
          i3=index(ncol2,j,ncol2)
          step=(grid2(i1)-grid2(i2))/(ncol2-ncol+1)
          do 20 i=ncol+1,ncol2
            ij=index(i,j,ncol2)
20          grid2(ij)=grid2(i2)+step*(i-ncol)
      else
        pause 'EXPAND:  ncol2.le.ncol'
      end if
```

Subroutine B.29. Subroutine to expand the dimensions of a grid. Rows and columns are added to the east and north edges of the grid in order to eliminate discontinuities in discrete Fourier transforms. Elements added to each row change gradually from the value of the last element of the row to the value of the first element of the row, and likewise for each column.

```
      if(nrow2.gt.nrow)then
         do 30 i=1,ncol2
            j1=index(i,1,ncol2)
            j2=index(i,nrow,ncol2)
            j3=index(i,nrow2,ncol2)
            step=(grid2(j1)-grid2(j2))/(nrow2-nrow+1)
            do 30 j=nrow+1,nrow2
               ij=index(i,j,ncol2)
30             grid2(ij)=grid2(j2)+step*(j-nrow)
         else
            pause 'EXPAND:  nrow2.le.nrow'
            end if
      return
      end
```

Continuation of Subroutine B.29.

Appendix C
Review of Sampling Theory

Gridding embodies forgotten compromises made at an early stage of analysis from which we may suffer later on – an Achilles' heel.
(Lindrith Cordell)

The process of sampling a continuous function $f(x)$ can be represented mathematically as multiplication of $f(x)$ by a series of impulses,

$$f_D(x) = \sum_{m=-N}^{N} f(x)\,\delta(x - m\Delta x)$$

$$= f(x) \sum_{m=-N}^{N} \delta(x - m\Delta x)$$

$$= f(x)\,k_{N,\Delta x}(x)\,, \qquad\qquad (C.1)$$

where Δx is the sample interval and $2N + 1$ is the number of points in the sample sequence. The series of impulses $k_{N,\Delta x}(x)$ in equation C.1 is called the *sampling function* (Figure C.1). It may seem curious to represent $f(x)$ by a series of impulses, but the process of digitizing $f(x)$ at $x = x_0$ really amounts to finding an average value of $f(x)$ over a very

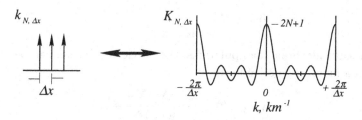

Fig. C.1. An example of the sampling function and its Fourier transform for $N = 1$ and $\Delta x = 1$.

413

narrow range of x centered about x_0, and equation 11.14 shows that the value of a function $f(x)$ at x_0 is given by the convolution of $f(x)$ with an impulse,

$$f(x_0) = \int_{-\infty}^{\infty} f(x)\,\delta(x - x_0)\,dx\,.$$

Note that $f_{\mathrm{D}}(x)$ and $f(x)$ do not have the same units because $\delta(x)$ has units of inverse distance.

We want to investigate the differences between $f(x)$ and $f_{\mathrm{D}}(x)$ in the Fourier domain, so we begin by Fourier transforming equation C.1 and then employing equation 11.20. Hence, the Fourier transform of $f_{\mathrm{D}}(x)$ will be given by the convolution of the Fourier transforms of $f(x)$ and $k_{N,\Delta x}(x)$,

$$F_{\mathrm{D}}(k) = \frac{1}{2\pi} \int_{-\infty}^{\infty} F(k')K_{N,\Delta x}(k - k')\,dk'\,.$$

The Fourier transform of the sampling function, therefore, will describe how the Fourier transform of a digitized function differs from the transform of its continuous form. From equation 11.15, it follows that the Fourier transform of an impulse at the origin is given by

$$\int_{-\infty}^{\infty} \delta(x)e^{-iks}\,dx = 1\,,$$

and the shifting theorem of Section 11.1.2 leads to

$$\int_{-\infty}^{\infty} \delta(x - m\Delta x)e^{-ikx}\,dx = e^{-ikm\Delta x}$$

and

$$K_{N,\Delta x}(k) = \sum_{m=-N}^{N} e^{-ikm\Delta x}\,.$$

After some algebraic manipulations, the previous equation can be written in a more instructive form,

$$K_{N,\Delta x}(k) = \frac{\sin k\Delta x(N + \frac{1}{2})}{\sin \frac{1}{2}k\Delta x}\,. \tag{C.2}$$

Equation C.2 shows that the Fourier transform of the sampling function is periodic and repeats itself over the interval $\frac{2\pi}{\Delta x}$. Figure C.1 shows an example for $N = 3$. The narrowness of the large peaks and the number of zero-crossings between the large peaks both increase with increasing N. In fact, it can be shown that as N becomes very large, the Fourier transform of the sampling function approaches an infinitely long series of impulses spaced along the k axis at intervals of $\frac{2\pi}{\Delta x}$,

$$\lim_{N \to \infty} K_{N,\Delta x}(k) = \frac{2\pi}{\Delta x} \sum_{j=-\infty}^{\infty} \delta\left(k - \frac{2\pi j}{\Delta x}\right).$$

Hence an infinite set of impulses spaced Δx apart Fourier transforms to an infinite set of impulses spaced $\frac{2\pi}{\Delta x}$ apart.

Suppose now that we were able to digitize an infinitely long continuous function $f(x)$. (This is not as impossible as it may seem; if we assume that $f(x) = 0$ for $|x| > N \Delta x$, then we may consider N to be infinite and $f(x)$ to be infinitely long.) According to equation 11.20, the Fourier transform of this infinite series would be

$$F_{\mathrm{D}}(k) = \frac{1}{\Delta x} \sum_{j=-\infty}^{\infty} F\left(k - \frac{2\pi j}{\Delta x}\right). \tag{C.3}$$

This summation describes *aliasing*, an important phenomenon of sampled data. We would like the Fourier transform of the digital function to equal the Fourier transform of the continuous function at any arbitrary wavenumber; that is, we would like $\Delta x F_{\mathrm{D}}(k_0) = F(k_0)$. Unfortunately, equation C.3 shows that the discrete Fourier transform is equal to $F(k_0)$ plus an infinite set of additional values:

$$F_{\mathrm{D}}(k) = \frac{1}{\Delta x}\left[\cdots + F\left(k_0 + \frac{2\pi}{\Delta x}\right) + F(k_0) + F\left(k_0 - \frac{2\pi}{\Delta x}\right) + \cdots\right].$$

In other words, the periodic nature of $F_{\mathrm{D}}(k)$ produces contamination at each value of k_0; instead of $F(k_0)$, we get $F(k_0)$ plus $F(k)$ evaluated at an infinite number of other values of k.

The amplitude spectrum for many kinds of physical phenomena (including potential-field anomalies) decreases rapidly with increasing $|k|$. Hence, the low-wavenumber part of $F_{\mathrm{D}}(k)$ may be a good approximation to $F(k)$ if $f(x)$ has been properly sampled. This should be clear from equation C.3; the contaminating terms in the summation will be small if $F(k)$ decreases rapidly with increasing $|k|$. Aliasing will be most severe at wavenumbers near $k = \frac{\pi}{\Delta x}$. This is the symmetry point in Figure C.1 and is called the *Nyquist* or *folding* wavenumber. Clearly one way to reduce aliasing is to increase the Nyquist wavenumber, and this can be done only by making the sample interval Δx smaller.

Note that the Nyquist wavenumber corresponds to a wavelength of twice the sample interval. We cannot find information at wavelengths smaller than this threshold. For some physical phenomena, it might be appropriate to assume that the energy-density spectrum or the power-density spectrum of a process is zero at wavenumbers greater than some critical wavenumber k_c. Such a function is said to be *band limited*. The *sampling theorem* states that if $f(x)$ is band limited with no energy at wavelengths less than λ_c, $F_D(k)$ will contain all the information of $F(k)$ so long as $\Delta x \leq \lambda_c/2$.

Appendix D
Conversion of Units

The only known antidotes to discussions of units are undisturbed silence in a dark room for 15 minutes or a brisk walk in the park.

(Robert F. Butler)

Two systems of electromagnetic units are in common usage in the geophysical literature: The venerable cgs system, also known as electromagnetic units (emu), and the more modern and internationally accepted *Système Internationale* (International System, abbreviated SI). Geophysical journals now require the use of SI units, and for the most part SI units are adhered to in this text. Geophysical journals published prior to 1980, however, employed emu, and even some recently published textbooks (e.g., Butler [47]) continue its usage. Consequently, reading the geophysical literature requires conversion between the two systems, a common source of frustration because the two systems are significantly different.

The following table summarizes these two important systems of units and the conversion factors between them. Additional discussions can be found in textbooks by Panofsky and Phillips [201, pp. 459–65], Butler [47, pp. 15–18], and the Society of Exploration Geophysicists [266], and in papers by Shive [255], Payne [213], and Moskowitz [189].

417

Electromagnetic Units (emu) International Standard (SI)

Quantity	Unit	Dimensions	Unit	Dimensions	Conversion
Force (F)	dyne	$g \cdot cm \cdot s^{-2}$	newton (N)	$kg \cdot m \cdot s^{-2}$	1 dyne $= 10^{-5}$ N
Current (I)	abampere	$10\ C \cdot s^{-1}$	ampere (A)	$C \cdot s^{-1}$	1 abampere $= 10$ A
Induction or flux density (B)	gauss (G)	$0.1\ g \cdot s^{-1} \cdot C^{-1}$	tesla (T)	$kg \cdot s^{-1} \cdot C^{-1}$	1 gauss $= 10^{-4}$ T
"	gamma	10^{-5} gauss	nanotesla (nT)	10^{-9} T	1 gamma $= 1$ nT
Field intensity (H)	oersted (Oe)	$0.1\ g \cdot s^{-1} \cdot C^{-1}$	$A \cdot m^{-1}$	$C \cdot s^{-1} \cdot m^{-1}$	1 oersted $= 10^3/4\pi$ $A \cdot m^{-1}$
Magnetization (M)	gauss[†] (G)	$0.1\ g \cdot s^{-1} \cdot C^{-1}$	$A \cdot m^{-1}$	$C \cdot s^{-1} \cdot m^{-1}$	1 gauss $= 10^3$ $A \cdot m^{-1}$
Moment (m)	gauss·cm³[‡]	$0.1\ g \cdot s^{-1} \cdot C^{-1} \cdot cm^3$	$A \cdot m^2$	$C \cdot s^{-1} \cdot m^2$	1 gauss $\cdot cm^3 = 10^{-3}$ $A \cdot m^2$
Susceptibility (χ)	—	dimensionless	—	dimensionless	1 (emu) $= 4\pi$ (SI)
Koenigsberger ratio (Q)	—	dimensionless	—	dimensionless	1 (emu) $= 1$ (SI)

[†] Also called emu·cm⁻³.
[‡] Also called emu.

Notes:

1. The units of field intensity **H** are different, both numerically and dimensionally, between the two systems. This is because of the different defining equations for **H**,

$$\mathbf{B} = \mathbf{H} + 4\pi\mathbf{M} \quad \text{(emu)},$$
$$\mathbf{B} = \mu_0(\mathbf{H} + \mathbf{M}) \quad \text{(SI)},$$

where $\mu_0 = 4\pi \times 10^{-7}$ $N \cdot A^{-2}$. Likewise, magnetization **M** differs both numerically and dimensionally in the two systems.

2. Although **B** and **H** are often considered equal fields outside magnetic material, in SI they differ by a factor μ_0, making them different both numerically and dimensionally.

3. Susceptibility, although dimensionless, differs by a factor of 4π between the two systems.

Bibliography

[1] Airy, G. B., "On the computation of the effect of the attraction of mountain-masses, as disturbing the apparent astronomical latitude of stations in geodetic surveys," *Philosophical Transactions of the Royal Society of London* **145** (1855), 101–4.

[2] Al-Chalabi, M., "Interpretation of gravity anomalies by non-linear optimisation," *Geophysical Prospecting* **20** (1972), 1–16.

[3] Alldredge, L. R., and Hurwitz, L., "Radial dipoles as the sources of the earth's main magnetic field," *Journal of Geophysical Research* **69** (1964), 2631–40.

[4] Ander, M. E., and Huestis, S. P., "Gravity ideal bodies," *Geophysics* **52** (1987), 1265–78.

[5] Arkani-Hamed, J., "Differential reduction-to-the-pole of regional magnetic anomalies," *Geophysics* **53** (1988), 1592–600.

[6] Arkani-Hamed, J., "Thermoviscous remanent magnetization of oceanic lithosphere inferred from its thermal evolution," *Journal of Geophysical Research* **94** (1989), 17, 421–36.

[7] Arkani-Hamed, J., and Strangway, D. W., "Lateral variations of apparent magnetic susceptibility of lithosphere deduced from Magsat data," *Journal of Geophysical Research* **90**, 2655–64, 1985.

[8] Backus, G. E., and Gilbert, F., "Uniqueness in the inversion of inaccurate gross Earth data," *Philosophical Transactions of the Royal Society of London, Series A* **226** (1970), 123–92.

[9] Baranov, V., "A new method for interpretation of aeromagnetic maps: pseudo-gravimetric anomalies," *Geophysics* **22** (1957), 359–83.

[10] Baranov, V., and Naudy, H., "Numerical calculation of the formula of reduction to the magnetic pole," *Geophysics* **29** (1964), 67–79.

[11] Barnett, C. T., "Theoretical modeling of the magnetic and gravitational fields of an arbitrarily shaped three-dimensional body," *Geophysics* **41** (1976), 1353–64.

[12] Barongo, J. O., "Euler's differential equation and the indentification of the magnetic point-pole and point-dipole sources," *Geophysics* **49** (1984), 1549–53.

[13] Bhattacharyya, B. K., "Magnetic anomalies due to prism-shaped bodies with arbitrary polarization," *Geophysics* **29** (1964), 517–31.

[14] Bhattacharyya, B. K., "Two-dimensional harmonic analysis as a tool for magnetic interpretation," *Geophysics* **30** (1965), 829–57.

[15] Bhattacharyya, B. K., "Continuous spectrum of the total-magnetic-field anomaly due to a rectangular prismatic body," *Geophysics* **31** (1966), 97–121.

[16] Bhattacharyya, B. K., "Some general properties of potential fields in space and frequency domain: a review," *Geoexploration* **5** (1967), 127–43.

[17] Bhattacharyya, B. K., and Chan, K. C., "Reduction of magnetic and gravity data on an arbitrary surface acquired in a region of high topographic relief," *Geophysics* **42** (1977), 1411–30.

[18] Bhattacharyya, B. K., and Leu, L.-K., "Analysis of magnetic anomalies over Yellowstone National Park; mapping of Curie point isothermal surface for geothermal reconnaissance," *Journal of Geophysical Research* **80** (1975), 4461–65.

[19] Blackwell, D. D.; Bowen, R. G.; Hull, D. A.; Riccio, J.; and Steele, J. L., "Heat flow, arc volcanism, and subduction in northern Oregon," *Journal of Geophysical Research* **87** (1982), 8735–54.

[20] Blakely, R. J., Documentation for Subroutine REDUC3, an Algorithm for the Linear Filtering of Gridded Magnetic Data, Open-File Report 77–784, U.S. Geological Survey, 1977.

[21] Blakely, R. J., "Statistical averaging of marine magnetic anomalies," *Journal of Geophysical Research* **88** (1983), 2289–96.

[22] Blakely, R. J., "The effect of topography on aeromagnetic data in the wavenumber domain," in *Proceedings of the International Meeting on Potential Fields in Rugged Topography, Bulletin 7*, pages 102–9, Institut de Géophysique de Université de Lausanne, Lausanne, Switzerland, 1985.

[23] Blakely, R. J., "Curie temperature isotherm analysis and tectonic implications of aeromagnetic data from Nevada," *Journal of Geophysical Research* **93** (1988), 11,817–32.

[24] Blakely, R. J., "Extent of magma chambers in the Cascade Range of western North America: constraints from gravity anomalies," *Journal of Geophysical Research* **99** (1994), 2757–73.

[25] Blakely, R. J., and Christiansen, R. L., "The magnetization of Mount Shasta and implications for virtual geomagnetic poles determined from seamounts," *Journal of Geophysical Research* **83** (1978), 5971–8.

[26] Blakely, R. J., and Cox, A., "Evidence for short geomagnetic polarity intervals in the Early Cenozoic," *Journal of Geophysical Research* **77** (1972), 7065–72.

[27] Blakely, R. J., and Cox, A., "Identification of short polarity events by transforming marine magnetic profiles to the pole," *Journal of Geophysical Research* **77** (1972), 4339–49.

[28] Blakely, R. J., and Grauch, V. J. S., "Magnetic models of crystalline terrane; accounting for the effect of topography," *Geophysics* **48** (1983), 1551–7.

[29] Blakely, R. J., and Hassanzadeh, S., "Estimation of depth to magnetic source using maximum entropy power spectra, with application to the Peru-Chile Trench," in *Nazca Plate; Crustal Formation and Andean Convergence*, 667–82, Geological Society of America Memoir 154, Boulder, CO (1981).

[30] Blakely, R. J., and Jachens, R. C., "Volcanism, isostatic residual gravity, and regional tectonic setting of the Cascade volcanic province," *Journal of Geophysical Research* **95** (1990), 19, 439–51.

[31] Blakely, R. J., and Jachens, R. C., "Regional study of mineral resources in Nevada: insights from three-dimensional analysis of gravity and magnetic anomalies," *Geological Society of America Bulletin* **103** (1991), 795–803.

[32] Blakely, R. J., and Schouten, H., "Comments on 'Filtering marine magnetic anomalies' by Hans Schouten and Keith McCamy," *Journal of Geophysical Research* **79** (1974), 773–4.

[33] Blakely, R. J., and Simpson, R. W., "Approximating edges of source bodies from magnetic or gravity anomalies," *Geophysics* **51** (1986), 1494–8.

[34] Booker, J. R., "Geomagnetic data and core motions," *Proceedings of the Royal Society of London, Series A* **309** (1969), 27–40.

[35] Bott, M. H. P., "The use of rapid digital computing methods for direct gravity interpretation of sedimentary basins," *Geophysical Journal of the Royal Astronomical Society* **3** (1960), 63–7.

[36] Bott, M. H. P., "Two methods applicable to computers for evaluating magnetic anomalies due to finite three dimensional bodies," *Geophysical Prospecting* **11** (1963), 292–9.

[37] Bott, M. H. P., "Solution of the linear inverse problem in magnetic interpretation with application to oceanic magnetic anomalies," *Geophysical Journal of the Royal Astronomical Society* **13** (1967), 313–23.

[38] Bott, M. H. P., and Hutton, M. A., "Limitations on the resolution possible in the direct interpretation of marine magnetic anomalies," *Earth and Planetary Science Letters* **8** (1970), 317–19.

[39] Bott, M. H. P., and Hutton, M. A., "A matrix method for interpreting oceanic magnetic anomalies," *Geophysical Journal of the Royal Astronomical Society* **20** (1970), 149–57.

[40] Bott, M. H. P., and Ingles, A., "Matrix methods for joint interpretation of two-dimensional gravity and magnetic anomalies with application to the Iceland–Faeroe Ridge," *Geophysical Journal of the Royal Astronomical Society* **30** (1972), 55–67.

[41] Bott, M. H. P., and Smith, R. A., "The estimation of the limiting depth of gravitating bodies," *Geophysical Prospecting* **6** (1958), 1–10.

[42] Bracewell, R., *The Fourier Transform and Its Applications,* McGraw-Hill, New York, 1965.

[43] Brozena, J. M., and Peters, M. F., "An airborne gravity study of eastern North Carolina," *Geophysics* **53** (1988), 245–53.

[44] Bullard, E. C., "The removal of trend from magnetic surveys," *Earth and Planetary Science Letters* **2** (1967), 293–300.

[45] Bullard, E. C.; Freedman, C.; Gellman, H.; and Nixon, J., "The westward drift of the earth's magnetic field," *Philosophical Transactions of the Royal Society of London, Series A* **243** (1950), 67.

[46] Bultman, M. W., and Gettings, M. E., *New Techniques of Geophysical Data Analysis; an Investigation of the Geometry, Structure, and Bedrock Lithology of the San Rafael Basin, Arizona,* U.S. Geological Circular 1103-A (1994), 14–15.

[47] Butler, Robert F., *Paleomagnetism, Magnetic Domains to Geologic Terranes,* Blackwell, Oxford, 1992.

[48] Cady, J. W., "Calculation of gravity and magnetic anomalies of finite-length right polygonal prisms," *Geophysics* **45** (1980), 1507–12.

[49] Cain, J. C.; Davis, W. M.; and Regan, R. D., "An $N = 22$ model of the geomagnetic field," *EOS, Transactions of the American Geophysical Union* **56** (1974), 1108.

[50] Cain, J. C.; Schmitz, D. R.; and Muth, L., "Small-scale features in the earth's magnetic field observed by Magsat," *Journal of Geophysical Research* **89** (1984), 1070–6.

[51] Cande, S. C., "A palaeomagnetic pole from Late Cretaceous marine magnetic anomalies in the Pacific," *Geophysical Journal of the Royal Astronomical Society* 44 (1976), 547–66.

[52] Cande, S. C., and Kent, D. V., "Constraints imposed by the shape of marine magnetic anomalies on the magnetic source," *Journal of Geophysical Research* 81 (1976), 4157–62.

[53] Carle, H. M., and Harrison, C. G. A., "A problem in representing the core magnetic field of the earth using spherical harmonics," *Geophysical Research Letters* 9 (1982), 265–8.

[54] Carmichael, R. S., "Magnetic properties of minerals and rocks," in *Handbook of Physical Properties of Rocks*, vol. 2, R. S. Carmichael (ed.), 229–87, CRC Press, Boca Raton, FL (1982).

[55] Chandler, V. W., and Malek, K. C., "Moving-window Poisson analysis of gravity and magnetic data from the Penokean orogen, east-central Minnesota," *Geophysics* 56 (1991), 123–32.

[56] Chapman, S., and Bartels, J., *Geomagnetism*, Oxford University Press, Oxford, 1940.

[57] Chikazumi, S., *Physics of Magnetism*, Wiley, New York, 1964; English edition prepared by S. J. Charap.

[58] Chovitz, B. H., "Modern geodetic earth reference models," *EOS, Transactions of the American Geophysical Union* 62 (1981), 65–7.

[59] Churchill, R. V., *Complex Variables and Applications*, 2nd ed., McGraw-Hill, New York (1960).

[60] Claerbout, J. F., *Fundamentals of Geophysical Data Processing with Applications to Petroleum Prospecting*, McGraw-Hill, New York (1976).

[61] Clarke, G. K., "Linear filters to suppress terrain effects on geophysical maps," *Geophysics* 36 (1971), 963–6.

[62] Coles, R. L., "A flexible iterative magnetic anomaly interpretation technique using multiple rectangular prisms," *Geoexploration* 14 (1976), 125–41.

[63] Connard, G.; Couch, R.; and Gemperle, M., "Analysis of aeromagnetic measurements from the Cascade Range in central Oregon," *Geophysics* 48 (1983), 376–90.

[64] Corbato, C. E., "A least-squares procedure for gravity interpretation," *Geophysics* 30 (1965), 228–33.

[65] Cordell, L., Iterative Three-Dimensional Solution of Gravity Anomaly Data, National Technical Information Service PB–196–979, U.S. Department of Commerce (1970).

[66] Cordell, L., "Gravimetric expression of graben faulting in Santa Fe country and the Española Basin, New Mexico," in *Guidebook to Santa Fe Country, 30th Field Conference*, R. V. Ingersoll (ed.), 59–64, New Mexico Geological Society (1979).

[67] Cordell, L., "Applications and problems of analytical continuation of New Mexico aeromagnetic data between arbitrary surfaces of very high relief," in *Proceedings of the International Meeting on Potential Fields in Rugged Topography*, Bulletin 7, 96–101, Institut de Géophysique de Université de Lausanne, Lausanne, Switzerland (1985).

[68] Cordell, L., and Grauch, V. J. S., "Mapping basement magnetization zones from aeromagnetic data in the San Juan Basin, New Mexico," Technical Program Abstracts and Biographies, Fifty-Second Annual International Meeting and Exposition, Dallas, Texas, October 17–21, 246–7, Society of Exploration Geophysicists (1982).

[69] Cordell, L., and Grauch, V. J. S., "Reconciliation of the discrete and integral Fourier transforms," *Geophysics* 47 (1982), 237–43.

[70] Cordell, L., and Grauch, V. J. S., "Mapping basement magnetization zones from aeromagnetic data in the San Juan basin, New Mexico," in *The Utility of Regional Gravity and Magnetic Anomaly Maps*, William J. Hinze (ed.), 181–97, Society of Exploration Geophysicists, Tulsa, OK (1985).

[71] Cordell, L., and Henderson, R. G., "Iterative three-dimensional solution of gravity anomaly data using a digital computer," *Geophysics* **33** (1968), 596–601.

[72] Cordell, L., and McCafferty, A. E., "A terracing operator for physical property mapping with potential field data," *Geophysics* **54** (1989), 621–34.

[73] Cordell, L.; Phillips, J. D.; and Godson, R. H., U.S. Geological Survey Potential-Field Software Version 2.0, Open-File Report 92–18, U.S. Geological Survey (1992).

[74] Cordell, L., and Taylor, P. T., "Investigation of magnetization and density of a North Atlantic seamount using Poisson's theorem," *Geophysics* **36** (1971), 919–37.

[75] Cordell, L.; Zorin, Y. A.; and Keller, G. R., "The decompensative gravity anomaly and deep structure of the region of the Rio Grande rift," *Journal of Geophysical Research* **96** (1991), 6557–68.

[76] Counil, J.-L.; Achache, J.; and Galdeano, A., "Long-wavelength magnetic anomalies in the Caribbean: plate boundaries and allochthonous continental blocks," *Journal of Geophysical Research* **94** (1989), 7419–31.

[77] Courtillot, V. E.; Ducruix, J.; and Le Mouël, J. L., "Le prolongement d'un champ de potentiel d'un contour quelconque sur un contour horizontal: une application de la méthode de Backus et Gilbert," *Ann. Géophys.* **29** (1973), 361–66.

[78] Courtillot, V.; Ducruix, J.; and Le Mouël, J. L., "A solution of some inverse problems in geomagnetism and gravimetry," *Journal of Geophysical Research* **79** (1974), 4933–40.

[79] Dampney, C. N. G., "The equivalent source technique," *Geophysics* **34** (1969), 39–53.

[80] Ducruix, J.; Le Mouël, J. L.; and Courtillot, V., "Continuation of three-dimensional potential fields measured on an uneven surface," *Geophysical Journal of the Royal Astronomical Society* **38** (1974), 299–314.

[81] Duff, G. F. D., and Naylor, D., *Differential Equations of Applied Mathematics*, Wiley, New York (1966).

[82] Dyson, F. W., and Furner, H., "The earth's magnetic potential," *Monthly Notices, Royal Astronomical Society of London, Geophysics Supplement 1* **3** (1923), 76–88.

[83] Emilia, D. A., "Equivalent sources used as an analytic base for processing total magnetic field profiles," *Geophysics* **38** (1973), 339–48.

[84] Emilia, D. A., and Bodvarsson, G., "Numerical methods in the direct interpretation of marine magnetic anomalies," *Earth and Planetary Science Letters* **7** (1969), 194–200.

[85] Emilia, D. A., and Massey, R. L., "Magnetization estimation for nonuniformly magnetized seamounts," *Geophysics* **39** (1974), 223–31.

[86] Evjen, H. M., "The place of the vertical gradient in gravitational interpretations," *Geophysics* **1** (1936), 127–36.

[87] Fabiano, E. B.; Peddie, N. W.; and Zunde, A. K., The Magnetic Field of the Earth, 1980, Miscellaneous Investigations Series, U.S. Geological Survey (1983). Five charts showing magnetic declination

(I-1457), inclination (I-1458), horizontal intensity (I-1459), vertical
intensity (I-1460), and total intensity (I-1461); scale 1:40,053,700.

[88] Fernie, J. D., "The shape of the earth," *American Scientist* **79** (1991),
108–10.

[89] Fernie, J. D., "The shape of the earth. Part II," *American Scientist*
79 (1991), 393–5.

[90] Fernie, J. D., "The shape of the earth. Part III," *American Scientist*
80 (1991), 125–7.

[91] Francheteau, J.; Harrison, C. G. A.; Sclater, J. G.; and Richards,
M. L., "Magnetization of Pacific seamounts: a preliminary polar curve
for the Northeastern Pacific," *Journal of Geophysical Research* **75**
(1970), 2035–61.

[92] Frost, B. R., and Shive, P. N., "Magnetic mineralogy of the lower
continental crust," *Journal of Geophysical Research* **91** (1986),
6513–21.

[93] Gass, S. I., *Linear Programming: Methods and Applications*,
McGraw-Hill, New York, 4th edition (1975).

[94] Gettings, M. E.; Bultman, M. W.; and Fisher, F. S., Detailed Profiles
of Southeastern Arizona Obtained by a Truck-Mounted Magnetometer
– A Step Toward Better Utilization of the Information Content of
Geophysical Data, U.S. Geological Survey Circular 1062 (1991), 31–2.

[95] Gettings, M. E.; Bultman, M. W.; and Fisher, F. S., "Geophysical
investigation in the search for covered mineral deposits," in *Integrated
Methods in Exploration and Discovery*, S. B. Romberger and D. I.
Fletcher (eds.), AB35–AB36, Society of Economic Geologists and
others (1993).

[96] Gibert, D., and Galdeano, A., "A computer program to perform
transformations of gravimetric and aeromagnetic surveys," *Computers
and Geosciences* **11** (1985), 553–88.

[97] Godson, R. H., and Plouff, D., BOUGUER Version 1.0, a
Microcomputer Gravity-Terrain-Correction Program, Open-File
Report 88–644, U.S. Geological Survey (1988).

[98] Goodacre, A. K., "Estimation of the minimum density contrast of a
homogeneous body as an aid to the interpretation of gravity
anomalies," *Geophysical Prospecting* **28** (1980), 408–14.

[99] Grant, F. S., and West, G. F., *Interpretation Theory in Applied
Geophysics*, McGraw-Hill, New York (1965).

[100] Grauch, V. J. S., TAYLOR: a FORTRAN Program Using Taylor
Series Expansion for Level-Surface or Surface-Level Continuation of
Potential Field Data, Open-File Report 84–501, U.S. Geological
Survey (1984).

[101] Grauch, V. J. S., "A new variable-magnetization terrain correction
method for aeromagnetic data," *Geophysics* **52** (1987), 94–107.

[102] Grauch, V. J. S.; Blakely, R. J.; Blank, H. R.; Oliver, H. W., Plouff,
D.; and Ponce, D. A., *Geophysical Delineation of Granitic Plutons in
Nevada*, Open-File Report 88–11, U.S. Geological Survey (1988).

[103] Grauch, V. J. S., and Cordell, L., "Limitations of determining density
or magnetic boundaries from the horizontal gradient of gravity or
pseudogravity data," *Geophysics* **52** (1987), 118–21.

[104] Gregotski, M. E.; Jensen, O. G.; and Arkani-Hamed, J., "Fractal
stochastic modeling of aeromagnetic data," *Geophysics* **56** (1991),
627–30.

[105] Griscom, A., "Klamath Mountains province," in *Interpretation of the
Gravity Map of California and Its Continental Margin*, Howard W.

Oliver (ed.), 34–6, California Division of Mines and Geology Bulletin 205, Sacramento, CA (1980).

[106] Gudmundsson, G., "Interpretation of one-dimensional magnetic anomalies by use of the Fourier-transform," *Geophysical Journal of the Royal Astronomical Society* **12** (1966), 87–97.

[107] Gudmundsson, G., "Spectral analysis of magnetic surveys," *Geophysical Journal of the Royal Astronomical Society* **13** (1967), 325–37.

[108] Hager, B. H., "Subducted slabs and the geoid: Constraints on mantle rheology and flow," *Journal of Geophysical Research* **84** (1984), 6003–15.

[109] Hammer, S., "Estimating ore masses in gravity prospecting," *Geophysics* **10** (1945), 50–62.

[110] Hanna, W. F., "Some historical notes on early magnetic surveying in the U.S. Geological Survey," in *Geologic Application of Modern Aeromagnetic Surveys*, W. F. Hanna (ed.), 63–73, U.S. Geological Survey Bulletin 1924, Denver, CO (1990).

[111] Hansen, R. O., and Miyazaki, Y., "Continuation of potential fields between arbitrary surfaces," *Geophysics* **49** (1984), 787–95.

[112] Hansen, R. O., and Pawlowski, R. S., "Reduction to the pole at low latitudes by Wiener filtering," *Geophysics* **54** (1989), 1607–13.

[113] Hansen, R. O., and Simmonds, M., "Multiple-source Werner deconvolution," *Geophysics* **58** (1993), 1792–800.

[114] Hansen, R. O., and Wang, X., "Simplified frequency-domain expressions for potential fields of arbitrary three-dimensional bodies," *Geophysics* **53** (1988), 365–74.

[115] Harrison, C. G. A., "A seamount with a nonmagnetic top," *Geophysics* **36** (1971), 349–57.

[116] Harrison, C. G. A., "Marine magnetic anomalies—the origin of the stripes," *Annual Reviews of Earth and Planetary Science* **15** (1987), 505–43.

[117] Harrison, C. G. A., and Carle, H. M., "Intermediate wavelength magnetic anomalies over ocean basins," *Journal of Geophysical Research* **86** (1981), 11,585–99.

[118] Harrison, C. G. A.; Carle, H. M.; and Hayling, K. L., "Interpretation of satellite elevation magnetic anomalies," *Journal of Geophysical Research* **91** (1986), 3633–50.

[119] Harrison, C. G. A.; Jarrard, R. D.; Vacquier, V.; and Larson, R. L., "Palaeomagnetism of Cretaceous Pacific seamounts," *Geophysical Journal of the Royal Astronomical Society* **42** (1970), 859–82.

[120] Hartman, R. R.; Teskey, D. J.; and Friedberg, J. L., "A system for rapid digital aeromagnetic interpretation," *Geophysics* **36** (1971), 891–918.

[121] Heiland, C. A., "Geophysical mapping from the air; its possibilities and advantages," *Engineering and Mining Journal* **136** (1963), 609–10.

[122] Heirtzler, J. R., and Le Pichon, X., "Crustal structure of the mid-ocean ridges. 3," Magnetic anomalies over the Mid-Atlantic Ridge," *Journal of Geophysical Research* **70** (1965), 4013–33.

[123] Heiskanen, W. A., and Moritz, H., *Physical Geodesy*, Freeman, San Francisco (1967).

[124] Henderson, R. G., "On the validity of the use of the upward continuation integral for total magnetic intensity data," *Geophysics* **35** (1970), 916–19.

[125] Henderson, R. G., and Zietz, I., "The upward continuation of anomalies in total magnetic intensity fields," *Geophysics* **14** (1949), 517–34.

[126] Henderson, R. G., and Zietz, I., "The computation of second vertical derivatives of geomagnetic fields," *Geophysics* **14** (1949), 508–16.

[127] Hildenbrand, T. G., FFTFIL: A Filtering Program Based on Two-Dimensional Fourier Analysis, Open-File Report 83–237, U.S. Geological Survey (1983).

[128] Hildenbrand, T. G.; Rosenbaum, J. G.; and Kauahikaua, James P., "Aeromagnetic study of the island of Hawaii," *Journal of Geophysical Research* **98** (1993), 4099–119.

[129] Hood, P., "Gradient measurements in aeromagnetic surveying," *Geophysics* **30** (1965), 891–902.

[130] Hubbert, M. K., "A line-integral method of computing the gravimetric effects of two-dimensional masses," *Geophysics* **13** (1948), 215–25.

[131] Huestis, S. P., and Ander, M. E., "IDB2–A Fortran program for computing extremal bounds in gravity data interpretation," *Geophysics* **48** (1983), 999–1010.

[132] Huestis, S. P., and Parker, R. L., "Bounding the thickness of the oceanic magnetized layer," *Journal of Geophysical Research* **82** (1977), 5293–303.

[133] Huestis, S. P., and Parker, R. L., "Upward and downward continuation as inverse problems," *Geophysical Journal of the Royal Astronomical Society* **57** (1979), 171–88.

[134] International Association of Geomagnetism and Aeronomy (IAGA) Division V, Working Group 8: Analysis of the main field and secular variation, R. A. Langel, Chairman, "International Geomagnetic Reference Field, 1991 revision," *Geophysics* **57** (1992), 956–9.

[135] Ishihara, T., "Gravimetric determination of the density of the Zenisu Ridge," *Tectonophysics* **160** (1989), 195–205.

[136] Jachens, R. C., and Griscom, A., "An isostatic residual map of California: a residual map for interpretation of anomalies from intracrustal sources," in *Technical Program Abstracts and Biographies, Fifty-Second Annual International Meeting and Exposition, Dallas, Texas*, October 17–21, 299–301, Society of Exploration Geophysicists (1982).

[137] Jachens, R. C., and Moring, B. C., Maps of Thickness of Cenozoic Deposits and the Isostatic Residual Gravity over Basement for Nevada, Open-File Report 90–404, U.S. Geological Survey (1990).

[138] Jachens, R. C., and Roberts, C. W., Documentation of a FORTRAN Program, 'Isocomp', for Computing Isostatic Residual Gravity, Open-File Report 81–574, U.S. Geological Survey (1981).

[139] Jacobs, J. A. (ed.), *Geomagnetism*, 4 volumes, Academic Press, London (1987–91).

[140] Jakosky, J. J., *Exploration Geophysics*, 2nd ed., Trija Publishing Company, Newport Beach, CA (1950).

[141] James, R. W., "More on secular variation," *Comments on Earth Sciences: Geophysics* **2** (1971), 28–9.

[142] James, R. W., "The inability of latitude-dependent westward drift to account for zonal secular variation," *Journal of Geomagnetism and Geoelectricity* **26** (1974), 359–61.

[143] Jensen, H., "The airborne magnetometer," *Scientific American* **204** (1961), 151–62.

[144] Johnson, W. M., "A least-squares method of interpreting magnetic anomalies caused by two-dimensional structures," *Geophysics* **34** (1969), 65–74.

[145] Kanasewich, E. R., and Agarwal, R. G., "Analysis of combined gravity and magnetic fields in wave number domain," *Journal of Geophysical Research* **75** (1970), 5702–12.

[146] Kellogg, O. D., *Foundations of Potential Theory*, Dover, New York (1953).

[147] Kis, K. I., "Transfer properties of the reduction of magnetic anomalies to the pole and to the equator," *Geophysics* **55** (1990), 1141–7.

[148] Klitgord, K. D., and Behrendt, J. C., "Basin structure of the U.S. Atlantic margin," in *Geological and Geophysical Investigations of Continental Margins*, J. S. Watkins, L. Montadert, and P. W. Dickerson (eds.), 85–112, American Association of Petroleum Geologists Memoir 29 (1979).

[149] Kodama, K., and Uyeda, S., "Magnetization of Isu Islands with special reference to Oshima Volcano," *Journal of Volcanology and Geothermal Research* **34** (1979), 65–74.

[150] Ku, C. C., and Sharp, J. A., "Werner deconvolution for automated magnetic interpretation and its refinement using Marquardt's inverse modeling," *Geophysics* **48** (1983), 754–74.

[151] Kucks, R. P., and Hildenbrand, T. G., Description of Magnetic Tape Containing Nevada State Magnetic Anomaly Data, Earth Resource Observation System Data Center Report D87–0270, U.S. Geological Survey, Sioux Falls, SD (1987).

[152] LaFehr, T. R., "The estimation of the total amount of anomalous mass by Gauss's theorem," *Journal of Geophysical Research* **70** (1965), 1911–19.

[153] LaFehr, T. R., "Gravity, isostasy, and crustal structure in the southern Cascade Range," *Journal of Geophysical Research* **70** (1965), 5581–97.

[154] LaFehr, T. R., "Standardization in gravity reduction," *Geophysics* **56** (1991), 1170–8.

[155] LaFehr, T. R. "An exact solution for the gravity curvature (Bullard B) correction," *Geophysics* **56** (1991), 1179–84.

[156] Langel, R. A., "Study of the crust and mantle using magnetic surveys by Magsat and other satellites," *Proceedings of the Indian Academy of Sciences* **99** (1990), 581–618.

[157] Langel, R. A., "International geomagnetic reference field: the sixth generation," *Journal of Geomagnetism and Geoelectricity* **44** (1992), 679–707.

[158] Langel, R. A., Chair, IAGA Division V, Working Group 8, "IGRF, 1991 revision," *EOS, Transactions of the American Geophysical Union* **73** (1992), 182.

[159] Langel, R. A.; Phillips, J. D.; and Horner, R. J., "Initial scalar magnetic anomaly map from Magsat," *Geophysical Research Letters* **9** (1982), 269–72.

[160] Langel, R. A., and Estes, R. H., "A geomagnetic field spectrum," *Geophysical Research Letters* **9** (1982), 250–3.

[161] Last, B. J., and Kubik, K., "Compact gravity inversion," *Geophysics* **48** (1983), 713–21.

[162] Lee, Y. W., *Statistical Theory of Communication*, Wiley, New York (1960).

[163] Lerch, F. J.; Klosko, S. M.; Laubscher, R. E.; and Wagner, C. A., "Gravity model improvement using Geos 3 (GEM 9 and 10)," *Journal of Geophysical Research* **84** (1979), 3897–916.

[164] Leu, L.-K., "Use of reduction-to-equator process for magnetic data interpretation," in *Fifty-First Annual International Meeting*, 445, Society of Exploration Geophysicists (1981).

[165] Lindsley, D. H.; Andreasen, G. E.; and Balsley, J. R., "Magnetic properties of rocks and minerals," in *Handbook of Physical Constants*, revised edition, S. P. Clark, Jr. (ed.), 543–52, Geological Society of America Memoir 97, New York (1966).

[166] Longman, I. M., "Formulas for computing the tidal accelerations due to the moon and the sun," *Journal of Geophysical Research* **64** (1959), 2351–5.

[167] Lowes, F. J., "Mean-square values on sphere of spherical harmonic vector fields," *Journal of Geophysical Research* **71** (1966), 2179.

[168] Lowes, F. J., "Spatial power spectrum of the main geomagnetic field, and extrapolation to the core," *Geophysical Journal of the Royal Astronomical Society* **36** (1974), 717–30.

[169] Lowes, F. J., "Do magnetometers measure B or H?" *Geophysical Journal of the Royal Astronomical Society* **37** (1974), 151–5.

[170] Macdonald, K. C., "Near-bottom magnetic anomalies, asymmetric spreading, oblique spreading, and tectonics of the Mid-Atlantic Ridge near lat 37°N," *Geological Society of America Bulletin* **88** (1977), 541–55.

[171] Macdonald, K. C.; Miller, S. P.; Luyendyk, B. P.; Atwater, T. M.; and Shure, L., "Investigation of a Vine–Matthews magnetic lineation from a submersible: the source and character of marine magnetic anomalies," *Journal of Geophysical Research* **88** (1983), 3403–18.

[172] MacMillan, W. D., *The Theory of the Potential*, Dover, New York (1958).

[173] Marion, J. B., *Classical Dynamics of Particles and Systems*, Academic Press, New York (1965).

[174] Marquardt, D. W., "An algorithm for least-squares estimation of nonlinear parameters," *Journal Soc. Indust. Appl. Math.* **11** (1963), 431–41.

[175] Marsh, J. G.; Koblinsky, C. J.; Zwally, H. J.; Brenner, A. C.; and Beckley, B. D., "A global mean sea surface based upon GEOS 3 and Seasat altimeter data," *Journal of Geophysical Research* **97** (1992), 4915–21.

[176] Marson, I., and Klingele, E. E., "Advantages of using the vertical gradient of gravity for 3-D interpretation," *Geophysics* **58** (1993), 1588–95.

[177] Mayhew, M. A., "Inversion of satellite magnetic anomaly data," *Journal of Geophysical Research* **45** (1979), 119–28.

[178] Mayhew, M. A., "An equivalent layer magnetization model for the United States derived from satellite altitude magnetic anomalies," *Journal of Geophysical Research* **87** (1982), 4837–45.

[179] Mayhew, M. A., "Magsat anomaly field inversion for the U. S.," *Earth and Planetary Science Letters* **71** (1984), 290–6.

[180] McGrath, P. H., and Hood, P. J., "An automatic least-squares multimodel method for magnetic interpretation," *Geophysics* **38** (1973), 349–58.

[181] McNutt, M., "Nonuniform magnetization of seamounts: a least squares approach," *Journal of Geophysical Research* **91** (1986), 3686–700.

[182] Menke, W., *Geophysical Data Analysis: Discrete Inverse Theory*, Academic Press, Orlando, FL (1984).

[183] Merrill, R. T., and McElhinny, M. W., *The Earth's Magnetic Field, Its History, Origin and Planetary Perspective*, Academic Press, London (1983).

[184] Milbert, D. G., and Dewhurst, W. T., "The Yellowstone–Hebgen Lake geoid obtained through the integrated geodesy approach," *Journal of Geophysical Research* **97** (1992), 545–57.

[185] Miyazaki, Y., "Analysis of potential field data over Long Valley, California; 2, mapping of Curie isothermal depth from aeromagnetic anomalies," *Butsuri-Tansa (Geophysical Exploration)* **44** (1991), 289–310.

[186] Morley, L. W., and Larochelle, A., "Paleomagnetism as a means of dating geological events," in *Geochronology in Canada*, F. F. Osborne (ed.), 39–51, Royal Society of Canada Special Publication Number 8, University of Toronto Press, Toronto (1964).

[187] Morrish, A. H., *The Physical Principles of Magnetism*, Wiley, New York (1965).

[188] Morse, P. M., and Feshbach, H., *Methods of Theoretical Physics*, McGraw-Hill, New York (1953).

[189] Moskowitz, B. M., "Fundamental physical constants and conversion factors," in *Handbook of Physical Constants*, T. J. Ahrens (ed.), American Geophysical Union, Washington, D. C., in press.

[190] Nabighian, M. N., "The analytic signal of two-dimensional magnetic bodies with polygonal cross section: its properties and use for automated anomaly interpretation," *Geophysics* **37** (1972), 507–17.

[191] Nabighian, M. N., "Additional comments on the analytic signal of two-dimensional magnetic bodies with polygonal cross section," *Geophysics* **39** (1974), 85–92.

[192] Nabighian, M. N., "Toward a three-dimensional automatic interpretation of potential field data via generalized Hilbert transforms: fundamental relations," *Geophysics* **49** (1984), 780–6.

[193] Nakatsuka, T., "Reduction of magnetic anomalies to and from an arbitrary surface," *Butsuri-Tanko* **34** (1981), 6–12.

[194] Needham, J., *Science and Civilization in China*, Vol. 4, Physics and Physical Technology, Part 1, Physics, Cambridge University Press, Cambridge (1962).

[195] Neidell, N. S., "Spectral studies of marine geophysical profiles," *Geophysics* **31** (1966), 122–34.

[196] O'Brien, D. P., "CompuDepth – a new method for depth-to-basement calculation," presented at the 42nd Meeting of the Society of Exploration Geophysicists, Anaheim, CA (1972).

[197] Okabe, M., "Analytical expressions for gravity anomalies due to homogeneous polyhedral bodies and translations into magnetic anomalies," *Geophysics* **44** (1979), 730–41.

[198] Okubo, Y.; Graf, R. J.; Hansen, R. O.; Ogawa, K.; and Tsu, H., "Curie point depths of the island of Kyushu and surrounding areas, Japan," *Geophysics* **50** (1985), 481–94.

[199] Okubo, Y.; Tsu, H.; and Ogawa, K., "Estimation of Curie point temperature and geothermal structure of island arcs of Japan," *Tectonophysics* **159** (1989), 279–90.

[200] Oldenburg, D. W., "The inversion and interpretation of gravity anomalies," *Geophysics* **39** (1974), 526–36.

[201] Panofsky, W. K. H., and Phillips, M., *Classical Electricity and Magnetism*, 2nd ed., Addison-Wesley, Reading, MA (1962).

[202] Papoulis, A., *The Fourier Integral and Its Applications*, McGraw-Hill, New York (1962).

430 *Bibliography*

[203] Parasnis, D. S., *Principles of Applied Geophysics*, 4th ed., Chapman and Hall, London (1986).
[204] Parker, R. L., "The rapid calculation of potential anomalies," *Geophysical Journal of the Royal Astronomical Society* **31** (1972), 447–55.
[205] Parker, R. L., "Best bounds on density and depth from gravity data," *Geophysics* **39** (1974), 644–9.
[206] Parker, R. L., "The theory of ideal bodies for gravity interpretation," *Geophysical Journal of the Royal Astronomical Society* **42** (1975), 315–34.
[207] Parker, R. L., "Understanding inverse theory," *Annual Reviews of Earth and Planetary Sciences* **5** (1977), 35–64.
[208] Parker, R. L., and Huestis, S. P., "The inversion of magnetic anomalies in the presence of topography," *Journal of Geophysical Research* **79** (1974), 1587–93.
[209] Parker, R. L., and Klitgord, K. D., "Magnetic upward continuation from an uneven track," *Geophysics* **37** (1972), 662–8.
[210] Parker, R. L.; Shure, L.; and Hildenbrand, J. A., "The application of inverse theory to seamount magnetism," *Reviews of Geophysics* **25** (1987), 17–40.
[211] Parkinson, W. D., *Introduction to Geomagnetism*, Scottish Academic Press, Edinburgh, Scotland (1983).
[212] Paterson, N. R., and Reeves, C. V., "Applications of gravity and magnetic surveys: the state-of-the-art in 1985," *Geophysics* **50** (1985), 2558–94.
[213] Payne, M. A., "SI and Gaussian CGS units, conversions and equations for use in geomagnetism," *Physics of the Earth and Planetary Interiors* **26** (1981), P10–P16.
[214] Pearson, W. C., and Skinner, C. M., "Reduction-to-the-pole of low latitude magnetic anomalies," In *Technical Program Abstracts and Biographies, Fifty-Second Annual International Meeting and Exposition, Dallas, Texas*, Society of Exploration Geophysicists, October 17–21, 356–8 (1982).
[215] Peddie, N. W., and Zunde, A. K., *The Magnetic Field in the United States, 1985*, Geophysical Investigations Map GP–986, U.S. Geological Survey (1988).
[216] Pedersen, L. B., "Wavenumber domain expressions for potential fields from arbitrary 2-, 2 1/2-, and 3-dimensional bodies," *Geophysics* **43** (1978), 626–30.
[217] Pedersen, L. B., "Relations between horizontal and vertical gradients of potential fields," *Geophysics* **54** (1989), 662–3.
[218] Pedersen, L. B., "On: 'Reduction of magnetic and gravity data on an arbitrary surface acquired in a region of high topographic relief' by B. K. Bhattacharyya and K. C. Chan, with reply by K. C. Chan," *Geophysics* **54** (1989), 664–5.
[219] Peters, L. J., "The direct approach to magnetic interpretation and its practical application," *Geophysics* **14** (1949), 290–320.
[220] Phillips, J. D., ADEPT: A Program to Estimate Depth to Magnetic Basement from Sampled Magnetic Profiles, Open-File Report 79–367, U.S. Geological Survey (1979).
[221] Pilkington, M., and Crossley, D. J., "Determination of crustal interface topography from potential fields. *Geophysics* **51** (1986), 1277–84.
[222] Pilkington, M., and Crossley, D. J., "Inversion of aeromagnetic data for multilayered crustal models," *Geophysics* **51** (1986), 2250–4.

[223] Pilkington, M.; Gregotski, M. E.; and Todoeschuck, J. P., "Using fractal crustal magnetization models in magnetic interpretation," *Geophysical Prospecting*, in press.

[224] Pilkington, M., and Roest, W., "Draping aeromagnetic data in areas of rugged topography," *Journal of Applied Geophysics* **29** (1992), 135–142.

[225] Pilkington, M., and Todoeschuck, J. P., "Naturally smooth inversions with a priori information from well logs," *Geophysics* **56** (1991), 1811–18.

[226] Pilkington, M., and Todoeschuck, J. P., "Fractal magnetization of continental crust," *Geophysical Research Letters* **20** (1993), 627–30.

[227] Plouff, D., Derivation of Formulas and FORTRAN Programs to Compute Gravity Anomalies of Prisms, National Technical Information Service PB–243–526, U.S. Department of Commerce (1975).

[228] Plouff, D., Derivation of Formulas and FORTRAN Programs to Compute Magnetic Anomalies of Prisms, National Technical Information Service PB–243–526, U.S. Department of Commerce (1975).

[229] Plouff, D., "Gravity and magnetic fields of polygonal prisms and application to magnetic terrain corrections," *Geophysics* **41** (1976), 727–41.

[230] Plouff, D., Preliminary Documentation for a FORTRAN Program to Compute Gravity Terrain Corrections Based on Topography Digitized on a Geographic Grid, Open-File Report 77–535, U.S. Geological Survey (1977).

[231] Pratt, J. H., "On the attraction of the Himalaya Mountains and of the elevated regions beyond them, upon the plumb-line in India," *Philosophical Transactions of the Royal Society of London* **145** (1855), 53–100.

[232] Pratt, J. H., "On the deflection of the plumb-line in India, caused by the attraction of the Himalaya Mountains and of the elevated regions beyond; and its modification by the compensating effect of a deficiency of matter below the mountain mass," *Philosophical Transactions of the Royal Society of London* **149** (1859), 745–78.

[233] Press, W. H.; Flannery, B. P.; Teukolsky, S. A.; and Vetterling, W. T., *Numerical Recipes, The Art of Scientific Computing*, Cambridge University Press, Cambridge (1986).

[234] Raff, A. D., and Mason, R. G., "Magnetic survey off the west coast of North America, 40°N latitude to 52°N latitude," *Geological Society of America Bulletin* **72** (1961), 1267–70 .

[235] Ramsey, A. S., *An Introduction to the Theory of Newtonian Attraction*, Cambridge University Press, Cambridge (1940).

[236] Rao, D. B., and Babu, N. R., "A rapid method for three-dimensional modeling of magnetic anomalies," *Geophysics* **56** (1991), 1729–37.

[237] Rasmussen, R., and Pedersen, L. B., "End corrections in potential field modeling," *Geophysical Prospecting* **27** (1979), 749–60.

[238] Ravat, D., "Use of fractal dimension to determine the applicability of Euler's homogeneity equation for finding source locations of gravity and magnetic anomalies," in *Proceedings of the Symposium of the Application of Geophysics to Engineering and Environmental Problems*, Boston, Massachusetts, March 1994, **1** (1994), 41–53; Environmental and Engineering Geophysical Society, Englewood, CO.

[239] Ravat, D. N.; Hinze, W. J.; and von Frese, R. R. B., "Lithospheric magnetic property contrasts within the South American plate derived from damped least-squares inversion of satellite magnetic data," *Tectonophysics* **192** (1991), 159–68.

[240] Reford, M. S., "Magnetic method," *Geophysics* **45** (1985), 1640–58.

[241] Regan, R. D.; Cain, J. C.; and Davis, W. M., "A global magnetic anomaly map," *Journal of Geophysical Research* **80** (1975), 794–802.

[242] Reid, A. B.; Allsop, J. M.; Granser, H.; Millett, A. J.; and Somerton, I. W., "Magnetic interpretation in three dimensions using Euler deconvolution," *Geophysics* **55** (1990), 80–91.

[243] Reynolds, R. L.; Rosenbaum, J. G.; Hudson, M. R.; and Fishman, N. S., "Rock magnetism, the distribution of magnetic minerals in the earth's crust, and aeromagnetic anomalies," in *Geologic Application of Modern Aeromagnetic Surveys*, W. F. Hanna (ed.), 24–45, U.S. Geological Survey Bulletin 1924, Denver, CO (1990).

[244] Ricard, Y., and Blakely, R. J., "A method to minimize edge effects in two-dimensional discrete Fourier transforms," *Geophysics* **53** (1988), 1113–17.

[245] Roest, W. R.; Arkani-Hamed, J.; and Verhoef, J., "The seafloor spreading rate dependence of the anomalous skewness of marine magnetic anomalies," *Geophysical Journal International* **109** (1992), 653–69.

[246] Roest, W. R., and Pilkington, M., "Identifying remanent magnetization effects in magnetic data," *Geophysics* **58** (1993), 653–9.

[247] Roest, W. R.; Verhoef, J.; and Pilkington, M., "Magnetic interpretation using the 3-D analytic signal," *Geophysics* **57** (1992), 116–25.

[248] Runcorn, S. K., "On the interpretation of lunar magnetism," *Physics of the Earth and Planetary Interiors* **10** (1975), 327–35.

[249] Sager, W. W.; Davis, G. T.; Keating, B. H.; and Philpotts, J. A., "A geophysical and geologic study of Nagata seamount, northern Line Islands," *Journal of Geomagnetism and Geoelectricity* **34** (1982), 283–305.

[250] Saltus, R. W., "Upper-crustal structure beneath the Columbia River Basalt Group, Washington: Gravity interpretation controlled by borehole and seismic studies," **105** (1993), 1247–59.

[251] Saltus, R. W., and Blakely, R. J., HYPERMAG, An Interactive, 2- and 2 1/2-Dimensional Gravity and Magnetic Modeling Program, Version 3.5, Open-File Report 93–287, U.S. Geological Survey (1993).

[252] Schouten, J. A., "A fundamental analysis of magnetic anomalies over oceanic ridges," *Marine Geophysical Researches* **1** (1971), 111–44.

[253] Schouten, H., and McCamy, K., "Filtering marine magnetic anomalies," *Journal of Geophysical Research* **77** (1972), 7089–99.

[254] Schouten, H., and Cande, S. C., "Palaeomagnetic poles from marine magnetic anomalies," *Geophysical Journal of the Royal Astronomical Society* **44** (1976), 567–75.

[255] Shive, P. N., "Suggestions for the use of SI units in magnetism," *EOS, Transactions of the American Geophysical Union* **67** (1986), 25.

[256] Shuey, R. T., and Pasquale, A. S., "End corrections in magnetic profile interpretation," *Geophysics* **38** (1973), 507–12.

[257] Shuey, R. T.; Schellinger, D. K.; Tripp, A. C.; and Alley, L. B., "Curie-depth determination from aeromagnetic spectra," *Geophysical Journal of the Royal Astronomical Society* **50** (1977) 75–101.

[258] Silva, B. C. J., "Reduction to the pole as an inverse problem and its application to low latitude anomalies," *Geophysics* **51** (1986), 369–82.

[259] Simpson, R. W., and Jachens, R. C., "Gravity methods in regional studies," in *Geophysical Framework of the Continental United States*, L. C. Pakiser and W. D. Mooney (eds.), 35–44, Geological Society of America Memoir 172, Boulder, CO (1989).

[260] Simpson, R. W.; Jachens, R. C.; and Blakely, R. J., AIRYROOT: A Fortran Program for Calculating the Gravitational Attraction of an Airy Isostatic Root Out to 166.7 km, Open-File Report 83–883, U.S. Geological Survey (1983).

[261] Simpson, R. W.; Jachens, R. C.; Blakely, R. J.; and Saltus, R. W., "A new isostatic residual gravity map of the conterminous United States with a discussion on the significance of isostatic residual anomalies," *Journal of Geophysical Research* **91** (1986), 8348–72.

[262] Skillbrei, J. R., "The straight-slope method for basement depth determination revisited," *Geophysics* **58** (1993), 593–5.

[263] Smith, R. A., "Some depth formulae for local magnetic and gravity anomalies," *Geophysical Prospecting* **7** (1959), 55–63.

[264] Smith, R. A., "A uniqueness theorem concerning gravity fields," *Proceedings of the Cambridge Philosophical Society* **57** (1961), 865–70.

[265] Smith, R. B.; Shuey, R. T.; Freidline, R. O.; Otis, R. M.; and Alley, L. B., "Yellowstone hot spot: new magnetic and seismic evidence," *Geology* **2** (1974), 451–5.

[266] Society of Exploration Geophysicists. *The SI Metric System of Units and SEG Tentative Metric Standard*, Published by the SEG Metrification Subcommittee, chaired by L. Lenz, Society of Exploration Geophysicists, Tulsa, OK.

[267] Spector, A., and Bhattacharyya, B. K., "Energy density spectrum and autocorrelation function of anomalies due to simple magnetic models," *Geophysical Prospecting* **14** (1966), 242–72.

[268] Spector, A., and Grant, F. S., "Statistical models for interpreting aeromagnetic data," *Geophysics* **35** (1970), 293–302.

[269] Spielman, J. B., and Ponce, D. A., Handtc, a Fortran Program to Calculate Inner-Zone Terrain Corrections, Open-File Report 84–777, U.S. Geological Survey (1984).

[270] Stacey, F. D., *Physics of the Earth*, 3rd ed., Brookfield Press, Brisbane, Australia (1992).

[271] Stacey, F. D.; Tuck, G. J.; Moore, G. I.; Holding, S. C.; Goodwin, B. D.; and Zhou, R., "Geophysics and the law of gravity," *Reviews of Modern Physics* **59** (1987), 157–74.

[272] Steenland, N. C., "Discussion on 'The geomagnetic gradiometer' by H. A. Slack, V. M. Lynch, and L. Langan (*Geophysics*, October 1967, pp. 877–992)," *Geophysics* **33** (1968), 680–3.

[273] Stewart, J. H., and Carlson, J. E., *Geologic Map of Nevada*, Technical Report, U.S. Geological Survey (1978).

[274] Strauss, W. A., *Partial Differential Equations, an Introduction*, Wiley, New York (1992).

[275] Talwani, M. "Computation with the help of a digital computer of magnetic anomalies caused by bodies of arbitrary shape," *Geophysics* **30** (1965), 797–817.

[276] Talwani, M., and Ewing, M., "Rapid computation of gravitational attraction of three-dimensional bodies of arbitrary shape," *Geophysics* **25** (1960), 203–25.

[277] Talwani, M., and Heirtzler, J. R., Computation of Magnetic Anomalies Caused by Two-Dimensional Structures of Arbitrary Shape, Stanford University Publications of the Geological Sciences, Computers in the Mineral Industries, Stanford University, Stanford, CA (1964), 464–80.

[278] Talwani, M.; Worzel, J. L.; and Landisman, M., "Rapid gravity computations for two-dimensional bodies with application to the Mendocino submarine fracture zone," *Journal of Geophysical Research* **64** (1959), 49–59.

[279] Telford, W. M.; Geldart, L. P.; and Sheriff, R. E., *Applied Geophysics*, 2nd ed., Cambridge University Press, Cambridge (1990).

[280] Thompson, D. T., "EULDPH: a new technique for making computer-assisted depth estimates from magnetic data," *Geophysics* **47** (1982), 31–7.

[281] Treitel, S.; Clement, W. G.; and Kaul, R. K., "The spectral determination of depths to buried magnetic basement rocks," *Geophysical Journal of the Royal Astronomical Society* **24** (1971), 415–28.

[282] Tsuboi, C., and Fuchida, T., "Relations between gravity values and corresponding subterranean mass distribution," *Earthquake Research Institute of the Tokyo Imperial University, Bulletin* **15** (1937), 636–49, 1937.

[283] Tsuboi, C., and Fuchida, T., "Relation between gravity anomalies and the corresponding subterranean mass distribution (II.)," *Earthquake Research Institute of the Tokyo Imperial University, Bulletin* **16** (1938), 273–84.

[284] Ueda, Y., "Geophysical study of two seamounts near Minami-Tori Sima (Marcus) Island, Western Pacific Ocean," *Journal of Geomagnetism and Geoelectricity* **40** (1988), 1481–501.

[285] Vacquier, V., "A machine method for computing the magnitude and the direction of magnetization of a uniformly magnetized body from its shape and a magnetic survey," in *Proceedings of the Benedum Earth Magnetism Symposium, 1962*, 123–37, University of Pittsburgh Press, Pittsburgh, PA (1963).

[286] Vine, F. J., and Matthews, D. H., "Magnetic anomalies over ocean ridges," *Nature* **199** (1963), 947–9.

[287] von Frese, R. R. B.; Hinze, W. J.; and Braile, L. W., "Spherical earth gravity and magnetic anomaly analysis by equivalent point source inversion," *Earth and Planetary Science Letters* **53** (1981), 69–83.

[288] Wang, X., and Hansen, R. O., "Inversion for magnetic anomalies of arbitrary three-dimensional bodies," *Geophysics* **55** (1990), 1321–26.

[289] Wasilewski, P.; Thomas, H.; and Mayhew, M. A., "The Moho as a magnetic boundary," *Geophysical Research Letters* **6** (1979), 541–4.

[290] Webring, M., SAKI: A Fortran Program for Generalized Linear Inversion of Gravity and Magnetic Profiles, Open-File Report 85–122, U.S. Geological Survey (1985).

[291] Werner, S., "Interpretation of magnetic anomalies at sheet-like bodies," *Sveriges Geologiska Undersok.*, Årsbok 43, no. 6, series C, no. 508 (1953).

[292] Yukutake, T., "The westward drift of the magnetic field of the earth," *Bulletin of the Earthquake Research Institute* **40** (1962), 1–65.

[293] Yukutake, T., "Complexity of the geomagnetic secular variation," *Comments on Earth Sciences: Geophysics* **1** (1970), 55–64.

[294] Yukutake, T., and Tachinaka, H., "Separation of the earth's magnetic field into the drifting and the standing parts," *Bulletin of the Earthquake Research Institute* **47** (1969), 65–97.

[295] Zoback, M. L., and Thompson, G. A., "Basin and Range rifting in northern Nevada: clues from a mid-Miocene rift and its subsequent offsets," *Geology* **6** (1978), 111–16.

[296] Zorin, Y. A.; Pismenny, B. M.; Novoselova, M. R.; and Turutanov, E. K., "Decompensative gravity anomaly," *Geol. Geofiz.* **8** (1985), 104–5 (in Russian).

[297] Zumberge, M. A.; Ander, M. E.; Lautzenhiser, T. V.; Parker, R. L., Aiken, C. L. V.; Gorman, M. R.; Nieto, M. M.; Cooper, A. P. R., Ferguson, J. F.; Fisher, E.; Greer, J.; Hammer, P.; Hansen, B. L., McMechan, G. A.; Sasagawa, G. S.; Sidles, C.; Stevenson, J. M, and Wirtz, J., "The Greenland gravitational constant experiment," *Journal of Geophysical Research* **95** (1990), 15, 483–501.

Index